The Hive and Honey Bee Revisited

By
Roger Hoopingarner, Ph.D.

Published 2014 by

Wicwas Press LLC

Kalamazoo, Michigan

www.wicwas.com

ISBN 978-1-878075-36-9

Copyright 2006 by Roger Hoopingarner and Bee+ Books
All rights reserved. No part of this publication may be reproduced or trans-mitted in any form, or by any means, or stored in any information storage and retrieval system, without the prior written permission of Bee+ Books, Holt, Michigan

The first edition of **The Hive and Honey Bee Revisited** was published by Bee+ Books, Holt, Michigan in 2006.

The Hive and Honey Bee Revisited

The Remarkable Langstroth 150 Years Later

By

Roger Hoopingarner, Ph.D.
Professor Emeritus
Michigan State University

Preface To "The Hive and Honey Bee Revisited"

It has been over a 100 years since the death of Rev. L.L. Langstroth, and 150 years since his discovery of the movable-comb hive. There is enough "history buff" within me to think it was time to look at the beginnings of the beekeeping industry. What better way than to go back to the master beekeeper himself, and to compare his thoughts and ideas to our thinking about bees today. I know you will be surprised, as I was, when I re-read this edition just how much bee biology he knew! Some of it he read from other sources, but much of it was from just good, keen observations. Certainly when, and how, he learned beekeeping facts is different than how we would learn them today. He did not have the equipment or the background information that we have today—in many ways that makes his discoveries that much more remarkable. Beekeepers talk about their "Langstroth" hive but mostly do not know the depth of beekeeping knowledge that he possessed and passed on through his writings.

Langstroth's invention brought him almost no monetary gain as he was not able to pursue his claim against all the people that infringed on his patent. Yet fame he did receive. We read from the original introduction to the Third edition by the Reverend Dr. Robert Baird. "Not many years will pass away without seeing his important invention brought into extensive use, both in the Old and New World. Its great merits need only to be known; and this, Time will certainly bring about." How truly prophetic were these words.

The first reading that I did of Langstroth 50 years ago was from some of the later editions of "Hive and Honey-Bee" that were re-written by C.P. and M.G. Dadant. While these were good books, I think they took away some of the mastery of the subject that Langstroth displayed. In some places in this edition I am just amazed at how much he knew for the time period.

I have chosen to use the Third edition (1862; written in 1859) of his book, "The Hive and Honey-bee," for three reasons. The first is that this edition came out after his invention of the movable-comb hive had been in use for a few years and had gained some attention. Second, this is the last edition that Langstroth himself wrote. Finally, I did not have copies of the earlier editions in my library.

I have tried to keep the text as close to the original as possible as to format, style and spelling. For example, today in America we do not use honey-bee or honeybee but would write it as honey bee. This is because the North American Entomologists have a rule for the common names of insects. The rule is that the honey bee is a true bee, and thus is spelled as two words. If an insect's name does not fit it into its proper grouping it is spelled as one word. For example, a *butterfly* (Order Lepidoptera) is not a fly and thus is just one word, but a *honey bee* is a bee in the Order Hymenoptera.

Because of slight differences in type style and mode of printing most pages are not *exact* duplicates of the original Third edition, but I endeavored to keep them as close as possible to give you the "feel" of the original. On some occasions I was forced to make a page break, in order to keep page order as near as possible to the pagination of the original book.

The footnotes are divided into those given from the original book, by Langstroth, and those I have appended to give added, or new, meaning to the text. The original footnotes of Langstroth are in normal typeface whereas those of mine are *italicized* and numbered, and my additions to Langstroth's footnotes are italicized as well. The index also follows the same pattern, *i.e.*, the original is in normal font and my additions are in italics.

The original text was scanned into a computer using an optical-character recognition (OCR) program. Some errors occurred during the OCR process, and I have tried to read the text carefully to avoid misleading sentences. I am sure some escaped. Please do not blame Langstroth's writing, or his typesetter, but only my attempt to capture it for your reading. I am also sure you will recognize that spellings of some words have changed in the last 150 years, as well as the use of certain words that we rarely use.

Langstroth's writing reflects his university (theological) training and the nature of the way topics were written about 150 years ago. His style and long sentences may dismay you, but if you read them carefully many truths about bees will be found.

In the original book the full plate drawings were inserted throughout the text with a listing in the front. These original plates have been copied for your enjoyment. Since these engravings were inserted they have no page numbers themselves and have a blank page on the back side. In some of the drawings that illuminate the text I have used modern drawings to add clarity. They do not change the feeling of L.L. Langstroth's book. There is a full description of the plates at the end of the book with detailed measurements of each hive so that the reader could make the hive(s), if desired.

I have added a preface to many chapters to give the reader a little overview of what I think was the intention of Langstroth in his presentation, and to also give some historical perspective on the changes in apiculture since the 1850's.

I hope this book will allow many "Apiarians" a chance to read Langstroth's original masterwork and benefit from the knowledge modern science has clarified and amplified in the last 150 years.

This book is dedicated to the memory of my wife Barbara who gave me love, guidance and nurture for 45 years, and without which my career and this book would not have been possible.

Holt, Michigan
July, 2013

So work the Honey Bees,
Creatures, that by a rule in Nature, teach
The art of order to a peopled kingdom.—*Shakspeare.*

A PRACTICAL TREATISE

ON THE

HIVE AND HONEY-BEE

BY

L. L. LANGSTROTH

WITH

AN INTRODUCTION BY REV. ROBERT BAIRD, D. D.

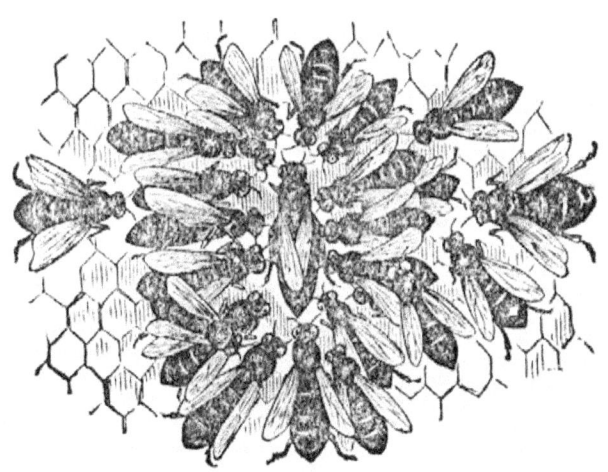

THIRD EDITION,

REVISED, AND ILLUSTRATED WITH SEVENTY-SEVEN ENGRAVINGS

NEW YORK
C. M. SAXTON, 25 PARK ROW
1862

Entered, according to Act of Congress, in the year 1859

By L. L. Langstroth,

In the Clerk's Office of the District Court of the Southern District of New York

Introduction.

———■———

I am happy to learn from my friend Mr. Langstroth, that a new edition of his work on the Hive and Honey-Bee is called for; I consider it by far the most valuable treatise on these subjects, which as come under my notice. Some years before it was published, I became acquainted with the main characteristics of his system of Bee-culture, and even then, I believed it to be incomparable superior to all others of which I had either read or heard. This conviction has been amply strengthened by the testimony of others, as well as by results, which have come under, by own observation.

In my early life I had no inconsiderable experience in the management of bees, and I am bold to say that the hive which Mr. Langstroth has invented, is in all respects greatly superior to any which I have ever seen, either in this or foreign countries. Indeed, I do not believe that any one who takes an intelligent interest in the rearing of bees, can for a moment hesitate to use it; or, rather, can be induced to use any other, when he becomes acquainted with its nature and merits.

At length the true secret has been discovered, of making these most industrious, interesting, and useful of insect-communities, work in habitations both comfortable to themselves and wonderfully convenient for their aggregation, division, and rapid increase; and all this without diminishing their productive labor, or resorting to the cruel measure of destroying them.

Mr. LANGSTROTH teaches us in his book, how bees can be taken care of without great labor, and without the risk of suffering from the weapon which the Creator has given them for self-defense. Even a delicate lady need not fear to undertake the task of cultivating this fascinating branch of Rural Economy. Nothing is easier for any family that resides in a favorable situation, than to have a number of colonies, and this at but little expense. I sincerely hope that many will avail themselves of the facilities now placed before them for prosecuting this easy branch of industry, not only for the sake of the large profit in proportion to its expense, which it may be made to yield, but also for the substantial pleasure which they may find in observing the habits of these wonderful little creatures. How remarkably does their entire economy illustrate the wisdom and skill of the Great Author of all things.

I cannot but believe that many Ministers of the Gospel, residing in rural districts, will accept Mr. LANGSTROTH'S generous offer to give them the free use of his Invention. With very little labor or expense, they can derive from bee-keeping considerable profit, as well as much pleasure.

INTRODUCTION

No industrial or material employment can be more innocent, or less inconsistent with their proper work.

There are few portions of our country which are not admirably adapted to the culture of the Honey-bee. The wealth of the nation might be increased by millions of dollars, if every family favorably situated for bee-keeping would keep a few hives. No other branch of industry can be named, in which there need be so little loss on the material that is employed, or which so completely derives its profits from the vast and exhaustless domains of Nature.

I trust that Mr. LANGSTROTH'S labors will contribute greatly to promote a department of Rural Economy, which in this country has hitherto received so little scientific attention. He well deserves the name of Benefactor; infinitely more so than many who in all countries and in all ages have received that honorable title. Not many years will pass away without seeing his important invention brought into extensive use, both in the Old and New World. Its great merits need only to be known; and this, Time will certainly bring about.

ROBERT BAIRD.

Preface

Encouraged by the favor with which the former editions of this work have been received, I submit to the public a Revised Edition, illustrated by additional woodcuts, and containing my latest discoveries and improvements. The information that it presents, is adapted not only to those who use the Moveable-Comb Hive, but all who aim at profitable bee-keeping, with any hive, or on any system of management.

Debarred, to a great extent, by ill-health, from the appropriate duties of my profession, and compelled to seek an employment calling me as much as possible into open air, I cherish the hope that my labors in an important department of Rural Economy, may prove serviceable to the community. Bee-keeping is regarded in Europe as an intellectual pursuit, and no one who studies the wonderful habits of his useful insect, will ever find the materials for new observations exhausted. The Creator has stamped the seal of his Infinity on all his works, so that it is impossible, even in the minutest, "by searching to find out the Almighty to perfection." In none of them, however, has he displayed himself more clearly than in the economy of the Honey-Bee:

> "What well-appointed commonwealth! where each
> Adds to the stock of happiness for all;
> Wisdom's own forums! whose professors teach
> Eloquent lessons in their vaulted hall!
> Galleries of art! and schools of industry!
> Stores of rich fragrance! Orchestras of song!
> What marvelous seats of hidden alchemy!
> How oft, when wandering far and erring long,
> Man might learn truth and virtue from the BEE!"
>   Bowring.

The attention of Ministers of the Gospel is particularly invited to this branch of Natural History. An intimate acquaintance with the wonders of the Bee-Hive, while beneficial to them in many ways, might lead them, in their preaching, to imitate more closely the example of Him who illustrated his teachings by "the birds of the air, and the lilies of the field," as well as the common walks of life, and the busy pursuits of men.

It affords me sincere pleasure to acknowledge my obligations to Mr. SAMUEL WAGNER, of York, Pennsylvania, for material assistance in the preparation of this Treatise. To his extensive and accurate acquaintance with Bee-keeping in Germany, my readers will find themselves indebted for much exceedingly valuable information.

L. L. LANGSTROTH.

OXFORD, BUTLER COUNTY, OHIO, March, 1859.

TABLE OF CONTENTS.

	Page.
List of Plates and Explanation of Wood-Cuts Illustrating the Natural History of Bees	11

Chapter.
- I. Facts connected with the invention of the Movable-Comb Bee-Hive ... 13
- II. The Honey-Bee capable of being tamed ... 24
- III. The Queen, or Mother-Bee –The Drones. –The Workers. –Facts in the Natural History ... 29
- IV. Comb ... 69
- V. Propolis ... 76
- VI. Pollen, or "Bee Bread." ... 80
- VII. Ventilation of the Bee-Hive ... 88
- VIII. Requisites of a Complete Hive ... 95
- IX. Natural Swarming, and Hiving of Swarms ... 109
- X. Artificial Swarming ... 143
- XI. Loss of the Queen ... 213
- XII. The Bee-Moth, and other Enemies of bees. –Diseases of Bees ... 228
- XIII. Robbing, and how Prevented ... 261
- XIV. Directions for Feeding Bees ... 267
- XV. The Apiary. —Procuring Bees to Stock it. —Transferring Bees from Common to Movable-Comb Hives ... 279
- XVI. Honey ... 285
- XVII. Bee-Pasturage. –Over-Stocking ... 292

TABLE OF CONTENTS

Chapter.	Page.
XVIII. The anger of Bees. —Remedies for their Stings	308
XIX. The Italian Honey-Bee	318
XX. Size, Shape, and Materials for Hives.–Observing –Hives	329
XXI. Wintering Bees	335
XXII. Bee-Keeper's Calendar. –Bee-Keeper's Axioms	362
Explanation of Wood-Cuts of Movable-Comb Hives, with Bills of Stock for making them	371
Copious Alphabetical Index	391

LIST OF PLATES

PAGE

Frontispiece
Movable-Comb Hive, with full
   glass Arrangement 12
Plate   I 20
    "   II 24
    "   III 28
    "   IV.. 36
    "   V... 44
    "   VI.. 48
    "   VII... 68
    "   VIII... 72
    "   IX 88
    "   X 96

PAGE

Plate XI...120
   " XII..128
   " XIII 144
   " XIV 168
   " XV 192
   " XVI..216
   " XVII...240
   " XVIII 264
   " XIX..288
   " XX 312
   " XXI..350
   " XXII...360
   " XXIII 368

EXPLANATION OF PLATES

PLATES I. to XI. Inclusive, show the various styles of Movable-comb
    Hives, and the Implements used in the Apiary. For explanation of these plates, see
    p. 371.

PLATE XII. –Figs. 31, 32. –Queen-Bee, of magnified and natural size. See p. 80.
    Figs. 33, 34. –Drone, of magnified and natural size. See p. 49.
    Figs. 35, 36. –Worker, of magnified and natural size. See p. 54.
    These Illustrations were copied (with some alternations) from *Bagster*.

PLATE XIII. Fig. 37. –Scales of wax, highly magnified. See p. 69.
    Fig. 38. –Abdomen of a worker-Bee, magnified, and showing the exuding scales of wax. See p. 69.
    Fig. 39. –Section of a Cell, magnified, and showing the usual position of the egg. See p. 44.
    Fig. 40. –Larvae of Bees, in various stages of development. See p. 44.
    Fig. 41. –Section of a Cell, magnified, and show larva. See p. 44.
    Fig. 42. –Worker-Larva, fully grown, and ready to spin its Cocoon. See p. 45.
    Fig. 43. –Worker-Nymph. See p. 45.

Fig. 49. –A Queen-Cell of natural size. See p. 62.
Fig. 50. –A Queen-Cell cut open, to show the unhatched queen. See p. 62.
Fig. 44. –Eggs of the Bee-Moth, of natural and magnified size. See. p. 234.
Fig. 45. –Larvae of the Bee-moth, fully grown. See p. 231.
Fig. 46. –Female Bee-Moth. See p. 229.
Fig. 59. –Female Bee-moth, with Ovipositor extruded, and eggs passing through it. See p. 230.
Fig. 60. –Male Bee-Moth. See p. 229.
Fig. 61. –Small Male Bee-Moth. See p. 229.
Fig. 62. –Head of Mexican Honey-Hornet, magnified. See p. 87.
Fig. 63. –Head of Honey-Bee, magnified. See p. 87.
Fig. 64, 65. –Jaws of Honey-Hornet and Honey-Bee magnified. See p. 87.
Some of these Illustrations were taken from Swammerdam, Reaumur, and Huber.

PLATE XIV. –For an explanation of this plate, which represents the different kinds of cells in the Honey-Comb, see p. 66.

PLATE XV. –For an explanation of Fig. 48, which represents Worker and Drone-Comb, of natural size, see p. 74.

Fig. 58. –A group of Queen Cells, drawn from a specimen found in the Author's hive. See p. 191.

PLATE XVI. –Fig. 51. –Proboscis of a Worker-Bee, highly magnified. See p. 56.

Fig. 68. – PLATE XIII., shows the Proboscis attached to the head.

Fig. 52. –Abdomen of a Worker-Bee, magnified.

PLATE XVII. –Fig. 53. –Sting of a Worker, highly magnified. See p. 56.

Fig. 54. –Honey-Bee, Intestines, Stomach, and Rectum of a Worker-Bee. See p. 56.

PLATE XVIII. –For an explanation of this plate, which represents the Ovaries (and adjacent parts) of a Queen-Bee, see p. 35.

PLATE XIX. –Fig. 56. –Cocoons spun by Larvae of the Bee-Moth. See p. 233.

PLATE XX. –Fig. 57. –Mass of Webs, Cocoons, and Excrements left in a hive destroyed by the Larvae of the Bee-Moth. See p. 235.

PLATE XXI. –Figs. 66, 67, 68, 69, and 70. –German method of Wintering Bees. See p. 348.

PLATE XXII. –Fig. 71 is the Frontispiece to the First Edition. See p. 331.

PLATE XXIII. –Shows the position in which a Frame is held when taken from the Movable-Comb Hive. –See p. 171

Movable-comb Hive with full Glass Arrangement

THE HIVE AND HONEY-BEE.

CHAPTER I.

FACTS CONNECTED WITH THE INVENTION OF THE MOVABLE-COMB BEE-HIVE.

PRACTICAL bee-keeping in this country is in a very depressed condition, being entirely neglected by the mass of those most favorably situated for its pursuit.[1] Notwithstanding the numerous hives which have been introduced, the ravages of the bee-moth [2] have increased, and success is becoming more and more precarious. While multitudes have abandoned the pursuit in disgust, many even of the most experienced are beginning to suspect that all the so called "Improved Hives" are delusions or impostures; and that they must return to the simple box or hollow log, and "take up" their bees with sulphur in the old-fashioned way.[3]

In the present state of public opinion, it requires no little confidence to introduce another patent hive, and a new system of management; but believing that a new era in bee-keeping has arrived, I invite the attention of Apiarians to the perusal of this Manual, trusting that it will convince them that there is a better way than any with which they have yet become acquainted. They will here find a clear explanation of many hitherto mysterious points in the physiology of the honey-

---

[1] *In the 1850's about 85% of the people were rural, and thus Langstroth considered them "favorably situated."*
[2] *Langstroth was very concerned about the bee-moth (wax-moth). His patent for the movable-comb hive remarked on how well the hive would control the pest.*
[3] *Since early times, beekeepers each year robbed about half of their hives by killing the colony with sulfur fumes. The following year the half that was not killed would swarm and the beekeeper would fill his empty skeps, or box hives, with these swarms. At the end of that season the process would be repeated with the alternate hives being killed, thus completing the cycle. This was in essence a robbing scheme, and not very much different than what the ancient cave dwellers did to secure their honey.*

bee, together with much valuable information never before communicated to the public.

It is now more than twenty years since I turned my attention to the keeping of bees. The state of my health [4] of late years having compelled me to live much in the open air, I have devoted a large portion of my time to a minute investigation of their habits, as well as to a series of careful experiments in the construction and management of hives.

Very early in my Apiarian studies I constructed a hive on the plan of the celebrated Huber,[5] and by verifying some of his most valuable discoveries became convinced that the prejudices existing against him were entirely unfounded. Believing that his discoveries laid the foundation or a more profitable system of bee-keeping, I began to experiment with hives of various construction.

Though the result of these investigations fell far short of my expectations, some of these hives now contain vigorous stocks fourteen years old, which without feeding have endured all the vicissitudes of some of the worst seasons [6] ever known for bees.

While I felt confident that my hive possessed valuable peculiarities, I still found myself unable to remedy many of the perplexing casualties to which bee-keeping is liable; and became convinced that no hive could do this, unless it gave the *complete control of the combs*, so that any or all of them might be removed at pleasure. The use of the Huber hive had satisfied me, that with proper precautions the combs might be removed without enraging the bees, and that these insects were capable of being tamed to a surprising degree.[7] Without a

---

[4] *Langstroth had several occurrences of bad health; see Florence Naile's book,* "The Life of Langstroth."

[5] *François Huber was a celebrated naturalist in France of the 1800's. His book on bees is regarded as one of the beginnings of scientific apiculture. Langstroth probably read the English translation of 1840,* "Observations on the Natural History of Bees."

[6] *It is hard to document the weather of the Eastern U.S. in the 1840-1860's. It may be that Langstroth was a typically pessimistic beekeeper as to honey producing seasons.*

[7] *I suspect that during this period Langstroth developed management skills, both in handling bees and also of selecting bees that were gentler. When a beekeeper did more than take up the bees, then a more docile bee was needed. It also should be noted that when a bee colony is completely dismantled they often sting less, as the colony is at a total loss of where to start defending the hive.*

knowledge of these facts I should have regarded a hive permitting the removal of the combs, as quite too dangerous for practical use. At first, I used movable slats or bars placed on rabbets in the front and back of the hive. The bees began their combs upon these bars, and then fastened them to the sides of the hive. By severing these attachments, the combs could be removed adhering to the bars. There was nothing new in the use of such bars the invention being probably a hundred years old [8] and the chief peculiarity in my hive was the facility with which they could be removed without enraging the bees, and their combination with my improved mode of obtaining the surplus honey.

With hives of this construction, I experimented on a larger scale than ever, and soon arrived at very important results. I could dispense entirely with natural swarming, and yet multiply colonies with greater rapidity and certainty than by the common methods. All feeble colonies could be strengthened, and those which had lost their queen furnished with the means of obtaining another.[9] If I suspected that any thing was wrong with a hive, I could quickly ascertain its true condition, and apply the proper remedies. In short, I felt satisfied that bee-keeping could be made highly profitable, and as much a matter of certainty, as most branches of rural economy.

One thing, however, was still wanting. The *cutting* of the combs from their attachments to the sides of the hive was attended with much loss of time both to myself and the bees. This led me to invent a method by which the combs were attached to MOVABLE FRAMES, so suspended in the hives as to touch neither the top, bottom, nor sides.[10] By this device the combs could be removed at pleasure, without any cutting, and speedily transferred to another hive. After exper-

---

[8] *The invention of such a slatted or top-bar hive probably dates to some very early Greek hives. The modern development of this technique of a top-bar hive is often used for African bees. The modern top-bar hive adds sloping sides to the hive to reduce, or eliminate, the fastening of the comb to the wall of the hive.*
[9] *This is truly the miracle of the movable-comb hive! While beekeepers are aided in the removal of honey from their hives, it is the ability to manage, divide or re-queen their hives that has lead to modern beekeeping.*
[10] *This sentence sums up the entirety and beauty of the modern hive that Langstroth first described in his diary on October 30, 1851, and then patented in 1852.*

imenting largely with hives of this construction, I find that they fully answer the ends proposed in their invention.

In the Summer of 1851 I ascertained that bees could be made to work in glass hives, exposed to the full light of day. This discovery procured me the pleasure of an acquaintance with Rev. Dr. Berg, then pastor of a Reformed Dutch Church, in Philadelphia. From him I first learned that a Prussian clergyman, of the name of Dzierzon[*] was attracting the attention of crowned heads by his discoveries in the management of bees. Before he communicated to me the particulars of these discoveries, I explained to Dr. Berg my own system and showed him my hive. He expressed great astonishment at the wonderful similarity in our methods of management, neither of us having any knowledge of the labors of the other.

Our hives he found to differ in some very important respects. In Dzierzon's hive, the combs not being attached to movable frames but to bars cannot be removed without cutting. In my hive, any comb may be taken out without removing the others; whereas in the Dzierzon hive, it is often necessary to cut and remove many combs to get access to a particular one; thus if the tenth from the end is to be removed, nine must be taken out. The German hive does not furnish the surplus honey in a form the most salable in our markets, or admitting of safe transportation in the comb. Notwithstanding these disadvantages, it has achieved a great triumph in Germany, and given a new impulse to the cultivation of bees. The following letter from Samuel Wagner,[11] Esq., Cashier of The Bank of York, in York, Pennsylvania, will show the results obtained in Germany by the new system of management, and his estimate of the superior value of my hive to those there in use.

---

[*] Pronounced Tseertsone. *(Dr. Jan Dzierzon was an excellent scientific beekeeper. He was the first to write that drones come from unfertilized eggs. Dzierzon was also working on the development of a better bee hive. He was truly a leader in the international beekeeping community.)*

[11] *Samuel Wagner was a leader in U. S. beekeeping circles. He was responsible for starting the* American Bee Journal *in 1861, and was its first editor.*

"YORK, PA., Dec. 24, 1852.

"DEAR SIR :—The Dzierzon theory and the system of bee-management based thereon, were originally promulgated *hypothetically* in the 'Eichstadt Bienen-Zeitung,' or Bee-Journal, in 1845, and at once arrested my attention. Subsequently, when in 1848 at the instance of the Prussian Government, the Rev. Mr. Dzierzon published his 'Theory and Practice of Bee Culture,' I imported a copy which reached me in 1849, and which I translated prior to January, 1850. Before the translation was completed I received a visit from my friend the Rev. Dr. Berg, of Philadelphia, and in the course of conversation on bee-keeping, mentioned to him the Dzierzon theory and system as one which I regarded as new and very superior, though I had had no opportunity for testing it practically. In February following, when in Philadelphia, I left with him the translation in manuscript— up to which period I doubt whether any other person in this country had any knowledge of the Dzierzon theory; except to Dr. Berg, I had never mentioned it to any one save in very general terms.

"In September 1851, Dr. Berg again visited York, and stated to me your investigations, discoveries and inventions. From the account Dr. Berg gave me, I felt assured that you had devised substantially the *same system* [12] as that so successfully pursued by Mr. Dzierzon; but how far *your hive* resembled his I was unable to judge from description alone. I inferred, however, several points of difference. The coincidence as to system, and the principles on which it was evidently founded, struck me as exceedingly singular and interesting, because I felt confident that you had no more knowledge of Mr. Dzierzon and his labors, before Dr. Berg mentioned him and his book to you, than Mr. Dzierzon had of you. These circumstances made me very anxious to examine your hives, and induced me to visit your Apiary in the vil-

---

[12] *It seems that apiculturists throughout the world had all been moving to the same goal, that is, to be able to manipulate the combs of a hive without cutting out every comb or killing the bees. That similar systems developed is not surprising. The fact is that these experimenters were all reading the same books and articles, as well as the fact that some of them were corresponding with each other, all of which contributed to the development of similar systems of management. Like most inventions or discoveries, the invention of the movable-comb hive would have come along sooner or later. Langstroth was just the right person at the right time.*

lage of West Philadelphia, last August. In the absence of the keeper I took the liberty to explore the premises thoroughly, opening and inspecting a number of the hives and noticing the internal arrangement of the parts. The result was, that I came away converted that though your system was based on the same principles as Dzierzon's, your hive was almost totally different from his both in construction and arrangement; and that while the same objects *substantially* are attained by each, your hive is more simple, more convenient, and much better adapted for general introduction and use, since the mode of using it can be more easily taught. Of its ultimate and triumphant success I have no doubt. I sincerely believe that when it comes under the notice of Mr. Dzierzon, he will himself prefer it to his own. It in fact combines all the good properties which a hive ought to possess, while it is free from the complication, clumsiness, *vain whims* and decidedly objectionable features which characterizes most of the inventions which profess to be at all superior to the simple box, or the common chamber hive.

"You may certainly claim equal credit with Dzierzon for originality in observation and discovery in the natural history of the honey-bee, and for success in deducing principles and devising a most valuable system of management from observed facts. But in invention, as far as neatness, compactness, and adaptation of means to ends are concerned, the sturdy German must yield the palm to you.

"I send you herewith some interesting statements respecting Dzierzon, and the estimate in which his system is held in Germany.

Very truly yours,
Samuel Wagner

Rev. L. L. Lang troth"

The following are the statements to which Mr. Wagner refers:

"As the best test of the value of Mr. Dzierzon's system as the results which have been made to flow from it, a brief account of its rise and progress may be found interesting. In 1835, he commenced beekeeping in the common way with twelve colonies, and after various mishaps which taught him the defects of the common hives and the old mode of management, his stock was so reduced, that, in 1838, he had virtually to begin anew. At this period he contrived his improved hive, in its ruder form, which gave him the command over all the combs, and he began to experiment on the theory which observation and study had enabled him to devise. Thenceforward his progress was as rapid, as his success was complete and triumphant. Though he met with frequent reverses, about seventy colonies having been stolen from him, sixty destroyed by fire, and twenty-four by a flood, yet, in 1846, his stock had increased to three hundred and sixty colonies, and he realized from them that year six thousand pounds of honey,[13] besides several hundredweight of wax. At the same time, most of the cultivators in his vicinity who pursued the common methods had fewer hives than they had when he commenced.

"In the year 1848, a fatal pestilence, known by the name of foul brood,[14] prevailed among his bees, and destroyed nearly all his colonies before it could be subdued, only about ten having escaped the malady which attacked alike the old stocks and his artificial swarms. He estimates his entire loss that year at over five hundred colonies. Nevertheless, he succeeded so well in multiplying by artificial swarms, the few that remained healthy,[15] that, in the Fall of 1851, his

---

[13] *It is sometimes hard to realize how far beekeeping has come in 150 years. We certainly would not consider 16 pounds of honey per colony as a source of pride.*
[14] *It wasn't until the 20th century that the foulbrood diseases were separated into American foulbrood and European foulbrood. However, the total loss of colonies would point to the cause as AFB.*
[15] *You have to wonder how many times bees may have been selected for resistance to disease. Then we forget about the problem until the resistance characteristics have been diluted to the point the disease again strikes. Or in recent years we have relied upon antibiotics to control the disease. In earlier times the selection for disease resistance would have been one of the major roles of the beekeeper.*

stock consisted of nearly four hundred colonies. He must therefore have multiplied his stocks more than three-fold each year.

"The highly prosperous condition of his colonies attested by the Report of the Secretary of the Annual Apiarian Convention, which met in his vicinity last Spring. This Convention, the fourth which has been held, consisted of one hundred and twelve experienced and enthusiastic bee-keepers from various districts of Germany and neighboring countries, and among them were some which when they assembled, were strong opposers of his system.

"They visited and personally examined the Apiaries of Mr. Dzierzon. The report speaks in the very highest terms of his success, and of the manifest prosperity of his system of management. He exhibited and satisfactorily explained to his visitors his practice and principles and they remarked with astonishment the *singular docility* of his bees, and the thorough control to which they were subjected. After a full detail of the proceedings, the Secretary goes on to say:

"Now that I have seen Dzierzon's method practically demonstrated, I must admit that it is attended with fewer difficulties than I had supposed. With his hive and system of management, it would seem that bees become at once more docile than they are in other cases. I consider his system the simplest and best means of elevating bee-culture to a profitable pursuit, and of spreading it far and wide over the land; especially as it is adapted to district in which the bees do not readily and regularly swarm. His eminent success in re-establishing his stock after suffering so heavily from the devastating pestilence; in short the recuperative power of the system, demonstrates conclusively that it furnishes the best, perhaps the only means of re-instating bee-culture to a profitable branch of rural economy.

Fig. 1. PLATE I.

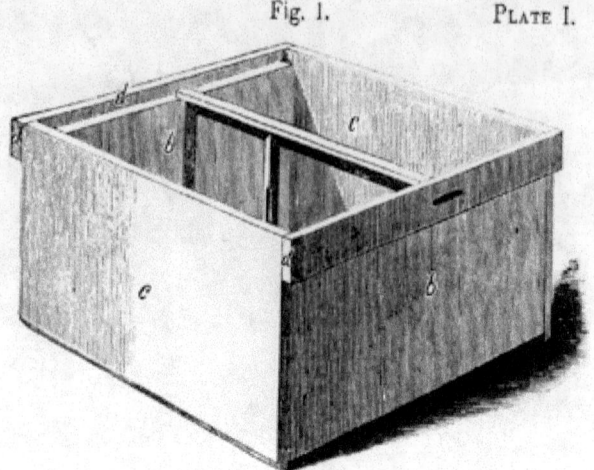

Fig. 2.

Fig. 3.

"'Dzierzon modestly disclaimed the idea of having attained perfection in his hive. He dwelt rather upon the truth and importance of his theory and system of management.'

"From the Leipzig Illustrated Almanac—Report on Agriculture for 1846:

"'Bee-culture is no longer regarded as of any importance in rural economy'"

"From the same for 1851 *and* 1853:

"'Since Dzierzon's system has been made known, and entire revolution in bee-culture has been produced. A new era has been created for it, and bee-keepers are turning their attention to it with renewed zeal. The merits of his discoveries are appreciated by the Government, and they are recommending his system as worthy the attention of teachers of common schools.

"Mr. Dzierzon resides in a poor, sandy district of Lower Silesia, which according to the common notions of Apiarians is unfavorable to bee-culture. Yet in spite of this and various other mishaps, he has succeeded in realizing nine hundred dollars as the product of his bees in one season!

"By his mode of management, his bees yield even in the poorest years from 10 to 15 per cent, on the capital invested; and where the colonies are produced by the Apiarian's own skill and labor, the cost him only about one-fourth the price at which they are usually valued. In ordinary seasons, the profit amounts to from 30 to 50 per cent., and in very favorable seasons from 80 to 100 per cent."

In communicating these facts to the public, I take an honest pride in establishing my claim to having matured by my own independent discoveries,[16] the system of bee-culture which has excited so much interest in Germany; I desire also to have the testimony to the merits of my hive, of Mr. Wagner, who is extensively known as an able German scholar. He has taken all the numbers of the Bee-Journal, which has

---

[16] *This edition was written seven or eight years after his discovery of the movable-comb hive and I suspect he was already having difficulties with people not recognizing his patent rights.*

been published monthly for more than nineteen years, in Germany; and he is undoubtedly more familiar than any other man in this country with the state of Apiarian culture abroad.

I wish, also, to show that the importance which I attach to my system of management, is amply justified by the success of those who, by the same system, even with inferior hives, have attained results which to common bee-keepers seem almost incredible. Inventors are prone to form exaggerated estimates of the values of their labors, and the public has been so often deluded by patent hives which have utterly failed to answer their professed objects, that they can scarcely be blamed for rejecting every new one as unworthy of confidence.

An American Bee Journal,[17] properly conducted, would have great influence in disseminating information, awakening enthusiasm, and guarding the public against the miserable impositions to which it has so long been subjected. Three such journals have been published monthly, in Germany; and their circulation has widely disseminated those principles which must constitute the foundation of any enlightened and profitable system of bee-culture.

While many of the principal facts in the physiology of the honey-bee were long ago discovered, it has unfortunately happened that some of the most important have been the most widely discredited. In themselves, they are so wonderful, and to those who have not witnessed them, often so incredible, that it is not strange that they have been rejected as fanciful conceits or bare-faced inventions.

---

[17] *It is interesting to note that Langstroth very possibly gave the name to the bee journal that his friend Samuel Wagner started, and edited, two years after this was written—*American Bee Journal.

For more than half a century, hives have been in use containing only one comb enclosed on both sides by glass. These hives are darkened by shutters, and when opened the queen is as much exposed to observation as the other bees. I have discovered that, with proper precautions, colonies can be made to work in observing-hives exposed continually to the full light of day; so that observations may be made at all times, without interrupting by any *sudden* admission of light the ordinary operations of the bees. In such hives, many intelligent persons from various States in the Union have seen the queen-bee depositing her eggs in the cells, while surrounded by an affectionate circle of her devoted children. They have also witnessed with astonishment and delight, all the mysterious steps in the process of raising queens from eggs, which with the ordinary development would have produced only the common bees. Often for more than three months, there has not been a day in my Apiary in which some colonies were not engaged in rearing new queens to supply the place of those taken from them; and I have had the pleasure of exhibiting these facts to beekeepers who never before felt willing to credit them.[18]

As all my hives are made so that each comb can be taken out and examined at pleasure, those who use them can obtain all the information which they need without taking anything upon trust. May I be permitted to express the hope, that the time is now at hand when the number of practical observers will be so multiplied, and the principles of bee-keeping so thoroughly understood, that ignorant and designing men will not be able to impose their conceits and falsehoods upon the public, by depreciating the discoveries of those who have devoted years of observation to the advancement of Apiarian knowledge!

---

[18] *Rev. Langstroth is comparing his movable-comb hive to that of an observation hive as to the ability to see such bee biology as rearing of queens. The comparison to an observation hive is quite valid since it is the ability of observing of the whole colony that makes the movable-comb hive the marvel that it is.*

Preface to Chapter II

This chapter is interesting in its location within the book. Most modern beekeeping books would cover the fact that bees sting much later in the book. Though at the time that this book was written bees were considered very vicious, and in some respects by modern standards they were as the bees were Apis mellifera mellifera, *the German black bee, that are quite defensive of their hive. Langstroth was anxious to allay most fears. First, he was trying to preach the "gospel" of beekeeping as an avid practitioner. Secondly, he also was pushing an invention, which the sales depended upon the acceptance of bees as capable of being tamed. Later in the book (Chapter XVIII) he covers what to do about stinging and the remedies for stings.*

Here, he also talks about bees and beekeeping as a practical enterprise and not just a box of wild things. Mainly bees could be handled and "tamed" because of a remarkable new invention, the movable-comb hive.

The movable-comb hive does allow many manipulations without a lot of disturbance to the bees. I think that Langstroth recognized this fact and was trying to convey that feeling within his title for the chapter —"The Honey-Bee Capable of Being Tamed." Beekeeping prior to the movable-comb hive would involve cutting combs or other major manipulations that could arouse the bees into stinging. Now you could remove covers (or inner covers) without much change within the hive, and thus the bees would not be as defensive. Today we take the existence of the movable-comb hive so much for granted that we don't even think of what beekeeping—and stinging—was like before Langstroth's invention.

CHAPTER II.

THE HONEY-BEE CAPABLE OF BEING TAMED

If the bee had not such a formidable weapon both of offense and defense, multitudes who now fear it might easily be induced to enter upon its cultivation. As my system of management takes the greatest possible liberties with this irascible insect, I deem it important to show in the very outset how all necessary operations may be performed without incurring any serious risk of exciting its anger.

Many persons have been unable to suppress their astonishment, as they have seen me opening hive after hive, removing the combs covered with bees, and shaking them off in front of the hives; forming new swarms, exhibiting the queen, transferring the bees with all their stores to another hive; and in short, dealing with them as if they were as harmless as flies. I have sometimes been asked, if the hives I was opening had not been subjected to a long course of training; when they contained swarms which had been brought only the day before to my Apiary.[1]

I shall, in this chapter, anticipate some principles in the natural history of the bee, to convince my readers that any one favorably situated may enjoy the pleasure and profit of a pursuit which has been appropriately styled, "the poetry of rural economy," without being made too familiar with a sharp little weapon which can speedily convert all the poetry into very sorry prose.

It must be manifest to every reflecting mind, that the Creator intended the bee, as truly as the horse or the cow, for the comfort of man.[2] In the early ages of the world, and indeed until quite modern

---

[1] *Honey bees can be made accustomed to disturbance, since a fluttering flag in the apiary will get the colonies accustomed to movement. Regular inspections of a colony while habituating it to disturbance, and thus less stinging, will also reduce its total honey yield if the visits are too often. Those colonies that are used to raise queens or royal jelly are usually examined three or four times a week. These colonies are uniformly gentle. They usually have to be fed, but are gentle.*

[2] *Rev. L. L. Langstroth's early 19th century theological leanings and training are often reflected in his writings.*

PLATE II.

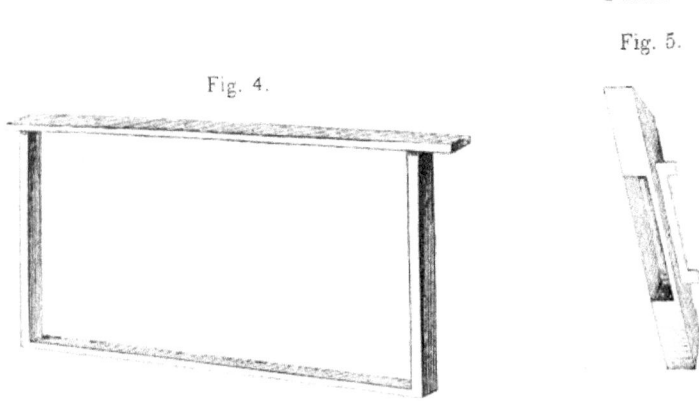

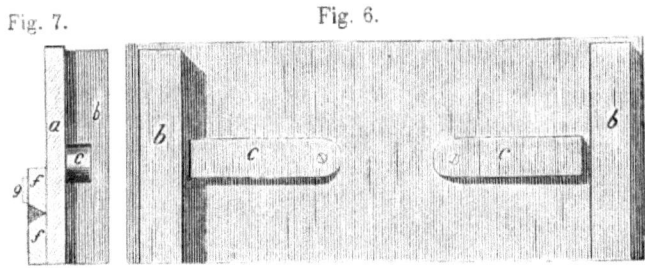

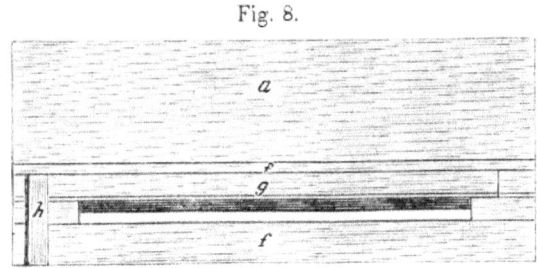

times, honey was almost the only natural sweet; and the promise of "a land flowing with milk and honey" had once a significance which it is difficult for us fully to realize.[3] The honey-bee, therefore, was created not merely to store up its delicious nectar for its own use, but with certain propensities, without which man could no more subject it to his control, than he could make a useful beast of burden of a lion or a tiger.

One of the peculiarities which constitutes the foundation of my system of management, and indeed of the possibility of domesticating at all so irascible, an insect, has never to my knowledge been clearly stated as a great and controlling principle. It may be thus expressed:

A honey-bee when filled with honey never volunteers an attack, but acts solely on the defensive.

This law of the honeyed tribe is so universal, that a stone might as soon be expected to rise into the air without any propelling power, as a bee well filled with honey to offer to sting, unless crushed or injured by some direct assault. The man who first attempted to hive a swarm of bees, must have been agreeably surprised at the ease with which he was able to accomplish the feat; for it is wisely ordered that bees, when intending to swarm, should fill their honey-bags to their utmost capacity. They are thus so peaceful that they can easily be secured by man,[4] besides having materials for commencing operations immediately in their new habitation, and being in no danger of starving if several stormy days should follow their emigration.

Bees issue from their hives in the most peaceable mood imaginable; and unless abused allow themselves to be treated with great familiarity. The hiving of them might always be conducted without risk, if there were not occasionally some improvident or unfortunate ones, who, coming forth without the soothing supply, are filled instead with the bitterest hate against any one daring to meddle with them.[5] Such

---

[3] *By the mid-1800's beet or cane sugar was well established as the general sweetener and thus the reduced importance of honey as the source of sugar.*
[4] *This trait is so universal that even the Africanized honey bee is extremely docile during swarming.*
[5] *While some swarms may be more inclined to sting than others, it is usually those swarms that have been out for several days without finding a new home that are likely to sting. This fact may underscore Langstroth's axiom that a bee filled with honey seldom stings.*

thriftless radicals are always to be dreaded, for they must vent their spleen on something, even though they perish in the act.

If a whole colony on sallying forth possessed such a ferocious spirit, no one could hive them unless clad in a coat of mail, bee-proof; and not even then, until all the windows of his house were closed, his domestic animals bestowed in some place of safety, and sentinels posted at suitable stations to warn all comers to keep at a safe distance. In short, if the propensity to be exceedingly good-natured after a hearty meal had not been given to the bee, it could never have been domesticated, and our honey would still be procured from the clefts of rocks or the hollows of trees.

A second peculiarity in the nature of the bee, of which we may avail ourselves with great success, may be thus stated:

Bees cannot under any circumstances resist the temptation to fill themselves with liquid sweets.[6]

It would be quite as difficult for them to do this, as for an inveterate miser to despise a golden shower of double eagles falling at his feet and soliciting his appropriation. If, then, when we wish to perform any operation which might provoke them, we can contrive to call their attention to a treat of flowing sweets, we may be sure that under its genial influence they will allow us to do what we please, so long as we do not hurt them.

Special care should be used not to handle them roughly, for they will never allow themselves to he pinched or hurt without thrusting out their sting to resent the indignity. If, as soon as a hive is opened, the exposed bees are gently sprinkled with water sweetened with sugar, they will help themselves with great eagerness, and in a few moments will he perfectly under control. The truth is, that bees thus managed are always glad to see visitors, for they expect at every call to receive an acceptable peace offering. The greatest objection to the use of sweetened water is, the greediness of bees from other hives, who, when there is any scarcity of honey in the fields, will often sur-

---

[6] *I think the behavior of bees to imbibe liquid honey is more a response to the threat of robbing from other bees and wasps. The bees store it in their honey stomach so that it cannot be readily taken by other bees. That is not to say that bees do not recognize sweet liquids, it is just not the reason for this behavior of removing any open honey within the hive.*

round the Apiarian as soon as he presents himself with his watering-pot, and attempt to force their way into any hive he may open, to steal if possible a portion of its treasures.[7]

A third peculiarity in the nature of bees gives an almost unlimited control over them, and may be expressed as follows:

Bees when frightened immediately begin to fill themselves with honey from their combs.

If the Apiarian only succeeds in frightening his little subjects, he can make them as peaceable as though they-were incapable of stinging. By the use of a little smoke from decayed wood,[*] the largest and most fiery colony may at once be brought into complete subjection. As soon as the smoke is blown among them, they retreat from before it, raising a subdued or terrified note; and, seeming to imagine that their honey is to be taken from them, they cram their honey-bags to their utmost capacity. They act either as if aware that only what they can lodge in this inside pocket is safe, or, as if expecting to be driven away from their stores, they are determined to start with a full supply of provisions for the way. The same result may be obtained by shutting them up in their hive and drumming upon it for a short time.[8] The various processes, however, for inducing bees to fill themselves with honey, are more fully explained in the chapter on Artificial Swarming.

By the methods above described, I can superintend a large Apiary, performing every operation necessary for pleasure or profit, without as much risk of being stung as must frequently be incurred in attempt-

---

[7] *Langstroth here recognizes robbing behavior. This behavior of stealing from other hives increases when nectar is no longer available in the field. Robbing can make the bees very mean and cause much discomfort to the beekeeper, and is one of the difficulties that a beekeeper can often get into without much warning.*

[*] Such wood is often called spunk, or touch-wood; it burns without any flame until consumed; and its smoke may easily be direct upon the bees, by the breath of the Apiarian.

(Smokers as we use them today were not yet invented, so the burning wood smoke had to be directed by blowing it onto the bees by the beekeeper.)

[8] *Old-time beekeepers used the drumming technique often to move bees out of an old box hive and into a modern hive. Beekeepers today often do not know how to use this trick. By regularly knocking (drumming) on the hive the bees will be induced to fill up with honey and move upward. If a new hive is placed over the old one and sufficient drumming is employed, the bees can be induced to move into a new box or super. Langstroth will describe this technique in full in the chapter on swarming.*

ing to manage a single hive in the ordinary way.

Let all your motions about your hives be gentle and slow. Accustom your bees to your presence: never crush or injure them, or breathe upon them in any operation;[9] acquaint yourself fully with the principles of management detailed in this treatise, and you will find that you have little more reason to dread the sting of a bee than the horns of a favorite cow, or the heels of your faithful horse.

Equipped with a bee-hat (Pl. XI., Figs. 25, 27) and India-rubber[10] gloves, even the most timid, by availing themselves of these principles, may open my hives and deal with their bees with a freedom astonishing to many of the oldest cultivators on the common plan: for in the management of the most extensive Apiary, no operation will ever be necessary, which, by exasperating a whole colony, impels them to assail with almost irresistible fury the person of the bee-keeper.

---

[9] *I recognize that my "breath* **can** *affect bees, and it is sometimes a useful tool. I will blow onto a cluster of bees on a frame to spread out the bees in case there might be a queen hiding under the large clump of workers.*
[10] *Heavy rubber gloves are quite good bee gloves as they can be cleaned very easily. Many bee inspectors use them as they can clean them between colonies and not be worried about accidentally transferring diseases from one colony to the next.*

PLATE III.

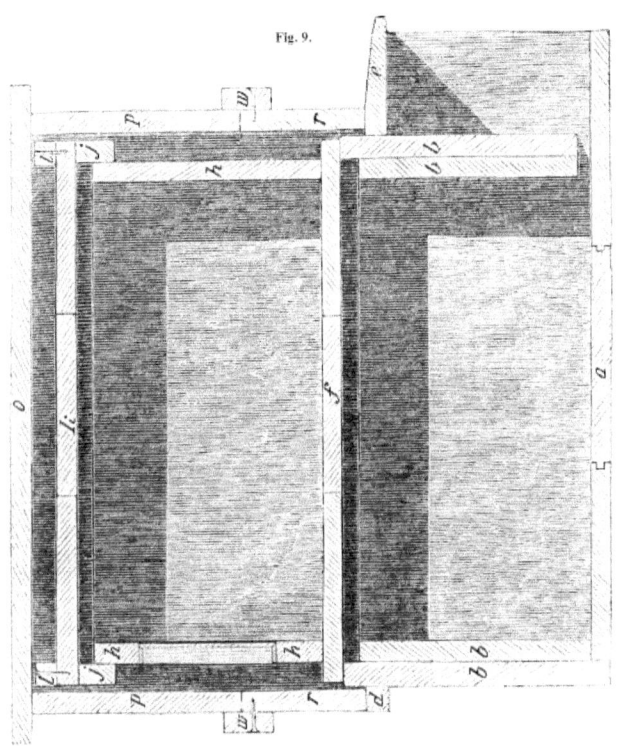

Fig. 9.

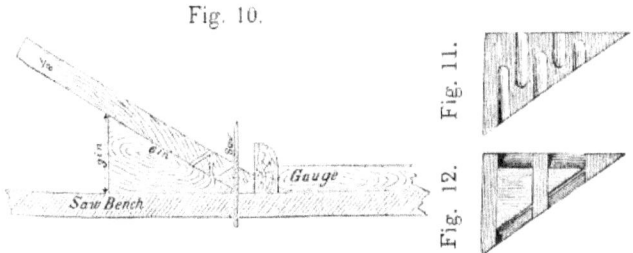

Fig. 10.   Fig. 11.   Fig. 12.

Preface to Chapter III

This is the one chapter that a person should read if they are just interested in the biology of the honey bee. It is the chapter that will give you some insights into what goes on in a hive so that you can use the information in beekeeping. It is in reading these writings that you realize that Langstroth was not only a very good observer he was also a learned man as well. Much of this material he gleaned from his reading, but then he put it all together very well.

There are a few facts about bees that we now know are different than have been described herein. Some of these facts we did not discover until just recently. Modern apicultural science has the advantage of many more measuring instruments with which to gather the information and to analyze the facts. Hopefully, I have been able to catch all of the technical errors that were described and have corrected them in the footnotes.

CHAPTER III.

THE QUEEN, OR MOTHER-BEE; THE DRONES; THE WORKERS; FACTS IN THEIR NATURAL HISTORY.

Honey-bees can flourish only when associated in large numbers, as in a colony. In a solitary state, a single bee is almost as helpless as a new-born child, being paralyzed by the chill of a cool Summer night.

If a strong colony preparing to swarm is examined, three kinds of bees will be found in the hive.

1st, One bee of peculiar shape, commonly called the *Queen-Bee*.

2d, Some hundreds and often thousands of large bees, called *Drones*.

3d, Many thousands of a smaller kind, called *Workers*, or common bees, such as are seen on the blossoms. Many of the cells will be found to contain honey and bee-bread; and vast numbers of eggs and immature workers and drones. A few cells of unusual size are devoted to the rearing of young queens. On Plate XII, the queen, drone, and worker are represented as magnified, and also of the natural size.

The *queen-bee* is the only *perfect female* in the hive, and all the eggs are laid by her. The *drones* are the *males*, and the *workers, females* whose ovaries, or "egg-bags," are so imperfectly developed that they are incapable of breeding; and which retain the instinct of females, only so far as to take care of the brood.

These facts have been demonstrated so repeatedly, that they are as well established as the most common laws in the breeding of our domestic animals. The knowledge of them in their most important bearings, is essential to all who would realize large profits from improved methods of rearing bees. Those who will not acquire the necessary information, if they keep bees at all, should manage them in the old-fashioned way, which demands the smallest amount of knowledge and skill.

I am well aware how difficult it is to reason with bee-keepers, who have been so often imposed upon, that they have no faith in statements made by any one interested in a patent hive; or who stigmatize all knowledge which does not square with their own, as mere "book knowledge" unworthy the attention of practical men.

If any such read this book, let me remind them that all my assertions may be put to the test. So long as the interior of a hive was to common observers a profound mystery, ignorant or designing men might assert what they pleased of what passed in its dark recesses; but now, when every comb can in a few moments be exposed to the full light of day, the man who publishes his own conceits for facts, will speedily earn the character both of a fool and an impostor.

The Queen-Bee, as she is the common mother of the whole colony, may very properly be called *the mother-bee*. She reigns most unquestionably by a divine right, for every good mother ought to be a queen in her own family. Her shape is widely different from that of the other bees. While she is not near so balky as a drone, her body is longer; and as it is considerably more tapering, or sugar-loaf in form than that of a worker, she has a somewhat wasp-like appearance. Her wings are much shorter in proportion than those of the drone, or worker; the under part of her body is of a golden color, and the upper part usually darker than that of the other bees. Her motions are generally slow and matronly, although she can, when she pleases, move with astonishing quickness. No colony can long exist without the presence of this all-important insect; but must as surely perish, as the body without the spirit must hasten to inevitable decay.

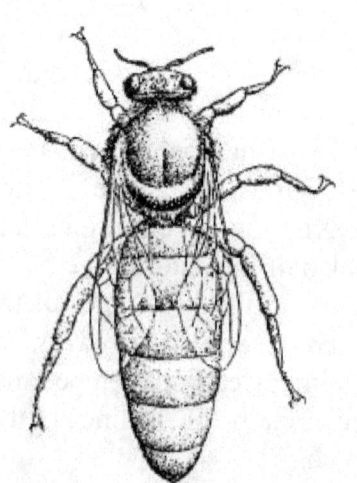

The queen is treated with the greatest respect and affection by the bees. A circle of her loving offspring constantly surrounds her,* testifying in various ways their dutiful regard; some gently embracing her with their antennae, others offering her honey from time to time, and all of them politely backing out of her way, to give her a clear path when she moves over the combs. If she is taken from them, the whole colony is thrown into a state of the most intense agitation as soon as they ascertain their loss;[1] all the labors of the hive are abandoned; the bees run wildly over the combs, and frequently rush from the hive in anxious search for their beloved mother. If they cannot find her, they return to their desolate home, and by their sorrowful tones reveal their deep sense of so deplorable a calamity. Their note at such times, more especially when they first realize their loss, is of a peculiarly mournful character; it sounds somewhat like a succession of wailings on the minor key,[2] and can no more be mistaken by an experienced bee-keeper, for their ordinary happy hum, than the piteous moanings of a sick child could be confounded by the anxious mother with its joyous crowings when overflowing with health and happiness.

I know that all this will appear to many much more like romance than sober reality; but, believing that it is a crime for any observer willfully to misstate or conceal important truths, I have determined, in writing this book to give facts, however wonderful, just as they are; confident that in due time they will be universally received and hoping that the many wonders in the economy of the honey-bee will not only excite a wider interest in its culture, but lead those who observe them to adore the wisdom of Him who gave them such admirable instincts.

---

* See the group of bees on the Title-Page.

[1] *In modern terms the loss of a queen, even for a few moments, means the loss of the "queen substance" or pheromone. This pheromone was the first of these communication substances to be chemically defined. It is produced primarily in the mandibular glands of the queen and distributed by the bees that feed and groom the queen—the "court" as seen on the title page.*

[2] *The sound that Langstroth is describing is caused by the fanning of many bees along with the exposure of the Nasonov gland on the tip of the abdomen. This gland produces the aggregation, or Nasonov, pheromone that might help a queen find her way back to the proper hive from a mating flight.*

The fertility of the queen-bee has been entirely under-estimated by most writers. During the height of the breeding season, she will often, under favorable circumstances lay from two to three thousand eggs a day! [3] In my observing-hives, I have seen her lay at the rate of six eggs a minute. The fecundity of the female of the white ant is, however, much greater than this, being at the rate of sixty eggs a minute; but her eggs are simply extruded from her body, and carried by the workers into suitable nurseries, while the queen-bee herself deposits her eggs in their appropriate cells.

It has been noticed that the queen-bee usually commences laying very early in the season, and always long before there are any males in the hive. How then, are her eggs impregnated? Francis Huber, of Geneva, by a long course of the most indefatigable observations, threw much light upon this subject. Before stating his discoveries, I must pay my humble tribute of gratitude and admiration to this wonderful man. It is mortifying to every naturalist and I might add, to every honest man acquainted with the facts, to hear such an Apiarian. as Huber, abused by the veriest novices and impostors; while others, who are indebted to his labors for nearly all that is of value in their works,

> "Damn with faint praise, assent with civil leer,
> And, without sneering, teach the rest to sneer."

Huber in early manhood lost the use of his eyes. His opponents imagine that to state this fact is to discredit all his observations. But to make their case still stronger they assert that his servant, Francis Burnens, by whose aid he conducted his experiments, was only an ignorant peasant. Now this so-called "ignorant peasant" was a man of strong native intellect, possessing the indefatigable energy and enthusiasm so indispensable to a good observer. He was a noble specimen of a self-made man, and rose to be the chief magistrate in the village

---

[3] *I feel that Langstroth is closer to the actual number of eggs laid in a day than some more modern estimates. While he may have arrived at the number by using some dubious means, the number is still correct. What is usually forgotten in research on number of eggs is that 7-10 percent die, and that there is an additional mortality of larvae and pupae. Thus, while she may lay 3,000 eggs in a day, not all make it to become adult bees.*

where he resided. Huber has paid an admirable tribute to his intelligence, fidelity, indomitable patience, energy and skill.*

It would be difficult to find in any language a better specimen of the inductive system of reasoning, than Huber's work on bees, and it might be studied as a model of the only way of investigating nature, so as to arrive at reliable results.

Huber was assisted in his researches, not only by Burnens, but by his own wife, to whom he was betrothed before the loss of his sight, and who nobly persisted in marrying him, notwithstanding his misfortune and the strenuous dissuasions of her friends. They lived longer than the ordinary term of human life in the enjoyment of great domestic happiness, and the amiable naturalist through her assiduous attentions scarcely felt the loss of his sight.

Milton is believed by many to have been a better poet in consequence of his blindness; and it is highly probable that Huber was a better Apiarian[4] from the same cause. His active yet reflective mind demanded constant employment; and he found in the study of the habits of the honey-bee, full scope for his powers. All the observations and experiments of his faithful assistants being daily reported, many inquiries and suggestions were made by him, which might not have suggested themselves, had he possessed the use of his eyes.

---

* A single fact will show the character of the man. It became necessary, in a certain experiment, to examine separately all the bees in two hives. "Burnens spent *eleven* days in performing this work, and during the whole time he scarcely allowed himself any relaxation but what the relief of his eyes required."

[4] *While we no longer use the word "Apiarian" to describe a beekeeper, the word would be proper as the noun derived from the genus of the honey bee,* Apis, *and as, similarly, we use the word Apiculture to describe the craft of beekeeping. It is just one of the words that never caught on, and has been dropped from our vocabulary.*

Few, like him, have such command of both time and money as to be able to prosecute on so grand a scale, for a series of years, the most costly experiments. Having repeatedly verified his most important observation, I take great delight in holding him up to my countrymen as the PRINCE OF APIARIANS.

To return to his discoveries on the impregnation of the queen-bee. By a long course of careful experiments, he ascertained that, like many other insects, she was fecundated in the open air and on the wing; and that the influence of this connection lasts for several years, and probably for life. He could, however, form no satisfactory conjecture how eggs were fertilized which were not yet developed in her ovaries. Years ago, the celebrated Dr. John Hunter, and others, supposed that there must be a permanent receptacle for the male sperm, opening into the oviduct. Dzierzon, who must be regarded as one of the ablest contributors of modern times to Apiarian science, maintains this opinion, and states that he has found such a receptacle filled with a fluid resembling the semen of the drones. He does not seem to have demonstrated his discoveries by any microscopic examinations.

In the winter of 1851-2, I submitted for scientific examination several queen-bees to Dr. Joseph Leidy, of Philadelphia, who has the highest reputation both at home and abroad, as a naturalist and microscopic anatomist. He found in making his dissections a small globular sac, about 1/38 of an inch in diameter, communicating with the oviduct, and filled with a whitish fluid; this fluid, when examined under the microscope, abounded in the spermatozoa which characterizes the seminal fluid.[5] A comparison of this substance, later in the season, with the semen of a drone, proved them to be exactly alike.

---

[5] *The drawing of the reproductive system of the queen, shown by Langstroth in this 3rd Edition, was taken from an original by Swammerdam. The reproductive system shown (next page) is a more modern dissection and drawing done by R.E. Snodgrass. The spermatheca is shown in the drawing. Either of the drawings shows the large number of ovarioles (egg tubes) that make up each ovary. There are approximately 300-350 in the two paired ovaries. I showed in my Ph.D. thesis that larger queens will have more ovarioles. Thus, it is important to have good queen rearing colonies to ensure that the young larvae are fed well to produce the largest queens.*

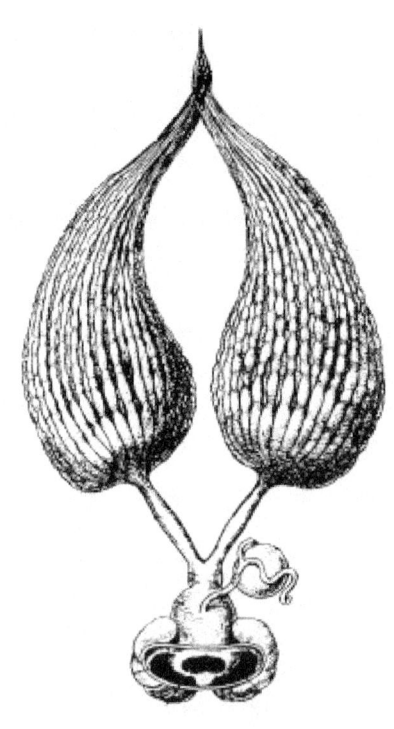

These examinations have settled, on the impregnable basis of demonstration, the mode in which the eggs of the queen are vivified. In descending the oviduct to be deposited in the cells, they pass by the month of this seminal sac, or "*spermatheca*," and receive a portion of its fertilizing contents. Small as it is, it contains sufficient to impregnate hundreds of thousands of eggs. In precisely the same way, the mother-wasps and hornets are fecundated. The females only of these insects survive the Winter, and often a single one begins the construction of a nest, in which at first only a few eggs are deposited. How could these eggs hatch, if the females had not been impregnated the previous season? Dissection proves that they have a spermatheca similar to that of the queen-bee. It never seems to have occurred to the opponents of Huber, that the existence of a permanently impregnated mother-wasp is quite as difficult to be accounted for, as the existence of a similarly impregnated queen-bee.

The celebrated Swammerdam, in his observations upon insects, made in the latter part of the seventeenth century, has given a highly magnified drawing of the ovaries of the queen-bee, a reduced copy of which I present (Plate XVIII), to my readers. The small globular sac (D), communicating with the oviduct (E), which he thought secreted a fluid for sticking the eggs to the base of the cells, is the seminal reservoir, or spermatheca. Anyone who will carefully dissect a queen-bee, may see this sac, even with the naked eye.[6] It will be seen that the

---

[6] *If you are willing to sacrifice a queen, or you find a queen recently killed, you can find the spermatheca and determine if the sac had sperm. With a pair of forceps, or tweezers, pull out the last segment of the abdomen of the queen. Carefully roll the contents out over your thumbnail and you should find the small, round sac. If it is freshly killed*

ovaries (G and H) are double, each consisting of an amazing number of ducts * filled with eggs, which gradually increase in size.† Huber, while experimenting to ascertain how the queen was fecundated, confined some young ones to their hive by contracting the entrances, so that they were more than three weeks old before they could go in search of the drones. To his amazement, the queens whose impregnation was thus retarded never laid any eggs but such as produced drones!

He tried this experiment repeatedly, but always with the same result. Bee-keepers, even from the time of Aristotle, had observed that all the brood in a hive were, occasionally drones. Before attempting to explain this astonishing fact, I must call the attention of the reader to another of the mysteries of the bee-hive.

It has already been stated, that the workers are proved by dissection to be females which under ordinary circumstances are barren. Occasionally, some of them appear to be sufficiently developed to be capable of laying eggs; but these eggs, like those of queens whose impregnations has been retarded, always produce drones! Sometimes when a colony which has lost its queen despairs of obtaining another, these drone-laying workers are exalted to her place, and treated with equal regard by the bees. Huber ascertained that fertile workers are usually reared in the neighborhood of the young queens, and thought that they received some particles of the peculiar food or jelly on which these

---

it will be silver white as it is surrounded with trachea filled with air. If you continue to roll the sac it will shed the trachea and expose the actual sac. If there are sperm within it will still be creamy white, but if there are no sperm it will be clear as water. If you have the equipment you can put the contents of a spermatheca into a haemocytometer and with the proper dilutions know exactly how many sperm were present.

* The ducts in this cut are represented as more numerous than those in Swammerdam's drawing.

† Since the first editions of this work were issued, I have ascertained that Posel describes the oviduct of the queen, the spermatheca and its contents, and the use of the latter in impregnating the passing egg. His work was published at Munich, in 1784. It seems also from his work (page 36), that before the investigations of Huber, Jansha, the bee-keeper royal of Maria Theresa, had discovered the fact that the young queens leave their hive in search of the drones.

PLATE IV.

Fig. 13.

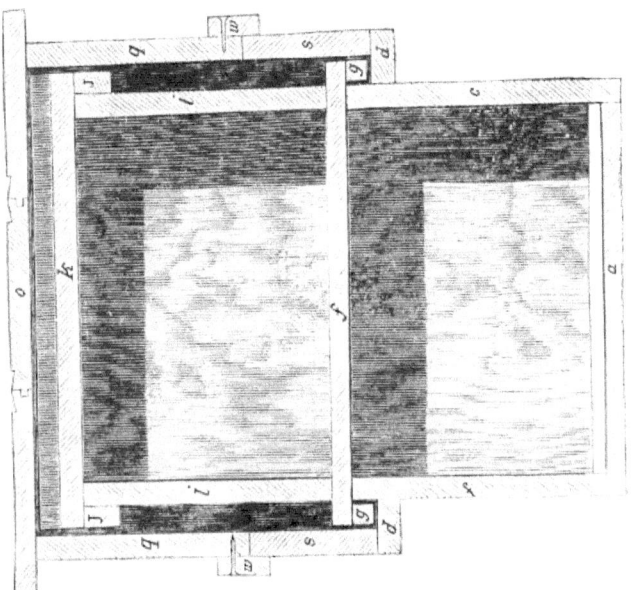

Fig. 14.

Fig. 15.

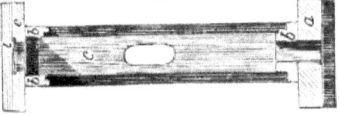

queens are fed. He did not pretend to account for the effect on the queen of retarded impregnation; and made no experiments on the fecundation of fertile workers.

Since the publication of Huber's work more than sixty years ago, no light has been shed upon the mysteries of drone-laying queens and workers, until quite recently. Dzierzon appears to have been the first to ascertain the truth on this subject; and his discovery must certainly be ranked among the most astonishing facts in all the range of animated nature. It seems at first view so absolutely incredible, that I should not dare mention it, if it were not supported by indubitable evidence, and if I had not determined to state all important and well-ascertained facts, however contrary to the prejudices of the ignorant and conceited. Dzierzon asserts, that all impregnated eggs produce females, either workers or queens; and all unimpregnated ones, males or drones! [7] He states that in several of his hives he found drone-laying queens, whose wings were so imperfect that they could not fly, and which on examination proved to be unfecundated. Hence, he concluded that the eggs laid by the queen-bee and fertile worker [8] had, from the previous impregnation of the egg from which they sprung, sufficient vitality to produce the drone, which is a less highly organized insect than the queen or worker. It had long been known that the queen deposits drone-eggs in the large or drone-cells, and worker-eggs in the small or worker-cells, and that she makes no mistakes. Dzierzon inferred, therefore, that there was some way in which she was able to decide the sex of the egg before it was laid, and that she must have such a control over the mouth of the seminal sac as to be able to extrude her eggs, allowing them at will to receive or not a portion of its fertilizing contents. In this way he thought she determined their sex, according to the size of the cell in which she laid them.

---

[7] *Parthenogenesis, or laying of unfertilized eggs, which produce males, is found throughout the order Hymenoptera. The Hymenoptera includes the bees, wasps and ants. Parthenogenesis is also found in some groups in other insect orders, but not as inclusive as in the Hymenoptera.*

[8] *The only "fertile workers" known would be the laying workers of such strains as the Cape bee. In these strains the polar bodies in the egg produced in the meiotic (chromosome reduction) division sometimes fuses with the egg nucleus producing a diploid egg that develops into a female, either queen or worker. I suspect that Langstroth is here using "fertile" to mean the ability to lay eggs—laying workers.*

My friend, Mr. Samuel Wagner, of York, Pennsylvania, has advanced a highly ingenious theory, which accounts for all the facts, without admitting that the queen has any special knowledge or will on the subject. He supposes that when she deposits her eggs in the worker-cells, her body is slightly compressed by their size, thus causing the eggs as they pass the spermatheca to receive its vivifying influence. On the contrary, when she is laying in drone-cells, as this compression cannot take place, the mouth of the spermatheca is kept closed and the eggs are necessarily unfecundated.[9]

In the Autumn of 1852, my assistant found a young queen whose progeny consisted entirely of drones. The colony had been formed by removing a few combs containing bees, brood, and eggs, from another hive, and had raised a new queen. Some eggs were found in one of the combs, and young bees were already emerging from the cells, all of which were drones. As there were none but worker-cells in the hive, they were reared in them, and not having space for full development, they were dwarfed in size, although the bees had pieced the cells to give more room to their occupants

I was not only surprised to find drones reared in worker cells, but equally so that a young queen, who at first lays only the eggs of workers, should be laying drone-eggs; and at once conjectured that this was a case of an unimpregnated drone-laying queen, sufficient time not having elapsed for her impregnation to be unnaturally retarded. All necessary precautions were taken to determine this point. The queen was removed from the hive, and although her wings appeared to be perfect, she could not fly.[10] It seemed probable, therefore, that she had never been able to leave the hive for impregnation.

---

[9] *I have heard this theory about how queens are able to lay either fertilized or unfertilized stated since I first started keeping bees more than 65 years ago. Most likely from the beekeepers reading this or other such accounts. The most compelling argument against this theory follows in Langstroth. Drones are usually reared in groups because the drone comb is usually found in large patches. Thus the queen does not turn off and on from one cell to the next and back again. The queen, however, responds to cell size when she inspects them prior to oviposition.*

[10] *Langstroth is not clear at this point whether he allowed the queen to cease laying a few days before he tried to see if she could fly. Even if she was laying drone eggs she still could not fly. Queens that are in colonies that are preparing to swarm stop laying eggs, or are forced to stop laying by the workers. This is the only way a mated queen is*

To settle the question beyond the possibility of doubt, I submitted this queen to Professor Leidy for microscopic examination. The following is an extract from his report. "The ovaries were filled with eggs, the poison-sac full of fluid; and the spermatheca distended with a perfectly colorless, transparent, viscid liquid, *without a trace of spermatozoa.*"

This examination demonstrates Dzierzon's theory that queens do not need impregnation to lay the eggs of males.

Considerable doubt seemed to rest on the accuracy of Dzierzon's statements on this subject, chiefly because of his having hazarded the unfortunate conjecture that the place of the poison-bag in the worker is occupied in the queen by the spermatheca. Now this is so completely contrary to fact (Pl. XVIII., A, D) that it was a natural inference that this acute and thoroughly honest observer made no microscopic dissections of the insects which he examined. I consider myself peculiarly fortunate, in having obtained the aid of a naturalist so celebrated for microscopic dissections as Dr. Leidy.

On examining this same colony a few days liter, I found satisfactory evidence that these drone-eggs were laid by the queen which had been removed. No fresh eggs had been deposited in the cells, and the bees on missing her had begun to build royal calls, to rear, if possible, another queen; this they would not have done, if a fertile worker had been present, by which the drone-eggs had been deposited.

---

able to leave and fly with a swarm. So it doesn't matter if a queen is laying drone eggs or worker eggs, if her ovaries are expanded with developing eggs she is unable to fly.

Another interesting fact proves that *all* the eggs laid by this queen were drone-eggs. Two of the royal cells were in a short time discontinued; while a third was sealed over in the usual way, to undergo its changes to a perfect queen. As the bees had only a drone-laying queen, whence came the female egg from which they were rearing a queen?

At first I imagined that they might have stolen it from another hive; but on opening this cell it contained only *a dead drone*! Huber had described a similar mistake made by some of his bees. At the base of this cell was on unusual quantity of the peculiar jelly fed to develop young queens. One might almost imagine that the bees had dosed the unfortunate drone to death; as though they hoped by such liberal feeding to produce a change in his sexual organization.

In the Summer of 1854, I found another drone-laying queen in my Apiary, with wings so *shrivelled* that she could not fly. I gave her successively to several queen-less colonies, in all of which she deposited only drone-eggs.

On the 14th of July, 1855, a queen in one of my observation-hives began to lay, when nine days old, a few eggs on the edges of the combs, instead of in the cells. She persisted in this for some days, until I transferred her to a colony which had been queenless for some weeks, hoping that she might, if unimpregnated, make an excursion from their hive to meet the drones. The observing-hive in which she was hatched was exposed to the full light of day; the entrance small, and difficult to find; and I had noticed on several occasions, that when the drones left the hive in the greatest numbers, the queen seemed unable to find her way out. At such times she manifested unusual excitement, and the whole colony were almost as much agitated as though they were swarming. After she had been in the second hive a short time, I found that she had laid a number of drone-eggs. They were deposited near the bottom and edge of the comb, in cells a little larger than the worker-size, and which the bees had begun to lengthen, to adapt them to the growth of their occupants. There was no other brood in the hive. On the 9th of August, I found the combs nearly filled with worker-brood, in a state considerably *less advanced* than the drones. Is there any reason to doubt that these drone-eggs were laid by the queen before, and the worker- eggs after, her impregnation?[11]

In Italy there is a variety of the honey-bee differing in size and color from the common kind. If a queen of this variety is crossed with the common drone her drone-progeny will be *Italian*, and her worker brood a cross between the two; thus showing the kind of drones she will produce has no dependence on the male by which she is fecundated.

It appears from recent discoveries in physiology, that to impregnate the ovum of an animal it is necessary that that spermatozoa should not simply come in contact with it, but actually enter into it through a small opening. In applying this discovery to bees, Prof. Siebold, of Germany, dissected a number of worker-eggs, and found in each from one to three spermatozoa; while he found none in dissecting drone-eggs.

Dr. Dönhoff, of Germany, in the Summer 1855, reared a worker-larvae from a drone-egg,[*] which be had artificially impregnated.

Aristotle noticed, more than 2,000 years ago, that the eggs which produce drones are like the worker-eggs. With the aid of powerful microscopes we are still unable to detect any difference in the size or appearance of the eggs of the queen.

---

[11] *Young laying queens often will lay drone eggs for a few days, or some lay egg patterns that look like laying workers for a few days. Eggs are laid in the wrong places in the cell, yet the queen has been mated. Oviposition is just a physiological adjustment that some accomplish earlier and easier than others. I think that all newly mated queens will lay a few drone eggs, whereas a little later in their development they would not lay drones under the same conditions.*

[*] I attempted to do this in 1852; but to my great disappointment, the bees removed or devoured all the eggs thus treated; owing as I then supposed to their unwillingness to raise workers in drone-cells. If some of the eggs just deposited in a piece of drone-comb are touched with a fine brush dipped in the diluted semen of drones and given to bees which have neither queen nor brood of any kind, I believe that queens, workers, and drones, may be raised from them.

(This idea has been tried, and reported several times in the scientific literature, with little or no success. It seems that the micropyle (egg opening) on the top of the egg is covered with a sealing substance before it is laid which prohibits the easy introduction of sperm after it is laid. It may be possible that with proper treatment to the egg and sperm that this system could be made to work.)

These facts taken in connection, appear to constitute a perfect demonstration that unfecundated queens are not only able to lay eggs, but that their eggs have sufficient vitality to produce drones.[12]

It seems to me probable, that after fecundation has been delayed for about three weeks, the organs of the queen-bee are in such a condition that it can no longer be effected; just as the parts of a flower, after a certain time, wither and shut up, and the plant becomes incapable of fruitification. Perhaps, after a certain time, the queen loses all desire to go in search of the male. The fertile drone-laying workers would seem to be physically incapable of impregnation.

There is something analogous to these wonders in the "*aphides*" or green lice, which infest plants. We have undoubted evidence that a fecundated female gives birth to other females, and they in turn to others, all of which without impregnation are able to bring forth young; until, after a number of generations, perfect males and females are produced, and the series starts anew!

However improbable it may appear that an unimpregnated egg can give birth to a living being, or that sex can depend on impregnation,[13] we are not at liberty to reject facts because we cannot comprehend the reasons of them. He who allows himself to be guilty of such folly, if he aims to be consistent, must eventually be plunged into the dreary gulf of atheism. Common sense, philosophy, and religion alike teach us to receive, with becoming reverence, all undoubted facts, whether in the natural or spiritual world; assured that however mysterious they

---

[12] *The term given for this ability to produce offspring from unfertilized eggs is parthenogenesis, which means "virgin birth."*

[13] *Modern studies have shown that it is not exactly the fact that drones are haploid that determines sex. There are sex alleles (genes) found on a pair of chromosomes. When these alleles are heterozygous (different alleles on each chromosome) the egg produces a female; when homozygous (alike genes for the sex allele) it produces a diploid male egg that 'dies'. Woyke in the 1960's clearly showed that you could produce diploid drones by hatching them in an incubator. What happens in nature when an egg is homozygous at the sex locus is that the workers destroy the diploid larva by eating them! I don't think anyone knows how the bees can determine that the larva has sex alleles that are homozygous rather than heterozygous.*

If queens are closely mated to brothers, or relatives, then half the young larvae that are eaten and thus the colony becomes very weak and usually dies—a consequence of inbreeding.

may appear to us, they are beautifully consistent in the sight of Him whose "understanding is infinite."

All the leading facts in the breeding of bees ought to be as familiar to the Apiarian, as the same class of facts as the rearing of his domestic animals.[*] A few crude and half-digested notions, however satisfactory to the old-fashioned bee-keeper, will no longer meet the wants of those who desire to conduct bee-culture on an extended and profitable system.

The extraordinary fertility of the queen-bee has already been noticed. The process of laying has been well described by the Rev. W. Dunbar, a Scotch Apiarian.

"When the queen is about to lay, she puts her head into a cell, and remains in that position for a second or two, to ascertain its fitness for the deposit she is about to make. She then withdraws her head, and curving her body downwards,[†] she inserts the lower part of it into the cell: in a few seconds she turns half round upon herself and withdraws, leaving an egg behind her. When she lays a considerable number, she does it equally on each side of the comb, those on the one side being as exactly opposite to those or the other as the relative position of the cells will admit. The effect of this is to produce the utmost possible concentration and economy of heat for developing the various changes of the brood!"

Here, as at every step in the economy of the bee, we behold, in the perfect adaptation of means to ends, a sagacity which seems scarcely inferior to that of man.

"The eggs of bees [‡] are of a lengthened, oval shape (Pl. XIII, Fig. 39), with a slight curvature, and of a bluish white color: being besmeared, at the time of laying, with a glutinous substance, they adhere

---

[*] "If it were possible," said an able German Apiarian, in 1846, "to ascertain the reproductive process of bees with as much certainty as that of our domestic animals, bee-culture might unquestionable be pursued with positive assurance of profit; and would assume a high rank among the various branches of rural economy."

(The fact that honey bee queens leave the hive and mate some distance from the parent hive has plagued bee breeding programs for many years. The development of instrumental insemination was an attempt to answer this problem. "Isolated" mating yards have also been used to help control the mating of various stocks.)

[†] She is thus sure to deposit the egg in the selected cell.
[‡] "Bevan on the Honey-Bee."

to the bases of the cells, and remain unchanged in figure or situation for three or four days; they are then hatched, the bottom of each cell presenting to view a small white worm. On its growing (Pl. XII, Figs. 40, 41), so as to touch the opposite angle of the cell, it coils itself up, to use the language of Swammerdam, like a dog when going to sleep; and floats in a whitish transparent fluid, which is deposited in the cells by the nursing-bees, and by which it is probably nourished; it becomes gradually enlarged in its dimensions, till the two extremities touch one another, and form a ring. In this state it is called a larva, or worm. So nicely do the bees calculate the quantity of food which will be required, that none remains in the cell when it is transformed to a nymph. It is the opinion of many eminent naturalists, that farina [14] does not constitute the sole food of the larva, but that it consists of a mixture of farina, honey, and water, partly digested in the stomachs of the nursing-bees.

"The larva having derived its support, in the manner above described, for four, five, or six days, according to the season,[15] continues to increase during that period, till it occupies the whole breadth, and neatly the length of the cell. The nursing-bees now seal over the cell with a light brown cover, externally more or less convex (the cap of a drone cell being more convex than that of a worker), and thus differing from that of a honey cell, which is paler and somewhat concave."

---

[14] *It was not until later that this brood food would be called either "worker jelly" or "royal jelly". Worker larvae receive not only this jelly but also some pollen after about the third day of larval life, whereas the queen appears to only receive royal jelly.*

[15] *Apparently Langstroth was influenced by the temperature dependence of other insects that develop faster when the weather is warmer. He did not know about the precise temperature regulation that occurs in the brood nest of honey bees. With the broodnest temperature normally constant at 94-95° F. the immature stages of the honey bee are quite constant, approximately 21 days for workers and 24 days for drones, and 16 days for a queen.*

Fig. 16. PLATE V.

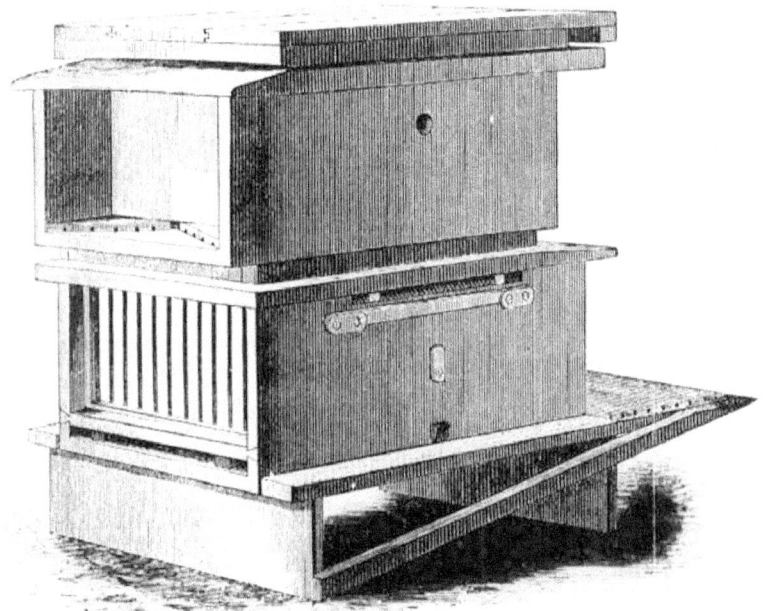

Fig. 17.

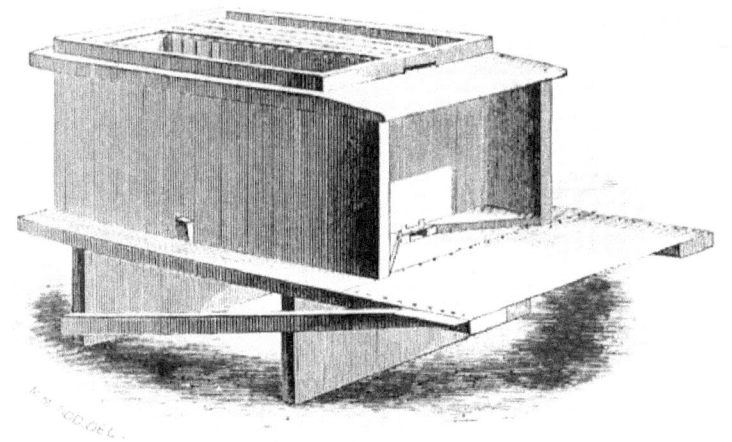

The cap of the brood-cell is not made of pure wax, but of a mixture of bee-bread and wax;[16] and appears under the microscope to be full of fine holes, to give air to the enclosed insect. From a texture and shape it is easily thrust off by the bee when mature, whereas if it consisted wholly of wax, the insect would either perish for lack of air, or be unable to free its way into the world. Both the material and shape of the lids which close the honey-cells are different: they are of pure wax, and thus air-tight, to prevent the honey of souring or candying in the cells; and are slightly concave, the better to resist the pressure of their contents.

To return to Bevan. "The larva is no sooner perfectly inclosed than it begins to line the cell by spinning round itself, after the manner of the silk-worm (Pl. XIII., Fig. 42), a whitish silky film, or cocoon, by which it is encased, as it were, in a pod. When it has undergone this change, it has usually borne the name of *nymph*, or *pupa*. It has now attained its full growth, and the large amount of nutriment which it has taken serves as a store for developing the perfect insect.

"The *working bee-nymph* spins its cocoon in thirty-six hours. After passing about three days in this state of preparation for a new existence, it gradually undergoes so great a change (Pl. XIII., Fig. 43) as not to wear a vestige of its previous form.

"When it has reached the twenty-first day of its existence, counting from the time the egg is laid, it comes forth a perfect winged insect. The cocoon is left behind, and forms a closely attached and exact lining to the cell in which it was spun; by this means the breeding cells become smaller, and their partitions stronger, the oftener they change their tenants; and may become so much diminished in size, as not to admit of the perfect development of full-sized bees.

---

[16] *The light brown color to the wax covering brood cells probably comes from the mixing of wax from the nearby darker brood combs. If a queen lays eggs in a newly drawn comb of beeswax, the cell capping will be lighter than those found on an old comb. There is little pollen within this wax.*

While the capping that covers honey is not as porous as that of a brood cell, air and moisture can move through the wax covering as well. The capping on honey may slow the movement of moisture into and out of honey, but it does not stop the process of uptake of water by honey. Combs of honey will ferment in regions of high humidity if they have been removed from the bees for a period of time. Honey picks up moisture at relative humidity above 50 percent, and loses water at relative humidity below 50 percent.

"Such are the respective stages of the working-bee; those of the royal bee are as follows: she passes three days in the egg, and is five a worm; the workers then close her cell, and she immediately begins spinning her cocoon, which occupies her twenty-four hours. On the tenth and eleventh days, and a part of the twelfth, as if exhausted by her labor, she remains in complete repose. Then she passes four days and a part of the fifth as a nymph. It is on the sixteenth day, therefore, that the perfect state of queen is attained.

"The drone passes three days in the egg, and six and a half as a worm, and changes into a perfect insect on the twenty-fourth or twenty-fifth day after the egg is laid.

"The development of each species likewise proceeds more slowly when the colonies are weak, or the air cool. Dr. Hunter has observed that the eggs, worms, and nymphs all require a heat at above 70° of Fahrenheit for their evolution.[17] Both drones and workers, on emerging from the cell, are at first gray, soft, and comparatively helpless, so that some time elapses before they take wing.

"The workers and drones spin *complete cocoons*, or inclose themselves on every side, while the royal larvae construct only *imperfect cocoons*, open behind, and enveloping only the head, thorax, and first ring of the abdomen; and Huber concludes, without any hesitation, that the final cause of this is, that they may he exposed to the mortal sting of the first hatched queen, whose instinct leads her instantly to seek the destruction of those who would soon become her rivals.

"If the royal larva spun complete cocoons, the sting of the queens seeking to destroy their rivals might be so entangled in their meshes that they could not be disengaged. 'Such,' says Huber, 'is the instinctive enmity of young queens to each other, that I have seen one of them, immediately on its emergence from the cell, rush to those of its sisters, and tear to pieces even the imperfect larvae. Hitherto, philosophers have claimed our admiration of nature for her care in preserv-

---

[17] *The broodnest temperature is kept fairly constant at 94-95° F. (35° C.) as long as brood is present. When brood rearing starts in early January the area maintained at this temperature may not be very large, within the winter cluster, but the bees will maintain this high temperature as long as there is brood present. This constant temperature will be maintained until brood rearing stops in September or October. (See note 15).*

ing, and multiplying the species. But from these facts we must now admire her precautions in exposing certain individuals to a mortal hazard.'"

The cocoon of the royal larvae is very much stronger and coarser than that of the drone or worker,— its texture considerably resembling that spun by the silk-worm. The young queen does not ordinarily leave her cell until she is quite mature; and as its great size allows the free exercise of her wings, she is usually capable of flying as soon as she quits it.[18] While still in her cell, she makes the fluttering and piping noises so familiar to observant bee-keepers.[19]

When the eggs of the queen are fully developed like those of the domestic hen, they must be extruded; but some Apiarians believe that she can regulate their development so that few or many are produced, according to the necessities of the colony. That this is true to a certain extent, seems highly probable; for if a queen is taken from a feeble colony, her abdomen seldom appears greatly distended; and yet if put in a strong one, she speedily becomes very prolific. Mr. Wagner says, "I conceive that she has the power of regulating or repressing the development of her eggs, so that *gradually* she can diminish the number maturing, and finally cease laying and remain inactive, as long as circumstances require. The old queen appears to qualify herself for accompanying a first swarm by repressing [*] the development of eggs, and as this is done at the most genial season of the year, it does not seem to be the result of atmospheric influence."

It is certain that when the weather is uncongenial, or the colony too feeble to maintain sufficient heat, fewer eggs are matured, just as unfavorable circumstances diminish the number of eggs laid by the hen; and when the weather is very cold, the queen stops laying in weak colonies.

In the latitude of Northern Massachusetts, I have found that the queen ordinarily ceases to lay some time in October; and begins

---

[18] *Queens still take a few days in which to harden their exoskeleton sufficiently to be able to fly.*

[19] *Most piping occurs outside of the cell, however they will also pipe within. The queen presses her abdomen to the surface and then by vibrating her wing bases very rapidly makes the sound.*

[*] Huber attributes her reduced size before swarming to a wrong cause.

again, in strong stocks, in the latter part of December.[20] On the 14th of January, 1857 (the previous month having been very cold, the thermometer sometimes sinking to 17° below zero), I examined three hives, and found that the central combs in two contained eggs and unsealed brood; there were a few cells with sealed brood in the third. Strong stocks even in the coldest climates usually contain some brood ten months in the year.

It is amusing to see how the supernumerary eggs of the queen are disposed of. If the workers are too few to take charge of all her eggs, or there is a deficiency of bee-bread to nourish the young; or if, for any reason, she does not judge best to deposit them in the cells, she stands upon a comb, and simply extrudes them from her oviduct, the workers devouring them as fast as they are laid. I have repeatedly witnessed in observing-hives the sagacity of the queen in thus economising her necessary work, instead of depositing her eggs in cells where they are not wanted. What a difference between her and the stupid hen, which so obstinately persists in sitting upon addled eggs, pieces of chalk, and on nothing at all.

The workers devour also all eggs which are dropped or deposited out of place by the queen; thus, even a tiny egg, instead of being wasted, is turned to good account.

One who carefully watches the habits of bees will often feel inclined to speak of his little favorites as having an intelligence almost if not quite akin to reason; and I have sometimes queried, whether the workers who are so fond of a tit-bit in the shape of a newly laid egg, ever experience a struggle between appetite and duty; so that they must practice self-denial to refrain from breakfasting on the eggs so temptingly deposited in the cells.

---

[20] *Langstroth observed what has escaped most beekeepers over the years. Honey bees start rearing brood soon after the day-length starts to increase; almost regardless of the outside temperature. The size of the colony will have an effect, as a small cluster would not have the ability to heat the area to the required 95° F. (35° C.) until later in the winter when the outside temperatures rise. If the colony is strong the brood rearing starts soon after the winter solstice. The bees know that Spring is coming!*

PLATE VI.

Fig. 18.

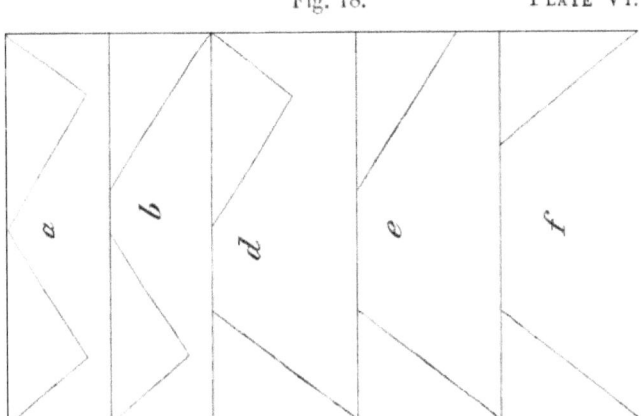

Fig. 72.

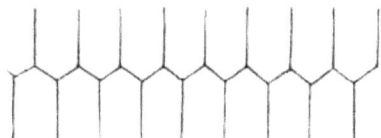

Fig. 19.

It is well known to breeders of poultry, that the fertility of a hen decreases with age, until at length she may become entirely barren. By the same law, the fecundity of the queen-bee ordinarily diminishes after she has entered her third year. An old queen sometimes ceases to lay worker-eggs; the contents of her spermatheca becoming exhausted, the eggs are no longer impregnated, and produce only drones.

The queen-bee usually dies of old age in her fourth year, although she has been known to live much longer. There is great advantage, therefore, in hives which allow her, when she has passed the period of her greatest fertility, to be easily removed.

Before proceeding farther in the natural history of the queen-bee, I shall describe more particularly the other inmates of the hive.

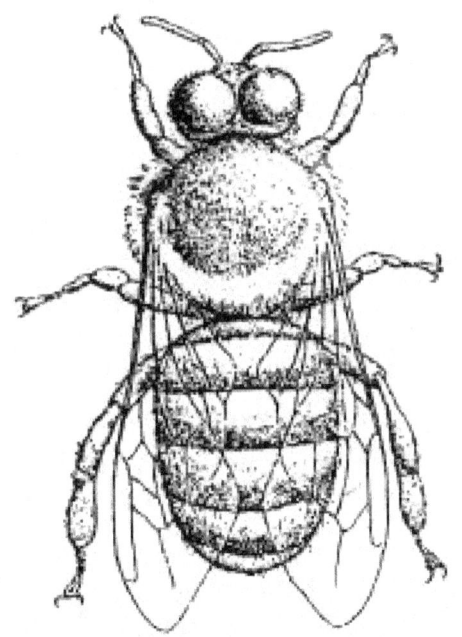

The DRONES are, unquestionable, the male bees; dissection proving that they have the appropriate organs of generation. They are much larger and stouter than either the queen or workers; although their bodies are not quite so long as that of the queen. They have no sting with which to defend themselves; and no suitable proboscis for gathering honey from the flowers; no baskets on their thighs for holding bee-bread, and no pouches on their abdomens for secreting wax. They are, therefore, physically disqualified for the ordinary work of the hive. Their proper office is to impregnate the young queens, and they are usually destroyed by the bees soon after this is accomplished.

Dr. Evans, an English physician and the author of a beautiful poem on bees, thus appropriately describes them:

"Their short proboscis sips
No luscious nectar from the wild thyme's lips,
From the lime's leaf no amber drops they steal,
Nor bear their grooveless thighs the foodful meal:
On other's toils, in pamper'd leisure thrive
The lazy fathers of the industrious hive."

The drones begin to make their appearance in April or May; earlier or later, according to the forwardness of the season, and the strength of the stock. In colonies too weak to swarm, none as a general rule are reared; for in such hives, as no young queens are raised, drones would be only useless consumers.

The number of drones in a hive is often very great, amounting not merely to hundreds, but sometimes to thousands. As a single one [21] will impregnate a queen for life, it would seem that only a few should be reared. But as sexual intercourse always takes place high up in the air, the young queens must necessarily leave the hive; and it is very important to their safety that they should be sure to find a drone without being compelled to make frequent excursions; for being larger than workers, and less active on the wing, queens are more exposed to be caught by birds, or destroyed by sudden gusts of wind.

In a large Apiary, a few drones in each hive, or the number usually found in one, would suffice. But under such circumstances bees are not in a state of nature, like a colony living in a forest, which often has no neighbors for miles.[22] A good stock, even in our climate, sometimes sends out three or more swarms, and in the tropical climates, of which the bee is probably a native, they increase with astonishing ra-

---

[21] *It was not until recent years that it was known that queens mate with about a dozen, or more, drones over a few days of mating activity. This multiple mating provides the subsequent colony greater genetic diversity, and prevents the degeneration of the colony. If a drone happens to have a sex allele in common with the queen these homozygous sex alleles then produce diploid drones that are killed by the nurse bees. If only one drone was to mate with the queen then half of the eggs would be lost and the colony would probably die. This has been described as inbreeding depression. In this case it is probably more than inbreeding per se, but a real genetic reduction in viable brood.*

[22] *I cannot be certain of the number of wild, or feral, colonies that existed in the 1850's but I suspect that Langstroth has under estimated the population greatly. Until the parasitic mites reduced this population they accounted for 50%, or more, of the bees.*

pidity.* Every new swarm except the first, is led off by a young queen; and as she is never impregnated until she has been established as the head of a separate family, it is important that each should be accompanied by a goodly number of drones: this requires the production of a large number in the parent hive.[23]

As this necessity no longer exists when the bee domesticated, the breeding of so many drones should be discouraged. Traps † have been invented to destroy them, but it is much better to save the bees the labor and expense of rearing such a host of useless consumers. This can readily be done, when we have the control of the combs; for by removing the drone-comb, and supplying its place with worker-cells, the over production of drones may be easily prevented.[24] Those who object to this, as interfering with nature, should remember that the bee is not in a state of nature; and that the same objection might, with equal force, be urged against killing off the supernumerary males of our domestic animals.

When a new swarm is building its combs, if the honey-harvest is abundant, the bees will frequently construct an unusual amount of drone-combs, for storing it. In a state of nature, where bees have room, as in the hollow of a tree, or cleft of a rock, excess of drone-comb will be used another season for the same purpose, and new worker-comb made to meet the enlarged wants of the colony; but in hives of a limited capacity this cannot be done, and thus many stocks become so crowded with drones as to be of little value to their owner.[25]

---

* At Sydney, in Australia, a single colony is stated to have multiplied to 300, in three years.
[23] These "after swarms" are usually headed by virgin queens and as such probably never survive in the wild since the population of the swarm declines too much before the queen is able to mate and begin laying eggs. It takes 21 days to produce any new bees. This is the major reason that swarms leave with the old, mated queen as oviposition can begin immediately upon arrival at the new nest site.
† Such traps were used in Aristotle's time.
[24] Since the foundation press had only been invented in 1857 by Mehring, in Germany, Langstroth did not find out about this process until the book was essentially written, thus, he probably saved comb from other colonies.
[25] In nature the amount of drone comb found in wild colonies has been in the range of 15-20 percent. While this may be more than a modern beekeeper would like in his

In July or August, or soon after the swarming season is over, the bees usually expel the drones from the hive; though, when the honey-harvest is very abundant, they often allow them to remain much later. They sometimes sting them, or gnaw the roots of their wings, so that when driven from the hive, they cannot return. If not ejected in either of these summary ways, they are so persecuted and starved, that they soon perish. At such times they often retreat from the comb, and keep by themselves upon the sides or bottom-board of the hive. The hatred of the bees extends even to the unhatched young, which are mercilessly pulled from the cells and destroyed with the rest. How wonderful that instinct which, when there is no longer any occasion for their services, impels the bees to destroy those members of the colony reared but a short time before with such devoted attention!

None of the reasons previously assigned seem fully to account for the necessity of so many drones. I have repeatedly queried, why impregnation might not have taken place *in the hive*, instead of in the open air. A few dozen drones would then have sufficed for the wants of any colony, even if it swarmed, as in warm climates, half a dozen times, or oftener, in the same season; and the young queens would have incurred no risks by leaving the hive for fecundation.

For a long time I could not perceive the wisdom of the existing arrangement; although I never doubted that there was a satisfactory reason for this seeming imperfection. To have supposed otherwise, would have been highly unphilosophical, when we know that with the increase of knowledge many mysteries in nature, once inexplicable, have been fully cleared up.

The disposition cherished by many students of nature, to reject some of the doctrines of revealed religion, is not prompted by a true philosophy. Neither our ignorance of all the facts necessary to their full elucidation, nor our inability to harmonize these facts in their mutual relations and dependencies, will justify us in rejecting any truth which God has seen fit to reveal, either in the book of nature, or in His holy word. The man who would substitute his own speculations for the divine teachings, has embarked without rudder or chart, pilot or compass, on an uncertain ocean of theory and conjecture; unless he

---

hives, having a number of drones is not all that disastrous for the drones do provide metabolic heat to the colony.

turns his prow from its fatal course, storms and whirlwinds will thicken in gloom on his "voyage of life;" no "Sun of Righteousness" will ever brighten for him the expanse of dreary waters; no favoring gales will waft his shattered bark to a peaceful haven.

The thoughtful reader will require no apology for this moralizing strain, nor blame a clergyman, if sometimes forgetting to speak as the mere naturalist, he endeavors to find

> "Tongues in trees, books in running brooks,
> Sermons in 'bees,' and 'God' in every thing."

To return to the attempt to account for the existence of so many drones. If a farmer persists in what is called "breeding in and in," that is, without changing the blood, the ultimate degeneracy of his stock is the consequence. This law extends, as far as we know, to all animal life, man himself not being exempt from its influence. Have we any reason to suppose that the bee us an exception? or that degeneracy would not ensue, unless some provision were made to counteract the tendency to "in and in breeding?" If, fecundation had taken place in the hive, the queen would have been impregnated by drones from a common parent; and the same result must have taken place in each successive generation, until the whole species would eventually have "run out." By the present arrangement, the young queens when they leave the hive, often find the air swarming with drones, many of which belong to other colonies, and thus by crossing the breed provision is constantly made to prevent deterioration.

Experience has proved that impregnation may be effected not only when there are no drones in the colony of the young queen, but even when there are none in the immediate neighborhood.[26] Intercourse takes place very high in the air (perhaps that less risk may be incurred from birds), and this favors the crossing of stocks.[27]

---

[26] *Drone congregation areas (DCA's) are now known to provide places where mating takes place. Such DCA's are often a mile or more from the colony. Drones (and queens) fly out from a hive using visual cues. These may be an edge of a woods, or similar cues. When these cues stop, the drones circle to orient themselves. This circling flight pattern then forms the DCA.*

[27] *Through the use of radar and other means we know that "very high" is relative, and that mating occurs in the range of 30 to 50 meters (100 to 160 ft.) high.*

I am strongly persuaded that the decay of many flourishing stocks, even when managed with great care, may be attributed to the fact that they have become enfeebled by "close breeding," and are thus unable to resist injurious influences, which were comparatively harmless when the bees were in a state of high physical vigor. When a cultivator has but few colonies, or is remote from other Apiaries, he should guard against this evil by occasionally changing his stocks.

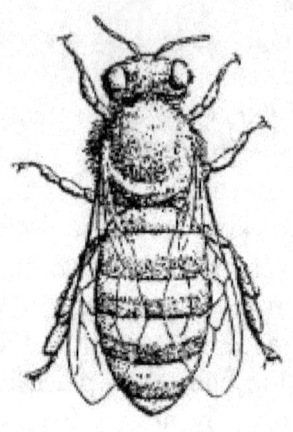

The WORKERS, or common bees, compose the bulk of the population of a hive. A good swarm ought to contain at least 20,000; and in large hives, strong colonies which are not reduced by swarming, frequently number two or three times as many during the height of the breeding season.[28] We are informed by Mr. Dobrogost Chylinski, that from the Polish hives, which often hold several bushels, swarms regularly issue so powerful that "they resemble a little cloud in the air."

It has already been stated, that the workers are all females whose ovaries are too imperfectly developed to admit of their laying eggs. Being for a long time regarded as neither males nor females, they were called Neuters; but careful microscopic examinations, by detecting the rudiments of their ovaries, have determined their sex. The accuracy of these examinations has been verified by the well known facts respecting fertile workers.

Riem, a German Apiarian, first discovered that workers sometimes lay eggs. Huber subsequently ascertained that such workers were bred in hives that had lost their queen, and near the royal cells in which young queens were being reared. He conjectured that small portions of the peculiar food of these infant queens were *accidentally* dropped

---

[28] *While early estimates of the number of bees in a colony were high, Langstroth was not as far off as some of his contemporaries. A good colony, at peak, is near 50,000 bees, and most colonies never reach that number.*

into their cells, by eating which their reproductive organs were more developed than those of other workers.[29]

In the Summer of 1854, I examined a brood-comb which had been given to a queenless colony. It contained eleven sealed queens; and numbers of the cells were capped with a round covering, as though they contained drones. Being opened, some contained drone, and others worker-nymphs. The latter seemed of a little more sugar-loaf shape than the common workers, and their cocoons were of a coarser texture than usual. I had previously noticed the same kind of cells in hives raising artificial queens, but thought they all contained drones. It is a well known fact, that bees often begin more queen-cells than they choose to finish. It seems to me probable, therefore, that when rearing queens artificially, they frequently give a portion of the royal jelly to larvae, which, for some reason, they do not develope as full grown queens; and that such larvae become fertile workers. Huber states that those fertile workers which lay only drone-eggs, prefer large cells in which to deposit them, resorting to small ones, only when usable to find those of greater diameter. A hive in my Apiary having much worker-comb, but only a small piece of drone size, a fertile worker filled the latter so entirely with eggs that some of the cells contained three or four each. Such workers have, in rare instances, been tolerated in hives containing a fertile, healthy queen.[30]

The worker is much smaller than either the queen or the drone. She is furnished with a tongue, or proboscis, so exceedingly curious and complicated, that a separate volume would hardly suffice to describe its structure and uses (Pl. XVI; Fig. 51). With this organ she obtains the honey from the blossoms, and conveys it to her honey-bag. This receptacle (PL XVI., Fig. 54, A), is not larger than a very small pea, and so perfectly transparent as to appear, when fined, of the same color with its contents; it is properly the first stomach, and is surrounded

---

[29] *This is conjecture without any scientific evidence. We know that laying workers are caused by the lack of queen pheromone within their body, and without this suppression their ovaries develop. The worker's ovaries have about two ovarioles (egg tubes) per ovary whereas the queen has about 150 in each. The ovarioles, when developed, are exactly like a queen's. It's just that they only have a very small number, and they do not mate so all of the eggs produce drones.*

[30] *There are estimates from current genetic studies that laying workers produce a small number of the drones from all hives.*

by muscles which enable the bee to compress it, and empty its contents through her proboscis into the cells.[31]

The hinder legs of the worker are furnished with a spoon-shaped hollow, or basket, to receive the pollen which she gathers from the flowers.

Every worker is armed with a formidable sting, and when provoked makes instant and effectual use of her natural weapon. When subjected to a microscopic examination (Pl. XVII, Fig. 53), it exhibits a very intricate mechanism. "It is moved by muscles[*] which, though invisible to the eye, are yet strong enough to force the sting, to the depth of one-twelfth of an inch, through the thick skin of a man's hand. At its root are situated two glands by which the poison is secreted; these glands uniting in one duct, eject the venomous liquid along the groove formed by the junction of the two piercers. There are four barbs on the outside of each piercer; when the insect is prepared to sting, one of these piercers, having its point a little longer than the other, first darts into the flesh, and being fixed by its foremost beard, the other strikes in also, and they alternately penetrate deeper and deeper, till they acquire a firm hold of the flesh with their barbed hooks, and then follows the sheath, conveying the poison into the wound. 'The action of the sting,' says Paley, 'affords an example of the union of *chemistry* and mechanism; of chemistry, In respect to the *venom* which can produce such powerful effects; of mechanism, as the sting is a compound instrument. The machinery would have been comparatively useless, had it not been for the chemical process by which, in the insect's body, *honey* is converted into *poison*; and on the other hand, the poison would have been ineffectual, without an instrument to wound, and a syringe to inject it.'

"Upon examining the edge of a very keen razor by the microscope, it appears as broad as the back of a pretty thick knife, rough, uneven, and full of notches and furrows, and so far from anything like sharp-

---

[31] *The honey stomach (morphologically the crop of insects) is probably responsible for the inversion of most of the sucrose and other complex sugars. Many nectars are mostly sucrose. As the bee carries the nectar to the hive in the honey stomach it has an invertase (enzyme) that breaks this down into the simple sugars of honey—fructose and glucose.*

[*] Bevan

ness, that an instrument as blunt as this seemed to be, would not serve even to cleave wood. An exceedingly small needle being also examined, it resembled a rough iron bar out of a smith's forge. The sting of a bee, viewed through the same instrument, showed everywhere a polish amazingly beautiful, without the least flaw, blemish, or inequality, and ended in a point too fine to be discerned."

As the extremity of the sting is barbed like an arrow, the bee can seldom withdraw it, if the substance into which she darts it is at all tenacious. In losing her sting she parts with a portion of her intestines, and of necessity soon perishes.

Although they pay so dearly for the exercise of their patriotic instincts, still, in defense of home and its sacred treasures, they

> "Deem life itself to vengeance well resign'd,
> Die on the wound, and leave their sting behind."

Hornets, wasps, and other stinging insects, are able to withdraw their stings from the wound. I have never seen the exception in the case of the honey-bee accounted for; [32] but as the Creator intended it for the use[*] of man, did He not give it this peculiarity, that it might be more completely subject to human control? Without a sting, it could not have defended its tempting sweets against a host of greedy depredators: while, if it had been able to sting a number of times, its thorough domestication would have been well nigh impossible.

The defense of the colony against enemies, the construction of the cells, and storing of them with honey and bee-bread, the rearing of the young, and in short, the whole work of the hive, the laying of eggs excepted, is carried on by the industrious little workers.

There may be *gentlemen* of leisure in the commonwealth of bees, but assuredly there are no such *ladies*, whether of high or low degree. The queen herself has her full share of duties, the royal office being no sinecure, when the mother who fills it must daily superintend the proper deposition of thousands of eggs.

The queen-bee will live four, and sometimes, though very rarely, five or more years. As the life of the drones is usually cut short by violence, it is difficult to ascertain its precise limit. Bevan estimates it not to exceed four months. The workers are supposed by him to live six or seven months; but their age depends very much upon their greater or less exposure to injurious influences, and severe labors. Those reared in the Spring and early part of Summer, upon whom the heaviest labors of the hive devolve, appear to live not more than two

---

[32] *The stinger of bees and wasps is derived from the ovipositor of an ancestral Hymenoperan. These ancient ovipositors, similar to those of the sawflies of today, were saw-like in appearance. The ovipositors are used to penetrate into wood and leaf tissue in order to lay the eggs. It happens that the honey bee retains more of this saw-like character than the other wasps and bees, thus the honey bee stinger's barbs (saw teeth) are caught in the flesh and the stinger and poison sac are pulled out of the bee's body.*

[*] Since the publication of the first edition of this treatise, I have had an opportunity during a visit to the Mexican frontier, of studying the habits of the honey-hornet, of that region. Its nest, in shape and material, resembles that of our common hornet; and some of them contain many pounds of delicious honey. This insect, which in those regions is so serviceable to man, like the honey-bee, is unable to withdraw its sting from the wound. It has also a queen, and lives in a colony state during the whole year.

or three months * while those bred at the close of Summer, and early in Autumn, being able to spend a large part of their time in repose, attain a much greater age.[33] It is very evident that "the bee" (to use the words of a quaint old writer), "is a Summer bird;" and that, with the exception of the queen, none live to be a year old.

Notched and ragged wings, instead of gray hairs and wrinkled faces, are the signs of old age in the bee, and indicate that its season of toil will soon be over. They appear to die rather suddenly; and often spend their last days, and sometimes even their last hours, in useful labors.

Place yourself before a hive, and see the indefatigable energy of these industrious veterans, toiling along with their heavy burdens, side by side with their more youthful compeers, and then judge if, while qualified for useful labor, you ought ever to surrender yourself to slothful indulgence. Let the cheerful hum of their busy old age inspire you with better resolutions, and teach you how much nobler it is to die with harness on, in the active discharge of the duties of life.

The age which individual members of the community may attain, must not be confounded with the colony. Bees have been known to occupy the same domicile for a great number of years. I have seen flourishing colonies more than twenty years old; the Abbe Della Rocca speaks of some over forty years old; and Stoche says, that he saw a colony, which he was assured had swarmed annually for forty-six years! "Such cases have led to the erroneous opinion, that bees are a long-lived race. But this, as Dr. Evans has observed, is just as if a stranger, contemplating a populous city, and personally unacquainted with its inhabitants, should, on paying it a second visit many years

---

* If an Italian queen be given, in the working season, to a swarm of common bees, in about three months only a few of the latter will be found in the colony.

[33] *Modern life history studies have given much shorter life spans to all of the members of the hive. Queens probably have an average life of two years, but may live a year longer. Drones live a month or two. Workers live about 30 to 35 days, on average, during the summer, and probably 120 to 150 days in the winter. Most stories on bees give worker bees much too long a life during the summer. There may be individual bees that live for six weeks or more, but most have a much shorter life.*

If Langstroth had calculated his 2,000 to 3,000 eggs per day along with several months' life, he would have had colonies with hundreds of thousands of bees.

after, and finding it equally populous, imagine that it was peopled by the same individuals, not one of whom might be living.

> 'Like leaves on trees, the race of bees is found,
> Now green in youth, now withering on the ground;
> Another race the Spring or Fall supplies,
> They droop successive, and successive rise.'"
> Evans.

The cocoons spun by the larvae are never removed by the bees; they adhere so closely to the sides of the cells that the labor of removal would cost more than it would be worth. As the breeding cells may eventually become too small for the proper development of the young, very old combs should be removed from the hive.[34] It is a great mistake, however, to imagine that the brood-combs ought to be changed every year. If it were desirable, this might easily be done in my hives; but to remove them oftener than once in five or six years, requires a needless consumption of honey to replace them, and injures the bees in Winter, as the new comb is much colder than the old.[35]

Inventors of hives have too often been "men of one idea:" and that one, instead of being a well established and important fact in the physiology of the bee, has frequently (like the necessity for a yearly change of the brood-combs), been merely a conceit of some visionary projector. This might be harmless enough, were no effort made to impose such crudities upon an ignorant public, either in the shape of a patented hive, or worse still, of an *unpatented* hive, the pretended *right* to use which is *fraudulently* sold to the cheated purchaser.*

---

[34] *This argument has raged up until this day. There is little scientific data to support the notion the cells will become so small as to affect the size of the bee. However, the replacement of combs, as Langstroth suggests, about every five or six years is valid. Diseases and pesticides will build up in the old combs and can affect the development of the larvae. Wax is a "sink" for many of the fat-soluble pesticides and these chemicals accumulate in the wax. Thus, periodic replacement cleans these out of the hive.*

[35] *The old combs do not provide warmth but generally have pollen in the combs. The pollen is needed in the winter cluster to produce new young bees.*

* Hives which have never been patented have been extensively sold as patent articles by men, who for years have been liable to prosecution for obtaining money under false pretenses. Others are disposed of, on the ground that the patent is still pending, when no application for a patent has ever been made, or has long ago been rejected. Often the

Apiarians unaware of the brevity of the bee's life, have often constructed huge "bee-palaces" and large closets, vainly imagining that the bees would fill them, being unable to see any reason why a colony should not increase until it numbers its inhabitants by millions or billions. But as the bees can never at one tune equal, still less exceed, the number which the queen is capable of producing in a season, these spacious dwellings have always an abundance of spare rooms. It seems strange that men can be thus deceived, when often in their own Apiary they have healthy stocks, which, though they have not swarmed for a year or more, are no more populous in the Spring, than those which have regularly parted with vigorous colonies.

It is certain that the Creator has wisely set a limit to the increase of numbers in a single colony; and I shall venture to assign a reason for this. Suppose he had given to the bee a length of life as great as that of the horse or the cow, or had made each queen capable of laying daily some hundreds of thousands of eggs; or had given several hundred queens to each hive; then a colony must have gone on increasing, until it became a scourge rather than a benefit to man. In the warm climates of which the bee is a native, it would have established itself in some cavern or capacious cleft in the rocks, and would soon have become so powerful as to bid defiance to all attempts to appropriate the avails of its labors.

It has already been stated that none, except the mother-wasps and hornets, survive the Winter. Had these insects, like the bee, been able to commence the season with the accumulated strength of a large colony, they would, long before its close, have proved an intolerable nuisance. If, on the contrary, the queen-bee had been compelled, solitary and alone, to lay the foundations of a new commonwealth, the honey-harvest would have disappeared long before she could become the parent of a numerous family.

---

patented part of a hive, being a worthless conceit, is carefully concealed, while much ingenuity is displayed, in exhibiting those features in the hive which any one has a right to use; and yet, which the vender, sometimes by implication, and sometimes by direct assertion, leads the purchaser to believe are essential parts of the patent.

No one should ever purchase a "patent hive" until he ascertains two things: 1st, that there is really a patent on the invention; and 2d, that the part patented is, in his opinion, worth to him the money asked for the right to use it.

The process of rearing Queen-Bees will now be more particularly described. Early in the season, if a hive becomes very populous, the bees usually make preparation for swarming. A number of royal cells are begun, being commonly constructed upon those edges of the combs (Pl. XIV., a,b,c,d), which are not attached to the sides of the hive. These cells somewhat resemble a small pea nut (Pl. XIII., Figs. 49, 50), and are about an inch deep, and one-third of an inch in diameter: being very thick, they require much wax for their construction. They are seldom seen in a perfect state after the swarming season, as the bees, after the queen has hatched, cut them down to the shape of a small acorn-cup. (Pl. XIV., c.) These queen-cells, while in progress, receive a very unusual amount of attention from the workers. There is scarcely a second in which a bee is not peeping into them; and as fast as one is satisfied, another pops in her head to report progress, or increase the supply of royal jelly. Their importance to the community might easily be inferred from their being the center of so much attraction.

While the other cells open sideways, the queen-cells always hang with their mouth *downwards*. Some Apiarians think that this peculiar position affects, in some way, the development of the royal larvae; while others, having ascertained that they are uninjured if placed in any other position, consider this deviation as among the inscrutable mysteries of the bee-hive. So it seemed to me, until convinced, by more careful observation, that they open downwards simply *to save room*. The distance between the parallel ranges of comb in the hive is usually too small for the royal cells to open sideways, without interfering with the opposite cells. To economize space, the bees put them on the unoccupied edges of the comb, where there is plenty of room for such very large cells.[36]

---

[36] *There is some speculation as to why the reproductive cells, queens and drones, are usually at the bottom of the combs. In an ancestral honey bee that hung the comb from a tree limb, much like* Apis dorsata *or* Apis florea *still do, the very important reproductive cells would be the farthest from the limb and invasion from ants. In the tropics ants are a major enemy of bees, and so you keep the reproductive castes as far from harm as possible. These ancient bees also probably coated the limb with propolis as well to help keep the ants away from their nest. Modern bees use this technique of coating limbs with propolis for the same reason. I suspect that honey bees use propolis more to keep out invaders than for sealing up the hive from wind or cold.*

The number of royal cells in a hive varies greatly; sometimes there are only two or three, ordinarily not less than five; and occasionally, more than a dozen. As it is not intended that the young queens should all be of the same age, the royal cells are not all begun at the same time. It is not fully settled how the eggs are deposited in these cells. In some few instances, I have thought that the bees transferred the eggs from common to queen-cells; and this may be their general method of procedure. I shall hazard the conjecture, that, in a crowded state of the hive the queen deposits her eggs in cells on the edges of the comb, some of which are afterwards changed by the workers into royal cells. Such is a queen's instinctive hatred to her own kind, that it seems improbable that she should be intrusted with even the initiatory steps for securing a race of successors.

The young queens are much more largely supplied with food than the other larvae so that they seem to lie in a thick bed of jelly, a portion of which may usually be found at the base of their cells, soon after they have hatched.[37] Unlike the food of the other larvae, it has a slightly acid taste; and when fresh, resembles starch; when old, a light quince jelly. The bees, if confined to their hive and supplied with water, can secrete it from the honey and bee-bread stored in their combs.

---

[37] *In the analysis of successful queen rearing colonies, or systems, the amount of remaining royal jelly left in the bottom of a queen cell has been used as a measure of how well the cells have been cared for during development of the queen larva.*

I submitted some royal jelly to Dr. Charles M. Wetherell, of Philadelphia; an interesting account of his analysis may be found in the Report of the Proceedings of the Philadelphia Academy of Natural Sciences for July, 1852. He speaks of the substance as being a "truly bread-containing, albuminous compound." A comparison of its elements with the food of the drone and worker-larvae, might throw some light on subjects now involved in obscurity.[38]

The effects produced upon the royal larvae by their peculiar treatment are so wonderful, that they have usually been rejected as idle whims, by those who have neither been eye-witnesses to them, nor acquainted with the opportunities enjoyed by others for accurate observation. They are not only contrary to all common analogies, but so marvelously strange and improbable, that many when asked to believe them, feel that an insult is offered to their common sense. The most important of these effects I shall briefly enumerate.

1st. The peculiar mode in which the worm designed for a queen is treated, causes it to arrive at maturity almost one-third earlier than if it had been reared a worker. And yet, as it is to be much more fully developed, according to ordinary analogy, it should have had a slower growth.

2d. Its organs of reproduction are completely developed, so that it can fulfill the office of a mother.

3d. Its size, shape, and color are greatly changed; its lower jaws are shorter, its head rounder, and its abdomen without the receptacles for secreting wax; its legs have neither brushes nor baskets, and its sting

---

[38] *It wasn't until 100 to 120 years after Langstroth that the chemical composition of royal jelly was finally determined, and then also the process by which queens were developed. Royal jelly is a highly nutritious food. Royal jelly's most unique substance is 10-hydroxy 2-decenoic acid (10 HDA), a 10-carbon fatty acid. While it is unique, it is not what makes a queen develop. It takes the stimulation of some sugar and a hormone to start the larva into feeding fast enough to become a queen. The sugar and hormone within the royal jelly start the process and keep in going until the queen is produced. Thus, a queen is produced most from the quantity of food. A larvae has to start the development before the third day of larval growth. The best, most fully developed, queens start earlier—the first or second day of larval life. Royal jelly has a fairly high protein level as well as many B vitamins. However it is the quantity of food and not some special substance that produces a queen.*

is more curved, and one-third longer (Pl. XVIII.) than that of a worker.

4th. Its instincts are entirely changed. Reared as a worker, it would have thrust out its sting at the least provocation; whereas now, it may be pulled limb from limb without attempting to sting. As a worker, it would have treated a queen with the greatest consideration; but now, if brought in contact with another queen, it seeks to destroy it as a rival. As a worker, it would frequently have left the hive, either for labor or exercise; as a queen, it never leaves it after impregnation except to accompany a new swarm.

5th. The term of its life is remarkably lengthened. As a worker, it would not have lived more than six or seven months; as a queen, it may live seven or eight times as long. All these wonders rest on the impregnable basis of demonstration, and instead of being witnessed only by a select few, may now, by the use of the movable-comb hive, be familiar sights to any bee-keeper who prefers an acquaintance with facts, to caviling and sneering at the labors of others.[*]

---

[*] A brief extract from the celebrated Dr. Boerhaave's memoir of Swammerdam, should put to blush the arrogance of those superficial observers, who are too wise in their own conceit to avail themselves of the knowledge of others.

"This treatise on Bees proved so fatiguing a performance, that Swammerdam never afterwards recovered even the appearance of his former health and vigor. He was almost continually engaged by day in making observations, and as constantly by night in recording them by drawings and suitable explanations.

"His daily labor began at six in the morning, when the sun afforded him light enough to survey such minute objects; and from that hour till twelve, he continued without interruption, all the while exposed in the open air to the scorching heat of the sun, bareheaded, for fear of intercepting his sight, and his head in a manner dissolving into sweat under the irresistible ardors of that powerful luminary. And if he desisted at noon, it was only because the strength of his eyes was too much weakened by the extraordinary afflux of light, and the use of microscopes, to continue any longer upon such small objects.

"He often wished, the better to accomplish his vast, unlimited views, for a year of perpetual heat and light to perfect his inquiries; with a polar night, to reap all the advantages of them by proper drawings and descriptions."

The process of rearing queens to meet some special emergency, is even more wonderful than the one already described. If the bees have worker-eggs, or worms not more than three days old, they make one large cell out of three, by nibbling away the partitions of two cells adjoining a third. Destroying the eggs or worms in two of these cells, they place before the occupant of the other, the usual food of the young queens; and by enlarging its cell, give it ample space for development. As a security against failure, they usually start a number of queen-cells, although often the work on all, except a few, is soon discontinued.

In from eleven to fourteen days, they are in possession of a new queen, in all respects resembling one reared in the natural way; while the eggs in the adjoining cells, which have been developed as workers, are nearly a week longer in coming to maturity.

The beautiful representation of comb, in Plate XVIII, is taken, with important alterations and additions of my own, from Cotton's "My Bee-Book," to which I am also indebted for the group of bees in the title-page. The royal cell (b), is a perfect queen-cell, from which the inmate has not yet emerged. The queen-cell (a), represents the cap or lid as it often appears just after-the young queen been hatched.[39] The queen-cell (d), which is open at the side, is one from which a young queen has been violently abstracted; the other (c), is one which the bees have nearly reduced to the acorn shape. It also resembles one only a few days old. On the face of the comb is a cell (n), just begun for the artificial rearing of a queen, this being the usual position of cells built to meet some unexpected emergency. To bring the points illustrated into a compact compass, the cells are drawn smaller than the natural size.

---

[39] *This hinged cap is one way to assess whether the young queen emerged. In my mating nucs, or divisions, I like to check the cells a day or so after they should have emerged. If I find the cap I usually look no further so as not to disturb the nuc very much. I then let the virgin mate and have time to begin laying eggs before I examine the nuc again. If you wait too long to examine a queen cell for this hinged cap the worker bees will have often torn the cell down to the point that you will not be able to determine if the queen emerged safely.*

I shall give, in this connection, a description of an interesting experiment.

A populous stock was removed, in the morning, to a new place, and an empty hive put upon its stand. Thousands of workers which were ranging the fields, or which left the old hive after its removal, returned to the familiar spot. It was truly affecting to witness their grief and despair; they flew in restless circles about the place where once stood their happy home, entering the empty hive continually, and expressing, in various ways, their lamentations over so cruel a bereavement. Towards evening, ceasing to take wing, they roamed in restless platoons, in and out of the hive, and over its surface, as if in search of some lost treasure. A small piece of brood-comb was then given to them, containing worker-eggs and worms. The effect produced by its introduction took place much quicker than can be described. Those which first touched it raised a peculiar note, and in a moment, the comb was covered with a dense mass of bees; as they recognized, in this small piece of comb, the means of deliverance, despair gave place to hope, their restless motions and mournful voices ceased, and a cheerful hum proclaimed their delight. If some one should enter a building filled with thousands of persons tearing their hair, beating their breasts, and by piteous cries, as well as frantic gestures, giving vent to their despair, and could by a single word cause all these demonstrations of agony to give place to smiles and congratulations, the change would not be more instantaneous than that produced when the bees received the brood-comb![40]

The Orientals call the honey-bee, "Deborah: She that speaketh." Would that this little insect might speak, in words more eloquent than those of man's device, to those who reject any of the doctrines of revealed religion, with the assertion that they are so improbable, as to labor under a fatal *a priori* objection. Do not all the steps in the development of a queen from a worker-egg, labor under the very same

---

[40] *This account points out the remarkable ability of olfaction by honey bees. They probably use detection of odors far more than we can imagine. If you think about this ability to detect the smallest odor you realize that they use this ability to detect minute odors to govern their foraging for flowers as well as to determine many different conditions within the dark hive.*

objection? and have they not, for this reason been always regarded, by many bee-keepers, as unworthy of belief? If the favorite argument of infidels will not stand the test, when applied to the wonders of the beehive, is it entitled to serious weight, when, by objecting to religious truths, they arrogantly take to task the Infinite Jehovah for what He has been pleased to do or to teach? With no more latitude than is claimed by such objectors, it were easy to prove that a man is under no obligation to believe any of the wonders of the beehive, even although he is himself an intelligent eyewitness to their substantial truth.

Preface to Chapters IV, V and VI

The next three chapters reflect, I think, on those topics that were (and are) the most unknown quantities in the beehive. Honeycomb not so much that it was unknown but that the average person had almost no idea on how it was formed and from what substance. Did the bees just find it, or was it something that they produced? Langstroth then spends a fair amount of time on the handling of the comb itself. He realized that this was an important product for the beekeeper and wanted to impart his wisdom on how to care and maintain this resource. The care of comb was certainly more important at that time as beeswax comb foundation had just been invented (in Europe) about two years before this edition was written, and was basically unknown to Langstroth.

Propolis was certainly an unknown to most people and even a beginning beekeeper would not necessarily understand all of the details of its origin and collection. It still may be the most unknown substance in the beehive. This is especially true since it has so many different trees and shrubs as its source.

Pollen is a vast resource for the colony and one that requires at least 25 percent of all foraging trips from the hive. In Langstroth's time its properties and importance was probably not as well known as today. Certainly the colony is a product of newly developed bees and they could not exist without the proteins from pollen.

Fig. 20. PLATE VII.

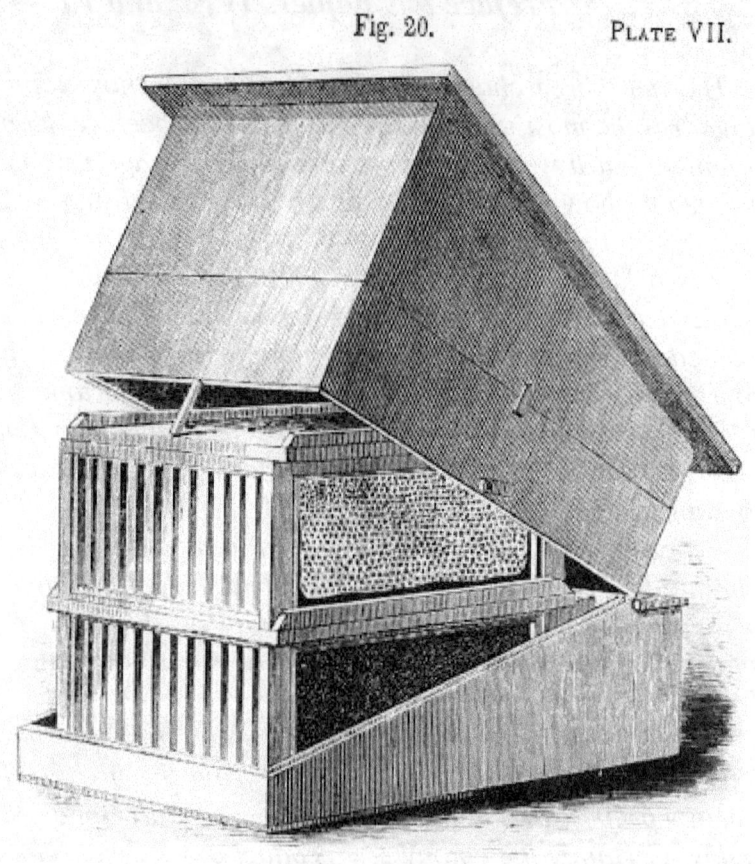

CHAPTER IV.

COMB.

Wax is a natural secretion of bees, and may be called their oil or fat. When gorged with honey, or any liquid sweet, if they remain quietly clustered together, it is secreted in the shape of delicate scales, in small pouches on their abdomen. (Pl. XIII., Figs. 37, 38.) Soon after a swarm is hived, the bottom-board will usually be covered with these scales. The bees seem to loosen them from their bodies by violently shaking themselves as they stand upon the combs.[1]

> "Thus, filter through yon flutterer's folded mail,
> Clings the cooled wax, and hardens to a scale.[2]
> Swift, at the well-known call, the ready train
> (For not a buz boon Nature breathes in vain)
> Spring to each falling flake, and bear along
> Their glossy burdens to the builder throng.
> These with sharp sickle, or with sharper tooth,
> Pare each excrescence, and each angle smoothe,
> Till now, in finish'd pride, two radiant rows
> Of snow white cells one mutual base disclose.
> Six shining panels gird each polish'd round;
> The door's fine rim, with waxen fillet bound;
> While walls so thin, with sister walls combined,
> Weak in themselves, a sure dependence find."
>   Evans

Most Apiarians before Huber's time supposed that wax was made from bee-bread, either in a crude or digested state. Confining a new swarm of bees to a hive in a dark and cool room, at the end of five days he found several beautiful white combs in their tenement; these

---

[1] *A certain number of the wax scales apparently fall to the bottom of a hive, though shaking is not required. When a colony produces wax there is not a general wholesale loss to the bottom of the hive.*

[2] *Whether the poet, Evans, knew that beeswax was secreted as a liquid from the wax glands or not, this is an accurate description of the process. The wax is extruded through the bee's inter-segmental membranes on the ventral abdomen. That is, between the segmental plates or "folded mail."*

being taken from them, and the bees supplied with honey and water, new combs were again constructed. Seven times in succession their combs were removed, and were in each instance replaced, the bees being all the time prevented from ranging the fields to supply themselves with bee-bread. By subsequent experiments, he proved that sugar-syrup answered the same end with honey. Giving an imprisoned swarm an abundance of fruit and bee-bread, he found that they subsisted on the fruit, but refused to touch the pollen; and that no combs were constructed, nor any wax-scales formed in their pouches.

Notwithstanding Huber's extreme caution and unwearied patience in conducting these experiments, he did not discover the whole truth on this important subject. Though he demonstrated that bees can construct comb from honey or sugar, without the aid of bee-bread, and that they cannot make it from bee-bread, without honey or sugar, he did not prove that when *permanently* deprived of bee-bread they can continue to work in wax or if they can, that the pollen does not aid in its elaboration.[3]

Some bee-bread is always found in the stomach of wax-producing workers, and they never build comb so rapidly as when they have free access to this article. It must, therefore, either furnish some of the elements of wax, or in some way assist the bee in producing it. Further investigations are necessary, before we can arrive at perfectly accurate results. Confident assertions are easily made, requiring only a little breath, or a few drops of ink; and those who like them best have often the profoundest contempt for observations and experiment. To establish any controverted truth on the solid foundation of demonstrated facts, usually requires severe and protracted labor.

Honey and sugar contain by weight about eight pounds of oxygen to one of carbon and hydrogen. When converted into wax, these proportions are remarkably changed, the wax containing only one pound of oxygen to more than sixteen of hydrogen and carbon.[4] Now as oxygen is the grand supporter of animal heat, the large quantity con-

---

[3] *Current knowledge of the biochemical synthesis of beeswax shows that Huber's conclusions were correct—wax is produced from sugars. Pollen, or bee-bread, is used in the production of proteins such as larval food.*

[4] *Modern chemistry would change these the ratios of molecules to $C_6 H_{10} O_5$, or $C_6 H_{12} O_6$, or $C_{12} H_{22} O_{11}$, depending on if the sugar was glucose, fructose or sucrose.*

sumed in secreting wax aids in generating that extraordinary heat which always accompanies comb-building, and which enables the bees to mould the softened wax into such exquisitely delicate and beautiful forms.* This interesting instance of adaptation, so clearly pointing to the Divine Wisdom, seems to have escaped the notice of previous writers.

Careful experiments prove that from thirteen to twenty pounds of honey are required to make a single pound of wax.[5] As wax is an animal oil, secreted chiefly from honey, this fact will not appear incredible to those who are aware how many pounds of corn or hay must be fed to cattle to have them gain a single pound of fat,

Many beekeepers are unaware of the value of empty comb. Suppose honey to be worth only fifteen cents per pound, and comb, when rendered into wax, to be worth thirty cents, the Apiarian who melts a pound of comb loses largely by the operation, even without estimating the time his bees have consumed in building it. It should, therefore, be considered a first principle in bee-culture never to melt good combs. A strong stock of bees, in the height of the honey-harvest, will fill them with very great rapidity.

Unfortunately, in the ordinary hives but little use can be made of empty comb, unless it is new, and can be put into the surplus honey-boxes; but by the use of bars, or movable frames, every good piece of worker-comb may be given to the bees.

When new, it may be easily attached to frames, or spare honey-receptacles, by dipping the edge into melted wax, and firmly holding it in place until it hardens; if it is old, or the pieces large and full of bee-bread, a mixture of melted wax and resin will secure a firmer ad-

---

* According to Dr. Dönhoff, the thickness of the sides of a cell in a new comb is only the one hundred and eightieth part of an inch!

[5] *Modern beekeepers are able to re-use comb by extracting honey and thus is able to produce much more honey as the bees do not need to expend the energy (honey) to produce wax. In earlier times the ratio of cost of honey to the cost of the wax may have been more favorable to wax since candles were more in need in the household. Though I doubt that beeswax ever was priced so well as to make it profitable to produce wax instead of honey.*

hesion. When comb is put into tumblers, or small receptacles,[6] it may be simply crowded in, so as to keep its place until fastened by the bees. As bees like "a good start in life," they prefer receptacles which contain some empty comb. All suitable drone-comb should be put into such receptacles, instead of being allowed to remain in the breeding apartment of the hive.

No one, to my knowledge, has ever attempted to imitate the delicate mechanism of the bee so closely, as to construct *artificial combs* for the ordinary uses of the hive. If store-combs could be made of gutta-percha,[7] they might be emptied of their contents, and returned to the hive.[8]

In the Summer of 1854, I ascertained that bees will, under some circumstances, use fine shavings of wax to build new comb. If this discovery can be made serviceable for practical purposes, it will both facilitate the cheap and rapid multiplication of colonies, and enable the bees to amass unusual quantities of honey. One pound of beeswax might be made to store nearly twenty pounds of honey; and the beekeeper would gain the difference in value between one pound of wax, and the honey which bees consume in making a pound of comb.[9] At times when no honey can be procured from the blossoms, strong stocks might be profitably employed in building spare comb, to strengthen feeble stocks, or for any other purpose.

The building of comb is usually carried on with the greatest activity

---

[6] *Beekeepers must have done this for many years. See the picture of the Langstroth "Patented" hive opposite Chapter I. There are many small tumblers for the storage of honey ready to be put on top of the hive.*
[7] *Gutta-percha is a resin from trees that would have been the nearest thing to plastic that was available in the 1850's. Modern plastic comb is the answer to this request.*
[8] *Langstroth seems to anticipate the invention of the extractor by about six years, as Van Hruschka, of Austria, did not invent the extractor until 1865.*
[9] *It would appear at this point that Langstroth did not have, or know about, comb foundation. Mehring had invented comb foundation in 1857 and this book was written in about 1859. Either the news had not yet reached him or he did not have access to comb foundation when this section of the book was written. Near the end of the book he mentions a "Wehring" and that comb foundation had been invented, but had not tried using it.*

Fig. 21. Plate VIII.

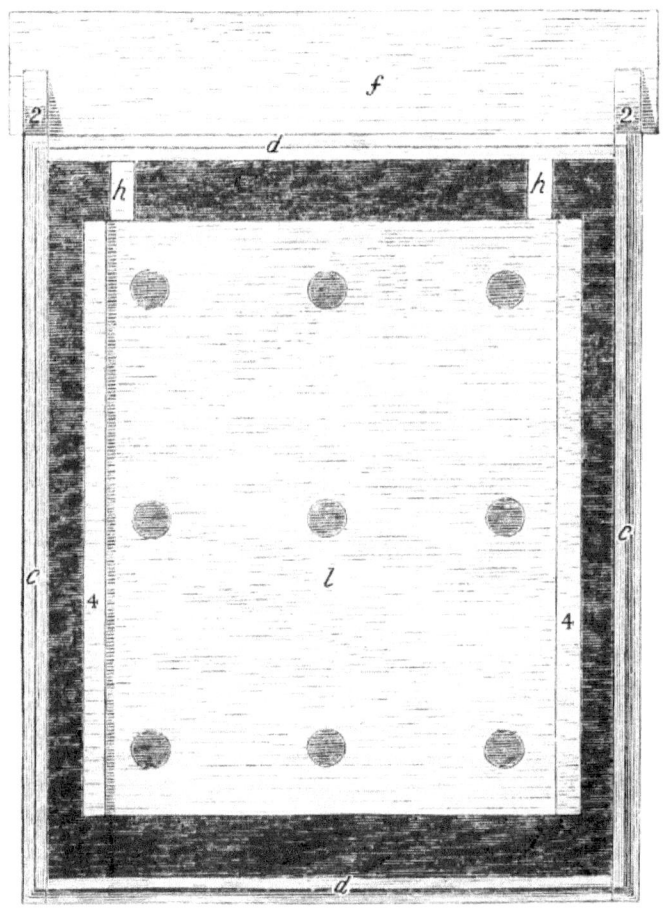

by night, while the honey is gathered by day.* Thus no time is lost. When the weather is too forbidding for out-door work, the combs are most rapidly constructed, the labor being vigorously carried on both by day and by night. On the return of a fair day, the bees, having plenty of room for its storage, gather unusual supplies. Thus, by their wise economy, they often lose no time, even if confined for several days to their hive,

"How doth the little busy bee improve each *shining* hour!"

The poet might, with equal truth, have described her as improving the gloomy days and dark nights in her useful labors.

It is an interesting fact, which seems hitherto to have escaped notice, that honey-gathering and comb-building go on simultaneously; so that when one stops, the other ceases also. As soon as the honey-harvest begins to fail, so that consumption is in advance of production, the bees cease to build new comb, even although large portions of their hive are unfilled. When honey no longer abounds in the fields, it is wisely ordered that they should not consume, in comb-building, the treasures which maybe needed for Winter use. What safer rule could have been given them?

As wax is a bad conductor, it can be more easily worked when warmed by the animal heat of the bees, than if it parted with its heat too readily. By this property, the combs aid in keeping the bees warm, and there is less risk of their cracking with frost, or of the honey candying in the cells. If wax were a good conductor of heat, the combs would often be icy cold, moisture would condense and freeze upon them, and they could not fulfill all their required ends.

---

* On very clear moonlight nights, I have known bees to gather honey from the tulip tree (*Liriodendron tulipifera*).

The size of the cells in which workers are reared never varies; the same may be substantially be said of the drone-cells, which are much larger; those in which honey is stored vary greatly in depth, while in diameter they are of all sizes, from that of worker to that of drone-cells. As five worker, or four drone-cells, will measure about one linear inch, a square inch of comb will contain, on each side, twenty-five worker, or sixteen drone-cells.

As bees in building their cells, cannot pass immediately from one size to another, they display an admirable sagacity in making the transition by a set of irregular intermediate cells. Plate XV. (Fig. 48), exhibits an accurate and beautiful representation of comb, drawn for this work from nature, by M. M. Tidd, and engraved by D. T. Smith, both of Boston, Mass. The cells are of the size of nature. The large ones are drone-cells, and the small ones, worker-cells. The irregular, five-sided cells between them, show how bees pass from one size to another.

The cells of bees are found to fulfill perfectly the most subtle conditions of an intricate mathematical problem. Let it be required to find what shape a given quantity of matter must take, in order to have the greatest *capacity* and *strength*, occupying, at the same time, the least *space*, and consuming the least *labor* in its construction. when this problem is solved by the most refined mathematical processes, the answer is the hexagonal or six-sided cell of the honey-bee, with its three four-sided figures at the base! [10]

The shape of these figures cannot be altered ever so little, except for the worse. In addition to the desirable qualities already enumerated, they serve as nurseries for rearing the young, and as small air-tight vessels for preserving the honey from souring or candying. Every prudent housewife who carefully stores her preserves in receptacles excluding the air, can appreciate the value of such an arrangement.

---

[10] *The simplest answer, though it also may not be the correct one, is that the facets of the compound eye of the honey bee are also six-sided. It may be that the bee just sees things in hexagonal shapes, thus builds the cells of the comb exactly as it sees them. Though the geometry of space saving is clearly in favor of the hexagonal cells. If you "stress" circles they become hexagonal as the shape has structural advantages.*

"There are only three possible figures of the cells," says Dr. Reid, "which can make them all equal and similar, without any useless spaces between them. These are the equilateral triangle, the square, and the regular hexagon. It is well known to mathematicians, that there is not a fourth way possible in which a plane may be cut into little spaces that shall be equal, similar, and regular, without leaving any interstices."

An equilateral triangle would have made a very uncomfortable tenement for an insect with a round body; and a square cell would have been but little better. A circle seems to be the best shape for the development of the larvae; but such a figure would have caused a needless sacrifice of space, materials, and strength; while the honey, which adheres so admirably to the many angles of the six-sided cell, would have been much more liable to run out.[11] The body of the immature insect, as it undergoes its changes, is charged with a superabundance of moisture, which passes off through the reticulated cover of its cell; may not a hexagon, therefore, while approaching so nearly to the shape of a circle, as not to incommode the young bee, furnish, in its six corners, the necessary vacancies for a more thorough ventilation?

Is it credible that these little insects can unite so many requisites in the construction of their cells, either by chance, or because they are profoundly versed in the most intricate mathematics? Are we not compelled to acknowledge that the mathematics by which they construct a shape so complicated, and yet the only one which can unite so many desirable requirements, must be referred to the Creator, and not to his puny creature?

---

[11] *Langstroth is here giving some properties to angles that are not justified. I really don't think that honey would run out of a circle any faster than the hexagonal cell of the honeycomb. The hexagonal cells provide lots of strength to the honeycomb as well as space saving. It probably is not necessary to lavish it traits that it does not possess.*

To an intelligent and candid mind, the smallest piece of honeycomb is a perfect demonstration that there is a "Great First Cause."

> " On books deep poring, ye pale sons of toil,
> Who waste in studious trance the midnight oil,
> Say, can ye emulate, with all your rules,
> Drawn from Grecian or from Gothic schools,
> This artless frame? Instinct her simple guide,
> A heaven-taught insect baffles all your pride.
> Not all yon marshall'd orbs, that ride so high,
> Proclaim more loud a present Diety,
> Than the nice symmetry of these small cells,
> Where on each angle genuine science dwells."
>
> Evans.

CHAPTER V.

PROPOLIS.

This substance is obtained by the bees from the resinous buds and limbs of trees; the different varieties of poplar yield a rich supply. When first gathered, it is usually of a bright golden color, and so adhesive that the bees never deposit it in cells, but apply it at once to the purposes for which they procured it. If a bee is caught while bringing in a load, it will be found to adhere very firmly to her legs.

"Huber planted in Spring some branches of the wild poplar, before the leaves were developed, and placed them in pots near his Apiary; the bees alighted on them, separated the folds of the large buds with their forceps, extracted the varnish in threads, and loaded with it, first one high and then the other; for they convey it like pollen, transferring it by the first pair of legs to the second, by which it is lodged in the hollow of the third." I have seen them thus remove the warm propolis from old bottom-boards standing in the sun.

PROPOLIS.

Propolis[1] is frequently gathered from the alder, horse-chestnut, birch, and willow; and as some think, from pines and other trees of the fir kind. Bees will often enter varnishing shops, attracted evidently by their smell;[2] and in the vicinity of Matamoras, Mexico where propolis seems to be scarce, I saw them using green paint from window blinds, and pitch from the rigging of a vessel. Bevan mentions the fact of their carrying off a composition of wax and turpentine from trees to which it had been applied. Dr. Evans says he has seen them collect the balsamic varnish which coats the young blossom-buds of the hollyhock, and has known them rest at least ten minutes on the same bud, moulding the balsam with their fore feet, and transferring it to the hinder legs, as described by Huber.

> "With merry hum the Willow's copse they scale,
> The Fir's dark pyramid, or Poplar pale;
> Scoop from the Alder's leaf its oozy flood,
> Or strip the Chestnut's resin-coated bud;
> Skim the light tear that tips Narcissus' ray,
> Or round the Hollyhock's hoar fragrance play;
> Then waft their nut-brown loads exulting home,
> That form of fret-work for the future comb;
> Caulk every chink where rushing winds may roar,
> And seal their circling ramparts to the floor."
>
> Evans.

A mixture of wax and propolis being much more adhesive than wax alone, serves admirably to strengthen the attachments of the combs to the top and sides of the hive. If the combs are not filled with honey or brood soon after they are built, they are varnished with delicate coating of propolis, which adds greatly to their strength; but as this natural varnish impairs their snowy whiteness, the bees ought not to be allowed access to combs in the surplus honey-receptacles, ex-

---

[1] *Propolis was so named by Aristotle. The name pro-polis, in Greek, means before the city. Aristotle saw this substance at the entrance to the hive. Sometimes in our modern hives the bees will form a curtain of propolis to seal off the opening before winter, and in this case re-creates its name. Propolis has some antibiotic and anti-fungal properties that aid the colony in some of its defense against these microorganisms.*

[2] *In the 1850's varnish was created from a natural resin, however, as seen from what follows in the text, bees will gather anything that will aid in the job of sealing the hive.*

cept when actively engaged in storing them with honey.

Bees make a very liberal use of propolis to fill any crevices about their premises; and as the natural summer heat of the hive keeps it soft, the bee-moth selects it as a place of deposit for her eggs.[3] Hives ought, therefore, to made of lumber entirely free from cracks. The corners, which the bees usually fill with propolis, may have a melted mixture run into them, consisting of three parts of resin and one of bees-wax; this remaining hard during the hottest weather, will bid defiance to the moth.

As bees find it difficult to gather propolis and equally so to work so sticky a material, they should be saved all unnecessary labor in amassing it. *To men, time is money; to bees, it is honey; and all the arrangements of the hive should he such as to economize it to the utmost.* [4]

Propolis is sometimes put to a very curious use by the bees. "A snail[*] having crept into one of M. Reaumur's hives early in the morning, after crawling about for some time, adhered, by means of its own slime, to one of the glass panes. The bees having discovered the snail, surrounded it, and formed a border of propolis round the verge of its shell, and fastened it so securely to the glass that it became immovable.

> 'Forever closed the impenetrable door;
> It naught avails that in its torpid veins
> Year after year, life's loitering spark remains.'
>   Evans

"Maraldi, another eminent Apiarian, states that a snail without a shell having entered one of his hives, the bees, as soon as they observed it, stung it to death; after which being unable to dislodge it,

---

[3] *Langstroth knew a great amount about wax-moths* (Galleria mellonella). *This observation is just a good example. The eggs are deposited in the cracks between supers and the 1st instar larvae crawl into the hive where they will infest combs. When the larvae grow large enough that the bees find them they kill and remove the larvae.*
[4] *I made this into italics as I used it in conjunction with a chapter on economics in the 1992 edition of "The Hive and Honey Bee" published by Dadant & Sons. I feel that if beekeepers used the thought of this phrase they would often be more prosperous.*
[*] Bevan

they covered it all over with an impervious coat of propolis.[5]

> 'For soon in fearless ire, their wonder lost,
> Spring fiercely from the comb the indignant host,
> Lay the pierced monster breathless on the ground,
> And clap in joy their victor pinions round;
> While all in vain concurrent numbers strive
> To heave the slime-girt giant from the hive-
> Sure not alone by force instinctive swayed,
> But blest with reason's soul-directing aid,
> Alike in man or bee, they haste to pour,
> Thick, hard'ning as it falls, the flaky shower;
> Embalmed in shroud of glue the mummy lies,
> No worms invade, no foul miasmas rise.'
>
> Evans.

"In these instances, who can withhold his admiration of the ingenuity and judgment of the bees? *In the first case*, a troublesome creature gained admission to the hive, which, from its unwieldiness, they could not remove, and which, from the impenetrability of its shell they could not destroy; here, then, their only resource was to deprive it of locomotion, and to obviate putrefaction; both which objects they accomplished most skillfully and securely, and, as is usual with these sagacious creatures, at the least possible expense of labor and materials. They applied their cement where alone it was required round the verge of the shell. *In the latter case*, to obviate the evil of decay, by the total exclusion of air, they were obliged to be more lavish in the use of their embalming material, and to case over the 'slime-girt giant,' so as to guard themselves from his noisome smell, What means more effectual could human wisdom have devised, under similar circumstance?"

When any member of a family dies, the bees are believed by many to know what has happened; and some are superstitious enough to put the hives in mourning, to pacify their sorrowing occupants; imagining that, unless this is done, the bees will never afterwards prosper! It has frequently been asserted that they sometimes take their loss so much

---

[5] *Bees will embalm most anything that they cannot remove from the hive—even a dead mouse.*

to heart, as to alight upon the coffin whenever it is exposed. A Clergyman told me, that he attended a funeral, where as soon as the coffin was brought from the house, the bees gathered upon it so as to excite much alarm. Some years after this occurrence, being engaged in varnishing a table, the bees alighted upon it in such numbers, as to convince him, that love of varnish, rather than sorrow or respect for the dead, was the occasion of their conduct at the funeral. How many superstitions, believed even by intelligent persons, might be as easily explained, if it were possible to ascertain as fully all the facts connected with them!

CHAPTER VI.

POLLEN, OR "BEE-BREAD."

POLLEN is gathered by the bees from blossoms, and is indispensable to the nourishment of their young—repeated experiments having proved that brood cannot be raised without it. It is very rich in the nitrogenous substances which are not contained in honey, and without which ample nourishment could not be furnished for the development of the growing bee. Dr. Hunter, on dissecting some immature bees, found that their stomachs contained pollen, but not a particle of honey.

We are indebted to Huber for the discovery, that pollen is the principal food of the young bees. As large supplies were often found in hives whose inmates had starved, it was evident that, without honey, it could not support the mature bees; and this led former observers to conclude that it served for the building of comb. Huber, after demonstrating that wax can he secreted from an entirely different substance, soon ascertained that pollen was used for the nourishment of the embryo bees. Confining some bees to their hive without any pollen, he supplied them with honey, eggs, and larvae. In a short time, the young all perished. A fresh supply of brood being given to them, with an ample allowance of pollen, the development of the larvae proceeded in the natural way.

I had an excellent opportunity of testing this substance, in the backward Spring of 1852. On the 5th of February, I opened a hive containing an artificial swarm of the previous year, and found many of the cells filled with brood. The combs being examined on the 23rd, contained neither eggs, brood, nor bee-bread; and the colony was supplied with pollen from another hive; the next day, a large number of eggs were found in the cells. When this supply was exhausted, laying again ceased, and was only resumed when more was furnished. During the time of these experiments, the weather was so unpromising, that the bees were unable to leave the hive.

Dzierzon is of opinion that bees can furnish food for their young, without pollen; although he admits that they can do it only for a short time, and at a great expense of vital energy; just as the strength of an animal nursing its young is rapidly reduced, if, for want of proper food, the very substance of the mother's body must be converted into milk. The experiment just described does not corroborate this theory, but confirms Huber's view, that pollen is indispensable to the development of brood.[1]

Gundelach, an able German Apiarian, says that if a colony with a fertile queen be confined to an empty hive, and supplied with honey,

---

[1] *Modern-day experiments confirm Dzierzon's view, that is, that bees can raise some brood from protein reserves in their body. These reserves are soon used up and if pollen, or some other source of protein, is not available to the bees then brood rearing will stop. This often happens in the winter cluster of the colony that runs out of pollen within the cluster area.*

comb will be rapidly built, and the cells filled with eggs, which in due time will be hatched; but the worms will all die within twenty-four hours. Some Apiarians believe that bees with an abundance of both pollen and honey, will secrete wax much faster than when supplied with honey alone; and that its secretion, without pollen, severely taxes their strength.

In September, 1856, I put a very large colony of bees into a new hive, to determine some points on which I was then experimenting. The weather was fine, and they gathered pollen, and built comb very rapidly; still, for ten days, the queen-bee deposited no eggs in the cells. during all that time, these bees stored very little pollen in the combs. One of the days being so stormy that they could not go abroad, they were supplied with rye flour, none of which, although very greedily appropriated, could be found in the cells. During all this time, as there was no brood to be fed, the pollen must have been used by the bees either for nourishment, or to assist them in secreting wax; or, as I believe, for both these purposes.

Bees prefer to gather *fresh* bee-bread, even when there are large accumulations of old stores in the cells. With [my type of] hives giving the control of the combs, the surplus of old colonies may be made to supply the deficiency of young ones; the latter, in Spring, being often destitute of this important article.[2]

If honey and pollen can both be obtained from the same blossoms, the industrious insect usually gathers a load of each. To prove this, let a few pollen-gatherers be dissected when honey is plenty; and their honey-sacs will ordinarily be full.[3]

The mode of gathering pollen is very interesting. The body of the bee appears to the naked eye to be covered with fine hairs, to which, when she alights on a flower, the farina [4] adheres. With her legs, she

---

[2] *Often in late Winter any pollen within the combs of the cluster area is used up and brood rearing stops. This is the reason that pollen substitute patties work so well at that time. The bees can not leave the cluster area for pollen that may be in other combs or areas of the hive, and the substitute patty is placed right over the bees.*

[3] *Unfortunately, Langstroth did not consider that a pollen-collecting bee needs energy with which to forage for the pollen. Bees often take a significant amount of honey with them when they forage for pollen. This does not mean that some bees do not collect both nectar and pollen, they do. The percentage that collects both is rather low.*

[4] *Langstroth is using a common word of the period for any ground flour.*

brushes it from her body, and packs it in the hollows, or *baskets*, one of which is on each of her thighs; [5] these baskets are surrounded by stouter hairs, which hold the load in its place. If from any cause the pollen cannot be readily gathered in balls, the bee will often roll herself in it, and return, all dusted over, to her hive.

When the bee brings home a load of pollen, she often shakes her body in a singular manner,[6] to attract the attention of other bees, who nibble from her thighs what they want for immediate use; the rest she stores away for future need, by inserting her body in a cell and brushing it from her legs; it is then carefully packed down, being often covered with honey, and sealed over with wax. Pollen is very rarely deposited in any except worker-cells.

Aristotle observed, that a bee, in gathering pollen, confines herself to the kind of blossom on which she begins, even if it is not so abundant as some others; thus a ball of this substance taken from her thigh, is found to be of a uniform color throughout; the load of one insect being yellow, of another, red, and of a third, brown; the color varying with that of the plant from which the supply was obtained.[7] They may prefer to gather a load from a single species of plant, because the pollen of different kinds does not pack so well together. Bees, by carrying the pollen or fertilizing substance of plants, on their bodies, from blossom to blossom, contribute essentially to their impregnation.

Though the importance of pollen has long been known, it is only of late that any attempts have been made to furnish a *substitute*. Dzierzon, early in the Spring, observed his bees bringing rye-meal to their hives from a neighboring mill, before they could procure any pollen from natural supplies.[8] The hint was not lost; and it is now a common

---

[5] *In entomological terms the pollen basket, or corbicula, is found on the tibia of the metathoracic (hind) legs. The comb that is used to clean the body hairs and then pack the pollen into the corbicula is on the inside of the next segment, called the basitarsus.*

[6] *Langstroth was observing the bee dances that communicate to the other bees a source of nectar or pollen. It took about 100 years before von Frisch sorted out the dance language. Bees will use the dances to indicate both nectar and pollen availability to nestmates*

[7] *Flower, or species, constancy is an important aspect of bee and flower evolution. If bees were not constant to one species on a single visit then cross pollination would not take place and the plants would not set fruit or seeds.*

[8] *Bees will often gather many different substances early in the Spring if pollen is not available— even dust particles from bird feeders. It is doubtful if they get much nutri-*

practice in Europe, where bee-keeping is extensively carried on, to supply the bees early in tile season with this article. Shallow troughs are set in front of the Apiaries, filled about two inches deep with *finely ground, dry, unbolted rye-meal*.[9] Thousands of bees, when the weather is favorable, resort eagerly to them, and rolling themselves in the meal, return heavily laden to their hives. In fine, mild weather, they labor at this work with great industry; preferring the meal to the old pollen stored in their combs. They thus breed early, and rapidly recruit their numbers. The feeding is continued till, the blossoms furnishing a preferable article, they cease to carry off the meal. The average consumption of each colony is about two pounds.

Mr. F. Sontag, a German Apiarian, says, that in the Spring of 1853, he fed one of his colonies with rye-meal, placed in the hive in an old comb; continuing the supply till they could procure fresh pollen abroad. This colony produced four strong swarms that Spring, and an adjoining stock not supplied with the meal, only one weak swarm.

Another German bee-keeper says, he has used wheat flour with very good results; the bees *forsaking the honey* furnished them, and engaging actively in carrying in the flour, which was placed about twenty paces in front of their hives.

The construction of my hives permits the flour to be easily placed where the bees can get it, without losing time in going abroad, or suffering for the want of it, when the weather confines them at home.

The discovery of this substitute removes a very serious obstacle to the culture of bees. In many districts, there is for a short time such an abundant supply of honey, that almost any number of strong colonies will, in a good season, lay up enough for themselves, and a large surplus for their owners. In many of these districts, however, the supply of pollen is often quite insufficient, and in Spring, the swarms of the previous year are so destitute, that unless the season is early, the pro-

---

tion from such ventures, but fungus spores within much of these materials may trigger the foraging. Ancient bees probably collected such spores for brood food long before there were flowering plants to supply pollen.

[9] *Modern pollen substitutes contain a much higher proportion of protein than either wheat or rye. They generally consist of various mixtures of soybean flour, and Brewer's yeast. Often the dry ingredients are then mixed with sugar syrup to form a patty of bread-dough-like consistency. The patty is then placed on the top bars of the brood chamber in late Winter.*

duction of brood is seriously checked, and the colony cannot avail itself properly of the superabundant harvest of honey.

While the honey-bee is regarded by the best informed horticulturists as a friend, a strong prejudice has been excited against it by many fruit-growers in this country; and in some communities, a man who keeps bees, is considered as bad a neighbor, as one who allows his poultry to despoil the gardens of others. Even the warmest friends of the "busy bee," may be heard lamenting its—propensity to banquet on their beautiful peaches and pears, and choicest grapes and plums.

In conversation with a gentleman, I once assigned three reasons, why the bees could not inflict any extensive injury upon his grapes. 1st, that as the Creator appears to have intended both the honey-bee and fruit for the comfort of man, it was difficult to conceive that he would have made one the natural enemy of the other. 2d, that as the supplies of honey from the blossoms had entirely failed, the season (1854) being exceedingly dry, if the numerous colonies in his vicinity had been able to help themselves to his sound grapes, they would have entirely devoured the fruit of his vines. 3d, that the jaws of the bee, being adapted chiefly to the manipulation of wax, were too feeble to enable it readily to puncture the skin even of his most delicate grapes.[10]

In reply to these arguments, being invited to go to his vines, and see the depredators in the very act, the result justified my anticipation's. Though many bees were seen banqueting on grapes, not one was doing any mischief to the *sound* fruit. Grapes which were bruised on the vines, or lying on the ground, and the moist stems, from which grapes had recently been plucked, were covered with bees; while other bees were observed to alight upon bunches, which, when found by careful inspection to be sound, they left with evident disappointment.

Wasps and hornets, which secrete no wax, being furnished with strong, saw-like jaws, for cutting the woody fibre with which they build their combs, can easily penetrate the skin of the toughest fruits. While the bees, therefore, appeared to be comparatively innocent, multitudes of these depredators were seen helping themselves to the best of the grapes. Occasionally, a bee would presume to alight upon a

---

[10] *Langstroth is correct in his observations on honey bees and fruit. They can not cut undamaged fruit and only take the sugar from the open or damaged fruit.*

bunch where one of these pests was operating for his own benefit, when the latter would turn and "show fight," much after the fashion of a snarling dog, molested by another of his species, while daintily discussing his own private bone.

After the mischief has been *begun* by other insects, or wherever a *crack*, or a spot of *decay* is seen, the honey-bee hastens to help itself, on the principle of "gathering up the fragments, that nothing may be lost." In this way, they undoubtedly do some mischief; but before war is declared against them, let every fruit-grower inquire if, on the whole, they are not more useful than injurious. As bees carry on their bodies the pollen, or fertilizing substance, they aid most powerfully in the impregnation of plants, while prying into the blossoms in search of honey or bee-bread. In genial seasons, fruit will often set abundantly, even if no bees are kept in its vicinity; but many Springs are so unpropitious, that often during the critical period of blossoming, the sun shines for only a few hours, so that those only can reasonably expect a remunerating crop whose trees are all murmuring with the pleasant hum of bees.

A large fruit-grower told me that his cherries were a very uncertain crop, a cold north-east storm frequently prevailing when they were in blossom. he had noticed, that if the sun shone only for a couple of hours, the bees secured him a crop.[11]

If the horticulturists who regard the bee as an enemy, could exterminate the race, they would act with as little wisdom as those who attempt to banish from their inhospitable premises every insectivorous bird, which helps itself to a small part of the abundance it has aided in producing. By making judicious efforts early in the Spring, to entrap the mother-wasps and hornets, which alone survive the Winter, an effectual blow may be struck at some of the worst pests of the orchard and garden. In Europe, those engaged extensively in the cultivation of fruit, often pay a small sum in the Spring for all wasps and hornets

---

[11] *With one or two good hives per acre of fruit trees a large amount of pollination can occur in a very short time—IF the weather cooperates. Often the weather is very cold during fruit bloom and bees are not able to forage in abundance. Bees do not do much foraging until the temperature reaches 60° F., from this temperature until about 95° F. there is almost a straight linear relationship to the number of foragers leaving the hive. It is the uncertainty of the weather that often requires more bee hives for a given crop than would be necessary in good weather.*

destroyed in their vicinity.

Fig. 62 (Pl. XIII.), shows the mangled head of a Mexican Honey-Hornet (p. 58). Fig. 63 shows the magnified head of the Honey-Bee. Fig. 64 shows the jaws of this Hornet, highly magnified. Fig. 65 shows the jaws of tile Honey-Bee highly mangled. A glance at these figures is enough to convince any intelligent horticulturist of the truth of Aristotle's remark made more than two thousand years ago that "bees hurt no kinds of fruit, but wasps and hornets are very destructive to them."

Preface to Chapter VII

There is no doubt in my mind that the hive designed by Langstroth was good for bees. His emphasis in this chapter on ventilation shows how much he understood about their needs as a colony of thousands of individuals. He was concerned about their well-being, both in the summer's heat and the winter's cold. He understood that if conditions were right then problems associated with the box hive would disappear--problems that contributed to swarming, diseases and to ripening of honey.

Why then did we stop making a hive much like his design with special ventilation ports? The problem is in the cost both of the original construction and in its maintenance. When a beekeeper has thousands of hives, the individual manipulation described here is just not practical. The ideas are right-it is just that we have to carry out the ventilation needs of the colony in different ways.

CHAPTER VII.

VENTILATION OF THE BEE-HIVE.

If a populous stock is examined on a warm day, a number of bees may be seen standing upon the alighting-board, with their heads turned towards the entrance of the hive, their abdomens slightly elevated, and their wings in such rapid motion, that they are almost as indistinct as the spokes of a wheel, in swift, rotation on its axis. A brisk current of air maybe felt proceeding from the hive; and if a small piece of down be suspended at its entrance, by a thread, it will be blown out from one part and drawn in at another. Why are these bees so deeply absorbed in their fanning occupation, that they pay no attention to the busy numbers constantly crowding in and out of the hive? And what is the meaning of this double current of air? To Huber, we owe the satisfactory explanation of these curious phenomena. The bees thus singularly plying their rapid wings, are ventilating the hive; and this double current is caused by pure air rushing in, to supply the place of the foul air which is forced out. By a series of beautiful experiments, Huber ascertained that the air of a crowded hive is almost as pure as the surrounding atmosphere. Now, as the entrance to such a hive is often very small, the air within cannot be renewed, without resort to artificial means. If a lamp is put into a close vessel, with only one small orifice, it will soon exhaust the oxygen, and cease to burn. If another small, orifice is made, the same result will follow; but if a current of air is by some device drawn out from one opening, an equal current will force its way into the other, and the lamp will burn until the oil is exhausted.

Fig. 22. PLATE IX.

It is on this principle of maintaining a double current by artificial means, that bees ventilate their crowded habitations. A file of ventilating bees stands inside and outside of the hive, each with head turned to its entrance, and while, by the rapid fanning of their "many twinkling" wings, a brisk current of air is blown out of the hive, an equal current is drawn in. As this important office demands unusual physical exertion, the exhausted laborers are, from time to time, relieved by fresh detachments.[1] If the interior of the hive permits inspection, many ventilators will be found scattered through it, in very hot weather, all busily engaged in their laborious employment. If its entrance is contracted, speedy accessions will be made to their numbers, both inside and outside of the hive; and if it is closed entirely, the heat and impurity quickly increasing, the whole colony will attempt to renew the air by rapidly vibrating their wings, and in a short time, if unrelieved, will die of suffocation.

Careful experiments show that pure air is necessary not only for the respiration of the mature bees, but for hatching the eggs, and developing the larvae; a fine netting of air-vessels enveloping the eggs, and the cells of the larvae being closed with a covering filled with air-holes.

In Winter, if bees are kept in a dark place, which is neither too warm nor too cold, they are almost dormant, and require very little air; but even under such circumstances, they cannot live entirely without it; and if they are excited by atmospheric changes, or in any way disturbed, a loud humming may be heard in the interior of their hives, and they need almost as much air as in warm weather.[2]

If bees are greatly disturbed, it will be unsafe, especially in warm weather, to confine them, unless they have a very free admission of air; and even then, unless it is admitted above, as well as below the mass of bees, the ventilators may become clogged with dead bees, and the colony perish. Bees under close confinement become excessively

---

[1] *The fanning of a bee's wings requires the same energy as flying and therefore once their metabolic reserves are used up they must rest to recharge the system.*

[2] *We have learned since this time in the designing of wintering houses and other metabolic studies that bees are very active during the winter. The bees in the center of the cluster generating the heat to keep the entire cluster warm. It is true that the bees on the outside of the cluster, where the temperature may be 45-50° F., will not use as much oxygen.*

heated, and their combs are often melted; if dampness is added to the injurious influence of bad air, they become diseased; and large numbers, if not the whole colony, may perish from dysentery. Is it not under precisely such circumstances that cholera and dysentery prove most fatal to human beings? The filthy, damp, and unventilated abodes of the abject poor, becoming perfect lazar-houses [3] to their wretched inmates.

I have several times examined the bees of new swarms which were brought to my Apiary, so closely confined, that they had died of suffocation. In each instance, their bodies were distended with a yellow and noisome substance, as though they had perished from dysentery. A few were still alive, and although the colony had been shut up only a few hours, the bodies of both the living and the dead were filled with this same disgusting fluid, instead of the honey they had when they swarmed.

In a medical point of view, these facts are highly interesting; showing as they do, under what circumstances, and how speedily, diseases may be produced resembling dysentery or cholera.[4]

In very hot weather, if thin hives are exposed to the sun's direct rays, the bees are excessively annoyed by the intense heat, and have recourse to the most powerful ventilation, not merely to keep the air of the hive pure, but to lower its temperature.

Bees, in such weather, often leave, almost in a body, the interior of the hive, and cluster on the outside, not merely to escape the close heat within, but to guard their combs against the danger of being dissolved. At such times, they are particularly careful not to cluster on new combs containing sealed honey, which from not being lined with cocoons, and from the extra amount of wax used for their covers, melt more readily than the breeding cells.

Apiarians have noticed that bees often leave their honey-cells almost bare us soon as they are sealed; but it seems to have escaped

---

[3] *Lazar-houses were the sick-houses for the poor. The name was also sometimes used for Leper-houses.*

[4] *There is no disease that kills the bees during the overheating and/or suffocation. First the bees probably regurgitate nectar or water to try to cool their bodies. This leaves the dead mass of bees somewhat wet looking. Secondly, when a bee dies it often defecates during the last throes of death, which adds to the appearance of dysentery.*

their observation, that this is absolutely necessary in very hot weather. In cool weather, they may frequently be found clustered among the scaled honey-combs, because there is then no danger of their melting.[5]

Few things are so well fitted to impress the mind with their admirable sagacity, as the truly scientific device by which they ventilate their dwellings. In this important matter, the bee is immensely in advance of the great mass of those who are called rational beings. It has, to be sure, no ability to decide, from an elaborate analysis of the chemical constituents of the atmosphere, how large a proportion of oxygen is essential to the support of life, and how rapidly the process of breathing converts it into a deadly poison: it cannot, like Liebig, demonstrate that God, by setting the animal and the vegetable world, the one over against the other, has provided that the atmosphere shall, through all ages, be as pure as when it first came from His creating hand. But shame upon us! That with all our boasted intelligence, most of us live as though pure air was of little or no importance; while the bee ventilates with a philosophical precision that should put to the blush our criminal neglect.

Is it said that ventilation, in our case, cannot be had without effort? Can it then be had for nothing, by the industrious bees? Those ranks of bees, so indefatigably plying their busy wings, are not engaged in idle amusement; nor might they, as some shallow utilitarian may imagine, be better employed in gathering honey, or superintending some other department in the economy of the hive. At great expense of time and labor, they are supplying the rest of the colony with the pure air so conducive to their health and prosperity.

Impure air, one would think, is bad enough; but all its inherent vileness is stimulated to still greater activity by air-tight, or rather lung-tight stoves,[*] which can economize fuel only by squandering

---

[5] *Honey filled comb has the additional problem in that it weighs much more than brood. However, old comb that has been used to rear brood is much stronger than new comb that has never had brood reared in the cells. This is because the bees add a "varnish" to the cells as well as the remnants of the silken cocoons of the larvae that get incorporated into the cell walls. These two added components keep strengthening the comb so that old, dark comb is very strong.*

[*] The beautiful open or Franklin stoves, for coal or wood, manufactured by Messrs. Treadwell, Perry & Norton, of Albany, New York, deserve the highest commendation as economizers of life, health, and fuel.

health and endangering life. Not only our private houses, but all our places of public assemblage, are either unproved with any means of ventilation, or to a great extent, supplied with those so deficient, that they only

> "Keep the word of promise to our ear,
> To break it to our hope."

That ultimate degeneracy must inevitably follow such gross neglect of the laws of health, cannot be doubted; and those who imagine that the physical stamina of a people may be undermined, and their intellectual, moral, and religious health suffer no decay, know little of the intimate connection which the Creator has established between body and mind.

Men may, to a certain extent, resist the injurious influences of foul air; as their employment's usually compel them to live more out of doors: but alas, alas! for the poor women! In the very land where they are treated with such merited deference and respect, often no provision is made to furnish them with that first element of health, cheerfulness and beauty, heaven's pure, fresh air.

The pallid cheek or hectic flush, the angular form and distorted spine, the enfeebled appearance of so large a portion of our women, who, to use the language of the lamented Downing, "in the signs of physical health, compare most unfavorably with all but the absolutely starving classes in Europe;" all these indications of debility, to say nothing of their care-worn faces and premature wrinkles, proclaim our violation of God's physical laws, and the dreadful penalty with which He is visiting our transgressions.

The man who shall convince the masses of the importance of ventilation, and whose inventive mind shall devise some simple, cheap, and efficacious way of furnishing a copious supply of pure air for our private dwellings, public buildings, and traveling conveyances, will be a greater benefactor than a Jenner or a Watt, a Fulton or a Morse.

(This, probably unsolicited, endorsement speaks to the advancement of heating stoves in an environment where heating stoves were still not well developed.)

VENTILATION.

In the ventilation of my hive, I have endeavored, as far as possible, to meet the necessities of the bees, under all the varying circumstances to which they are exposed in our uncertain climate, whose severe extremes of temperature forcibly impress upon the bee-keeper, the maxim of Virgil,

"Utraque vis pariter apihus metuenda."
"Extremes of heat or cold, alike are hurtful to the bees." [6]

To be useful to the majority of bee-keepers, *artificial* ventilation must be simple, and not as in Nutt's hive, and other labored contrivances, so complicated as to require almost as close supervision as a hot-bed or green-house.

By furnishing ventilation independent of the entrance,[7] we may improve upon the method which bees, in a state of nature, are often compelled to adopt, when the openings into their hollow trees are so small, that they must employ in hot weather, a larger force in ventilation, than would otherwise be necessary. By the use of my movable blocks (Pl. V., Fig. 17), the entrance may be kept so small, that only a single bee can go in at once, or it may be entirely closed, without the bees suffering for want of air. While the ventilators afford a sufficient supply, they pay be easily controlled, so as not to injure the brood by admitting too strong a current of chilly air. In the chapter on wintering bees, directions are given for ventilating the hives in cold weather, so as to carry off all superfluous moisture.

The construction of my hives allows of ventilation from above; and it should always be used, when bees are shut up for any length of

---

[6] *My feeling regarding winter honey stores is that an excess of actual colony need is often helpful because the honey acts as a buffer to sudden changes in temperature. That is, the honey holds some heat and therefore the bees are able to re-cluster when the temperature drops quite suddenly. These sudden changes from daylight to night can be quite great and the buffering effect of the honey is very helpful to the bees. The same argument can be made for honey in very warm days in the summer.*

[7] *Modern hive designs have abandoned Langstroth's ventilation scheme mostly because of the cost in making them. Such things as auger holes, though, are generally beneficial to the hive and should be encouraged. Many beekeepers want to keep their equipment "clean" and hesitate to bore holes into the hive bodies, yet such a simple addition will help the bees ventilate much more easily. Just don't put the auger holes in the handholds. Bore the auger holes below or to the side of the handholds.*

time, to be moved, that the colony may not be suffocated, by the lower ventilators becoming clogged by dead bees. As the entrance of the hive, may in a moment, be enlarged to any desirable extent, without perplexing the bees, any quantity of air which the bees may require, can be admitted; the ventilator on the back allowing a free current to sweep through the hive. The entrance may be fourteen inches and upwards in length; but as a general rule, in a large colony, it need not, in Summer, exceed four inches; while, during the rest of the year, one or two inches will suffice. In very hot weather, especially if the hive stands in the sun, the bees cannot have too much air; and the ventilators in the upper part of the main hive should all be kept open.

CHAPTER VIII.

REQUISITES OF A COMPLETE HIVE.

In this chapter, I shall enumerate certain advantages which seem essential to the idea of a complete hive. Instead of disparaging other hives, I prefer inviting the attention of bee-keepers to the importance of these requisites; some of which, I believe, are contained in no hive but my own. If, after careful scrutiny, they commend themselves to the judgment of practical cultivators, they will serve to test the comparative merits of the various hives in common use.[1]

1. A complete hive should give the Apiarian such perfect control of all the combs, that they may be easily taken out without cutting them, or enraging the bees.

2. It should permit all necessary operations to be performed without hurting or killing a single bee. Most hives are so constructed, that they cannot be used without injuring or destroying some of the bees; and the destruction of even a few, materially increases the difficulty of managing them.[2]

3. It should afford suitable protection against extremes of heat and cold, sudden changes of temperature, and the injurious effects of dampness.

The interior of a hive should be dry in winter, and free in Summer from a pent and almost suffocating heat.

4. It should permit every desirable operation to be performed, without exciting the anger of the bees.

5. Not one unnecessary motion should be required of a single bee.

---

[1] *We have such a complete standardization of one hive today that we have a hard time imagining so many different "patented" beehives. But in Langstroth's time there were dozens of hives to choose from, and this choice continued for many years after this book was written.*

[2] *The fact that an injured bee gives off alarm pheromone is often overlooked in most beekeeping management. A beekeeper will have less stinging behavior if the number of injured bees is kept to a minimum.*

As the honey-harvest, in most locations, is of short continuance, all the arrangements of the hive should facilitate, to the utmost, the work of the busy gatherers. Hives which compel them to travel with their heavy burdens through densely crowded combs, are very objectionable. Bees instead of forcing their way through thick clusters, can easily pass into the top surplus homey-boxes of my hives, from any comb in the hive, and into every box, without traveling at all over the combs.[3]

6. It should afford suitable facilities for inspecting, at all times, the condition of the bees.

7. It should be capable of being readily adjusted to the wants of either large or small colonies.

By means of a movable partition, my hive can be adjusted, in a few moments, to the wants of any colony how ever small; and with equal facility be enlarged, from time to time, or at once restored to its full dimensions.[4]

8. It should allow the combs to be removed without any jarring.

Bees manifest the utmost aversion to any motion which tends to loosen or detach their combs. The movable frames, however firmly fastened, can all be loosened in a few moments, without injuring or exciting the bees.

9. It should allow every good piece of comb to be given to the bees, instead of melting it into wax.

10. It should induce the bees to build regular combs. A hive containing too much comb suitable only for storing honey, or raising drones, cannot be expected to prosper.

11. It should furnish empty comb, to induce bees to occupy more readily the surplus honey-receptacles.

---

[3] *This can still be a problem with our "Langstroth" hives when we stack them up six or seven supers high. The simplest way to get around this is to have some entrances in the upper supers through auger holes or some other means of entrance for the incoming bees.*

[4] *"Follower boards" were much more common in earlier times than today. These thin boards, that have the dimensions of the inside of the hive, do provide a simple way to partition off a portion of the hive. I suspect that these boards are less used today simply because it is another piece of equipment to make and store when not in use.*

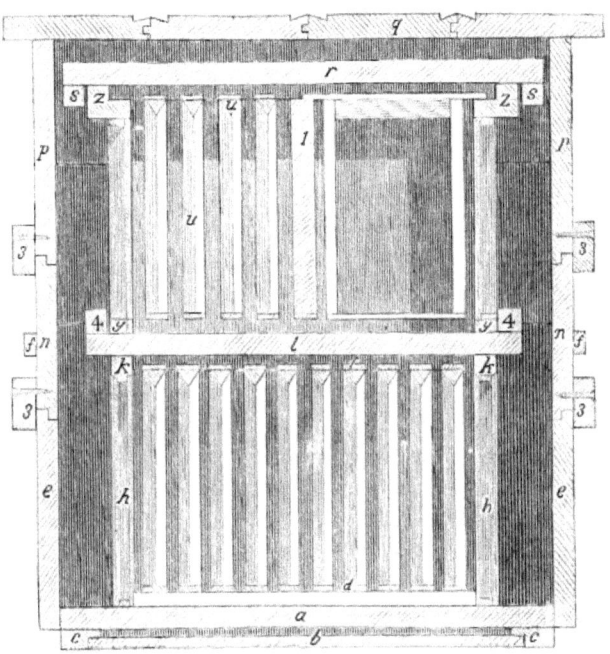

Fig. 23. PLATE X.

Fig 73.

REQUISITES OF A COMPLETE HIVE.

12. It should prevent the over-production of drones, by permitting the removal of drone-comb from the hive.

13. It should enable the Apiarian, if too many drones have been raised, to trap and destroy them, before they have largely consumed the honey of the hive.

This is effected, in my hives, by adjusting the blocks (Pl. III., Figs 11, 12) which regulate the entrance.

14: It should enable the Apiarian to remove such combs as are too old. [5]

The *upper* part of a comb, being generally used for storing honey, will last for many years.

15. It ought to furnish all needed security against the ravages of the bee-moth.

16. It should furnish to the Apiarian some accessible place, where the larvae of the bee-moth, when fully grown, may wind themselves in their cocoons.[6]

17. It should enable the Apiarian, by removing the combs, to destroy the worms, if they get the advantage of the bees.

18. The bottom-board should be permanently attached to the hive, for convenience in moving it, and to prevent the depredations of moths and worms.

Sooner or later, there will be crevices between every movable bottom-board and the sides of the hive, through. which moths will gain admission to lay their eggs, and under which worms, when fully grown, will retreat to spin their webs [7] In my hive, there is no place where the moth, can get in, except at the entrance for the bees, which may be contracted or enlarged, to suit the strength of the colony; and

---

[5] *The lacking of beeswax foundation would necessitate cutting the comb back to this acceptable upper edge. From that starting point the bees would be able to draw new, straight comb.*

[6] *The spinning of a wax moth cocoon should not happen in a strong hive as the bees will kill the larger larvae. When the larvae spin their cocoons they often chew out a depression in the hive body to protect the cocoon.*

[7] *Langstroth just did not appreciate, or know, how very small is the first instar of the wax moth larvae. They can enter most crevices where two hive bodies come together. The moth lays the eggs within these cracks and the small larvae crawl into the hive. Bees do recognize that the small larvae can enter through these openings and try to fill up these cracks with propolis.*

which, from its peculiar shape, the bees are easily enabled to defend. If, however, any prefer movable bottom-boards, they can be used in my hive.

19. The bottom board should slant toward the entrance, to facilitate the carrying out of dead bees, and other useless substances; to aid a colony in protecting itself against robbers; and to carry off moisture, and prevent rain from beating into the hive.

20. The bottom-board should admit of being easily cleared, in cold weather, of dead bees.

If suffered to remain, they often become mouldy, and injure the health of the colony. In dragging them out, when the weather moderates, the bees often fall with them on the snow, and are so chilled, that they never rise again; for a bee, in flying away with the dead, frequently retains its hold, until both fall to the ground.

21. No part of the interior of the hive should be below the level of the place of exit. If this principle is violated, the bees must at great disadvantage, drag, up all their dead, and all the refuse of the hive.[8]

22. It should afford facilities for feeding bees, both in warm and cold weather.

In this respect, the movable-comb hive has unusual advantages. In warm weather, sixty colonies may, in less than an hour, receive each a quart of food, without any feeder, and with no risk from robber-bees.

23. It should permit the easy hiving of a swarm, with out injuring any bees, or risking the destruction of the queen.

24. It should admit of the safe transportation of the bees to any distance whatever.

The permanent bottom-board, the firm attachment of each comb to a separate frame, and the facility with which air can be given to confined bees, admirably adapt my hive to this purpose.

25. It should furnish bees with air, when the entrance, for any cause, must be entirely shut.

---

[8] *This is one of the disadvantages of the anti-robbing screen that I have suggested (See footnote 4 on page 264). The bees do have some difficulty in removing dead bees and other debris. The simple answer to this problem is to occasionally remove the screen and clean the bottom. Good hygienic bees will still remove the debris from the hive even with an anti-robbing screen in place.*

26. It should furnish facilities for enlarging, contracting, and closing the entrance, to protect the bees against robbers, and the bee-moth; and when the entrance is altered, the bees ought not, as in most hives, to lose valuable time in searching for it.

27. It should give the requisite ventilation, without enlarging the entrance so much as to expose the bees to moths and robbers.

28. It should furnish facilities for admitting at once a large body of air, that the bees may be tempted to fly out and discharge their feces, on warm days in Winter, or early Spring.

If such a free admission of air cannot be given, the bees, by losing a favorable opportunity of emptying themselves, may suffer from diseases resulting from too long confinement.[9]

29. It should enable the Apiarian to remove the excess of bee-bread from old stocks. (See p. 82.)

30. It should enable the Apiarian to remove the combs, brood, and stores, from a common to an improved hive, so that the bees may be easily able to attach them again in their natural positions. A colony transferred to my hive will repair their combs, in a few days, so as to work as well as before their removal.

31. It should permit the safe and easy dislodgement of the bees from the hive. This requisite is especially important, when it becomes necessary to break up weak stocks, to join them to others.

32. It should allow the bees, together with the heat and odor of the main hive, to pass in the freest manner, to the surplus honey-receptacles.

In this respect, all other hives with which I am acquainted are more or less deficient: the bees being forced to work in receptacles difficult of access, and in which, in cool nights, they find it impossible to maintain the requisite heat for comb-building.[10] Bees cannot, in such hives; work to advantage in glass tumblers, or other small vessels. One of the most important arrangements of my hive, is that by which

---

[9] *Langstroth often refers to confinement or honey or lack of ventilation as contributing to dysentery. We now know that Nosema disease,* Nosema apis *(and to a lesser extent* Nosema ceranae*), is the major cause of dysentery or fecal spotting of the hive. Such things as confinement may exacerbate the problem, but are not the cause.*

[10] *The heat necessary for comb building is generally figured to be in the 105-106° F. range.*

the heat passes into the upper receptacles for storing honey, as naturally as the warmest air ascends to the top of a heated room.

33. It should permit the surplus honey to be taken away, in the most convenient, beautiful, and salable forms, and without risk of annoyance from the bees.

In my hives, it may be made on frames in an upper chamber, in tumblers, glass boxes, wooden boxes, small or large, earthen jars, flower-pots, in short, in any kind of receptacle which may suit the fancy or convenience of the beekeeper. Or these may all be dispensed with, and the honey taken from the interior of the main hive, by removing the full frames, and supplying their places with empty ones.

34. It should admit of the easy removal of good honey from the main hive, when its place can be supplied by the bees with an inferior article.

In districts where buckwheat is raised, any vacancies made by removing the choice honey from the hive will be rapidly filled.

35. When quantity and not quality is the object sought, it should allow the greatest yield, that the surplus of strong colonies may be given, in the Fall, to those which have an insufficient supply.[11]

By surmounting my hive with a box with the same dimensions, and transferring the combs to this box, the bees, when they build new comb, will descend and fill the lower frames, using, as fast as the brood hatches, the upper box for storing honey.[12] The combs in this box, containing a large amount of bee-bread, and being of a size adapted to the breeding of workers, will be very suitable for aiding weak colonies.

36. It should be able to compel the force of a colony to be may directed to raising young bees; that brood may be on hand to form new colonies, and strengthen feeble stocks.

---

[11] *Being able to add a comb of honey is a very powerful reason for using the movable-comb hive—just take a frame or two and add it to a weaker hive. Certainly, this would not have been possible with the old box hives.*

[12] *This, it seems to me, is the beginning of the concept of supering with boxes the same size as the brood chamber—the honey surplus management that we use today. While most beekeepers today practice the removal of whole supers of honey, in the days of Langstroth the removal of a single comb even would have been difficult since they were attached within the box hives. Thus, the killing of whole hives was practiced in order to collect the honey.*

37. It ought to be so constructed that, while well protected from the weather, the sun may be allowed in early Spring to encourage breeding, by warming up the hive.

38. The hive should be equally well adapted to be used as a swarmer, or non-swarmer.

In my hives, the bees may be allowed to swarm as in common hives, and be managed in the usual way. Even on this plan, the control of the combs will be found to afford unusual advantages. Non-swarming hives, managed in the ordinary way, are liable to swarm unexpectedly, in spite of all precautions. In my hives, the queen may be prevented from leaving, and a swarm will not depart without her.[13]

39. It should enable the Apiarian to prevent a new swarm from forsaking its hive. This vexatious occurrence can always be prevented, by so adjusting the entrance, for a few days, that the queen cannot leave the hive.

40. It should enable the Apiarian, if he allows his bees to swarm, and wishes to secure surplus honey, to prevent their swarming more than once in a season.[14]

41. It should enable the Apiarian, who relies on natural swarming, and wishes to multiply his colonies as fast as possible, to make vigorous stocks of all his small after-swarms.

Such swarms contain a young queen, and if they can be judiciously strengthened, usually make the best stock hives. My hives enable me to supply all such swarms at once with combs containing bee-bread, honey, and maturing brood.

42. It should enable the Apiarian to multiply his colonies with a certainty and rapidity which are impossible if he depends upon natural swarming.

43. It should enable the Apiarian to supply destitute colonies with the means of obtaining a new queen.

---

[13] *Langstroth makes a compelling argument here for his hive. In modern times we depend upon knowledge of the "condition" of the colony in order to manage it properly. Before the movable-frame hive the inner workings of the hive were mostly a mystery.*

[14] *It seems to me that here Langstroth is submitting to the common practices of the day, and not to his better judgment, as he will state later in the book, i.e., bees should not be allowed to swarm if you want the most honey.*

Every Apiarian, for this reason alone, would find it to his advantage to possess, at least, one such hive.

44. It should enable him to catch the queen, for any purpose; especially to remove an old one whose fertility is impaired by age.

45. While a complete hive is adapted to the wants of those who desire to manage their colonies on the most improved plans, it ought to be suited to the wants of those who, from timidity, ignorance or any other reason, prefer the common way.

46. It should enable a single bee-keeper to superintend the colonies of different individuals. [15]

Many persons would keep bees, if an Apiary, like a garden, could be superintended by a competent individual. No person can agree to do this with common hives. If the bees are allowed to swarm, he may be called in a dozen different directions at once, and if any accident, such as the loss of a queen, happens to the colonies of his customers, he can usually apply no remedy.

On my plan those who desire it, may witness the industry of this sagacious insect, and gratify their palates with its delicious stores harvested on their own premises, without incurring either trouble, or risk of annoyance.

47. All the joints of the hive should be water-tight, and there should be no doors or shutters liable to shrink, swell, or get out of order.

The importance of this requisite will be obvious to any one who has had the ordinary share of vexatious experience with such fixtures.

---

[15] *While the management of colonies of bees for other people is not unknown today, it is not common, yet maybe what is implied within this statement is the ability to manage a large number of colonies.*

48. It should enable the bee-keeper entirely to dispense with sheds or costly Apiaries; as the hive itself should alike defy heat or cold, rain or snow.[16]

49. It ought not to be liable to be blown down in high winds.

My hives may be made so low, for very windy situations, that it would require almost a hurricane to upset them.

50. A complete hive should have its alighting-board so constructed, as to shelter the bees against wind and wet, thus facilitating to the utmost their entrance with heavy burdens.

If this precaution is neglected, the colony cannot be encouraged to use, to the best advantage, the unpromising days which often occur in the working season.

51. A complete hive should be protected against the destructive ravages of mice in Winter.

When cold weather approaches, all my hives may have their entrances contracted by the movable blocks, so that a mouse cannot gain admission.

52. It should permit the bees to pass over their combs in the freest manner, both in Summer and Winter.

While such easy intercommunication facilitates the Summer work of the hive, it is often, in cold Winters, indispensable to the life of the colony.

53. It should permit the honey, after the gathering season is over, to be concentrated where the bees will most need it.

If the latter part of the season has been unpropitious, the centre combs, in which a colony usually winters, may have very little honey, while the others are well supplied. In hives where this cannot be remedied, it often causes the loss of the bees.[17]

54. It should permit a generous supply of honey to be left, in the Fall, in the hive, without detriment either to the bees, or to their owner.

---

[16] *Many beekeepers that used straw skeps had a covering over the hives. The Langstroth original hive was also double walled so it would have been better in the winter than our current models.*

[17] *This must have certainly been one of the most important features of the moveable-comb hive—the frames of honey, or brood, were interchangeable.*

If too much honey is taken, and the Winter prove very unfavorable, the bees may starve. In the common hives, if too much remains, it cannot be removed in the Spring, and it is thus worse than lost to the bee-keeper, by occupying the room needed for raising brood.

55. It should permit the Apiarian to remove such combs as cannot be protected by the bees, to a place of safety.

When a colony becomes greatly reduced in numbers, its empty combs may cause its destruction, by affording a harbor to the bee-moth; or its rich stores of honey may tempt robbing bees to despoil it. In the common hives, often nothing can be effectually done to prevent such casualties.

56. It should permit the space for spare honey-receptacles to be enlarged or contracted at will, without any alteration or destruction of existing parts of the hive.

Without the power to do this, the productive force of a colony is in some seasons greatly diminished.

57. It should be so compact as to economize, if possible, every inch of material used in its construction.[18]

58. The hive, while presenting a neat appearance, should admit, if desired, of being made highly ornamental.

59. It should enable an Apiarian to lock up his hives in some cheap and convenient way.

As my bottom-hoards are not movable, the contents of a hive, when it is locked, can only be reached by carrying it bodily away.

60. It should allow the contents of a hive bees, combs, and all, to be taken out when it needs any repairs.

As movable-comb hives can, at any time, be thoroughly overhauled and repaired, they should last for generations.

61. A complete hive, while possessing *all* these requisites, should, if possible, *combine* them in a *cheap* and *simple* form, adapted to the wants of all who are competent to cultivate bees.

---

[18] *The modern adaptation of the Langstroth hive is certainly more economical of lumber than the original patent. While all the extra flourishes may have used only scrap lumber it was still a lot of extra wood. Even our modern hives do not use standard dimensional lumber, For example, the standard hive body should be 7½ inches high, as this would use a standard 8-in. board.*

Few would imagine, in reading this long list of desirables, that any hive can combine them all, without being exceedingly complicated and expensive. On the contrary, the cheapness and simplicity with which the movable-comb hive effects this, is its most striking feature, and the one which has cost me more study than all the other points besides. Bees can work, in this hive, with even greater facility than in a simple box, as the frames being left rough by the saw, give them an admirable support while building their combs; and they can enter the spare honey-boxes with more ease than they could mount to an equal height in the upper part of a common box-hive.[19]

There are a few desirables to which my hive, even if it were perfect, could make no pretensions!

It promises no splendid results to those who are too ignorant or too careless to be entrusted with the management of bees. In bee-keeping, as in all other pursuits, a man must first understand his business, and then proceed upon the good old maxim, that "the hand of the diligent maketh rich."

It has no talismanic influence which can convert a bad situation for honey into a good one; or give the Apiarian an abundant harvest, whether the season is productive or otherwise. As well might a farmer seek for some kind of wheat which will yield an enormous crop, in any soil, and in every season.

It cannot enable the cultivator, while rapidly multiplying his stocks, to secure the largest yield of honey from his bees. As well might the breeder of poultry pretend, that in the same year, and from the same stock, he can both raise the greatest number of chickens, and sell the largest number of eggs.

Worse than all, it cannot furnish the many advantages enumerated, and yet be made in as little time, or quite as cheaply, as a hive which, in the end, proves to be a very dear bargain!

In the progress of my invention, while undoubtedly attaching undue importance to some points, I have steadily endeavored to avoid constructing a hive in accordance with crude theories, or mere conjectures. Having carefully studied the nature of the honey-bee for many

---

[19] *Langstroth may be accused of overselling at this point, as bees can walk up glass because of the pads on their tarsi (feet). While a well-used box hive may become rather smooth inside I do not think the bees would have any trouble walking up the walls.*

years, and compared my observations with that of writers and cultivators who have spent their lives in extending the sphere of Apiarian knowledge, I have endeavored to remedy the many difficulties with which bee-culture is beset, by adapting my invention to the actual habits and wants of the insect.[20] I have also tested the merits of this hive by long continued experiments, made on a large scale, so that I might not, by deceiving both myself and others, add another to the useless contrivances which have deluded and disgusted a too credulous public. I would, however, utterly repudiate all claims to having devised even a perfect bee-hive. Perfection belongs only to the works of Him, to whose omniscient eye were present all causes and effects, with all their relations, when he spake, and from nothing formed the Universe. For man to stamp the label of perfection upon any work of his own, is to show both his folly and presumption.

The culture of bees is confessedly at a low ebb in this country, when thousands can be induced to purchase hives which are in glaring opposition to the plainest dictates of common sense, as well as the simplest principles of Apiarian knowledge. Such have been the losses of deluded purchasers, that it is no wonder they turn from everything offered in the shape of a patent bee-hive, as a worthless conceit, if not an outrageous swindle.

So deleterious has been the influence of so-called "Improved Hives" that, as a general thing, only those who have used hives of the simplest form, have derived much profit from their bees. They have wasted neither time, money, nor bees, upon contrivances which can secure nothing in advance of a simple box-hive, with an upper chamber.

A *hive of the simplest possible construction*, is a close imitation of the abode of bees in a state of nature; being a mere hollow receptacle,

---

[20] *In many ways Langstroth is quite correct—the movable-comb hive was designed with the bee in mind. The hive was the culmination of many years of his testing, but also of many other minds working, over the centuries, on the problem of easy and successful bee culture. However, it was Langstroth who recognized the meaning of "bee space" and then designed a hive that used the space. Others, e.g., Dzierzon, may have had, or been close to, a movable-comb hive, but it was the recognition of the bee space AND the movable combs that made Langstroth the true Father of modern beekeeping.*

where, protected from the weather, they can lay up their stores.[21] *An improved hive*, is one which contains an additional, separate apartment, where bees can store their surplus honey for man. Most hives in common use are only modifications of this latter hive, and, as a general rule, are bad, exactly in proportion as they depart from it. While they tempt the common bee-keeper to ruinous departures from the beaten path, they furnish him no remedy for the loss of the queen, or the casualties to which bees are exposed. Such hives, therefore, form no reliable basis for any improved system of management; and hence, the cultivation c£ bees, in this country, has declined for the last fifty years, and the Apiarian is as dependent as ever upon the caprices of an insect, which more than any of his domestic animals, may be completely subjected to his control.

I would respectfully submit, that no hive which does not furnish a thorough control over every comb, can give that substantial advance over the simple improved or chamber hive, which the bee-keeper's necessities demand. Of such hives, the best are those which best unite *cheapness* and *simplicity*, with *protection in winter*, and *ready access* to the spare honey-boxes.

Having thus enumerated the tests to which all hives ought of be subjected, I submit them to the candid consideration of those, who, having the largest experience in the management of bees, are most conversant with the evils of the present system. If, on *full trial*, they find that the movable-comb hive can abide these tests, they may be willing to endorse the enthusiastic language of an experienced Apiarian, who, on examining its practical workings, declared that "it introduced not simply an improvement, but a *complete revolution* in bee-keeping." [22]

---

[21] *In many ways the modern beehive does resemble a cavity in a hollow tree, except for the fact that it might not have as much wood (insulation) as a tree. Recognizing the concept of a natural hive and adapting it for man's use is what Langstroth developed in the movable-comb hive.*

[22] *Since we generally credit Langstroth for the development of modern beekeeping, it can be said that he did start a "complete revolution" in beekeeping.*

Preface to Chapters IX and X

It is in these next two chapters that Langstroth gives us his great knowledge of bees and beekeeping. It was in his quest for knowledge about bees that he came across the need for examining bees on a regular basis and from such a quest came the movable-comb hive. Though you will see, as you read through these chapters, he recognized the fact that not everyone would buy (or had not yet been convinced to buy) his patented hive, and still he could give them information about the bees' habits that they could use.

Up until the movable-comb hive was invented the general mode of beekeeping was still unchanged from the Middle Ages...have bees swarm and then kill half each year to collect the honey and wax. It was a system that Langstroth did not think he could change easily so he often will give directions for any hive. In today's context the emphasis on swarming for swarming sake seems archaic. Today we do almost anything to keep bees from swarming. Yet, in Langstroth's beekeeping world, yearly swarming was the way people understood honey bees. Langstroth does mention that if his hives were used properly swarming (at least in the natural sense) was not necessary. In the long run (maybe not so long either) the box hives gave way to this new method of management that was made possible by the movable combs.

There is a lot of information regarding bees hidden here and there in these chapters. He just drops these wonderful little statements that carry lots of meaning. Maybe you do not catch the true importance the first time you read the statement, or maybe, because you are new to bees, you miss the essence of the words. I am impressed with how much he did know about honey bees.

CHAPTER IX.

NATURAL SWARMING, AND HIVING OF SWARMS.

The swarming of bees is one of the most beautiful sights in the whole compass of rural economy. Although many who use movable-comb hives prefer the artificial multiplication of colonies, few would be willing entirely to dispense with the pleasing excitement of natural swarming.

> "Up mounts the chief, and to the cheated eye
> Ten thousand shuttles dart along the sky;
> As swift through aether rise the rushing swarms,
> Gay dancing to the beam their sun-bright forms;
> And each thin form, still ling'ring on the sight,
> Trails, as it shoots, a line of silver light.
> High pois'd on buoyant wing, the thoughtful queen,
> In gaze attentive, views the varied scene,
> And soon her far-fetch'd ken discerns below
> The light laburnum lift her polish'd brow,
> Wave her green leafy ringlets o'er the glade,
> And seem to beckon to her friendly shade.
> Swift as the falcon's sweep, The monarch bends
> Her flight abrupt; the following host descends.
> Round the fine twig, like cluster'd grapes, they close
> In thickening wreaths, and court a short repose."
>
> Evans.

The multiplication of colonies by swarming, both guards the bee against the possibility of extinction,[1] and makes its labors in the highest degree useful to man. The laws of reproduction in insects not living in regular colonies, secure an ample increase of their numbers. The same is true of those which live in colonies during the warm weather only, as hornets, wasps and humble-bees.[2] In the Fall, the mates perish, while the impregnated females, retreating into Winter quar-

---

[1] *Swarming is reproduction in the honey bee. The division of a colony by binary fission allows the species to continue, as disease, fire or other calamities could kill the parent colony.*

[2] *Bumble bees were often called humble-bees until about the middle of the 20th century.*

ters, remain dormant till warm weather restores them to activity, that each may become the mother of a new family.[3]

The honey-bee, however is so organized that it must live in a community during the entire year; for while the balmy breezes of the Spring will quickly thaw the frozen body of a torpid wasp, the bee is chilled by a temperature no lower than 50°; and it would be as impossible to restore a frozen bee to animation, as to recall to life the stiffened corpses in the charnel-house[4] of the Convent of the Great St. Bernard. Bees, therefore, in cool weather, must associate in large numbers, to maintain the heat necessary for their preservation; and the formation of new colonies, after the manner of wasps and hornets, is out of the question. Even if the young queens, like the mother-wasps, were able, without any assistance, to found new colonies, they could not maintain the warmth requisite for the development of their young. And if this were possible, and they were furnished with a proboscis, for gathering honey, as long as that of a worker, baskets on their thighs for carrying bee-bread, and pouches on their abdomens for secreting wax, they would still be unable to amass treasures for our use, or even to lay up the stores requisite for their own preservation.

How admirably are all these difficulties obviated by the present arrangement! Their domicile being well supplied with all the requisite materials, the bees have added thousands, in the full vigor of youth, to their already numerous population, while such insects as depend upon the heat of the sun are still dormant. They can thus send off early colonies, strong enough to take full advantage of the honey-harvest, and to provision the new hive against the approach of Winter.[5] From these

---

[3] *All of these social insects survive by producing new queens in the fall, which mate, and then these mated queens find a protected place to hibernate; a place where they will not freeze, e.g., the leaf litter of the forest, in order to survive the winter. The old nests are not re-used.*

[4] *A charnel house is a house or place in which the bodies or bones of the dead are deposited.*

[5] *The "window of opportunity", or biological window, available for a swarm to survive the first year is rather narrow. The colony must produce enough bees early, and then the swarm must leave in time to store surplus honey to carry it through the following winter. Yet if the swarm were to leave too early, there is not sufficient new sources of nectar and pollen to allow it to survive, nor could the newly produced queen, in the parent colony, be able to fly out and mate if the weather was cold. Early swarms are essential but as the old verse ends, "...a swarm of bees in July, let them fly."*

considerations, it is evident that swarming, so far from being the forced or unnatural event which some imagine, is one, which could not possibly be dispensed with, in a state of nature.

Let us now inquire under what circumstances swarming ordinarily takes place.

The time when new swarms may be expected, depends, of course, upon the climate, the forwardness of the season, and the strength of the stocks. In our Northern and Middle States, they seldom issue before the latter part of May; and June may there be considered as the great swarming month. In Brownsville, Texas, on the lower Rio Grande, bees often swarm quite early in March.

In the Spring, as soon as a hive well filled * with comb, can no longer accommodate its teeming population,[6] the bees prepare for emigration, by building a number of royal cells. These cells are begun about the time that the drones make their appearance in the open air; and when the young queens arrive at maturity, the males are usually very numerous.

The first swarm is invariably led off by the old queen, unless she has died from accident or disease, then it is accompanied by one of the young ones reared to supply her loss. The old mother, unless delayed by unfavorable weather, usually leaves soon after one or more of the royal cells are sealed over. There are no signs from which the Apiarian can predict the certain issue of a *first* swarm. For years, I spent much time in the vain attempt to discover some *infallible* indications of first swarming; until facts convinced me that there can be no such indications. If the weather is unpleasant, or the blossoms yield an insufficient supply of honey, bees often change their minds, and refuse to swarm at all, even although their preparations have been

---

* In our Northern and Middle states, bees seldom swarm unless the hive is filled with comb; in Southern latitudes, however, the swarming instinct seems to be much more powerful. In Matamoras and Brownsville, I have seen many colonies issue from hives only partially filled with comb.

[6] *One of the major triggers of swarming is that the queen pheromone becomes so diluted within at least a certain population of workers that the bees start queen cells. Older queens may have less pheromone and thus these colonies are more likely to swarm. Though, in all probability it is poor distribution of the queen pheromone that is the major cause of queen cells being built and the bees swarming.*

so fully completed, that, like the traveler whose trunks are packed, they have filled their honey-sacs for their intended journey.

If, in the swarming season, but few bees leave a strong hive, on a clear, calm, and warm day, when other colonies are busily at work, we may look with great confidence for a swarm, unless the weather prove suddenly unfavorable. As the old queens which accompany the 'first swarm' are heavy with eggs, they fly with such difficulty, that they are shy of venturing out, except on fair, still days. If the weather is very sultry, a swarm will sometimes issue as early as seven o'clock in the morning; but from ten, A.M., to two, P.M., is the usual time; and the majority of swarms come off when the sun is within an hour of the meridian. Occasionally, a swarm ventures out as late as five, P.M.; but an old queen is seldom guilty of such an indiscretion.[7]

I have repeatedly witnessed, in my observing-hives, the whole process of swarming. On the day fixed for their departure, the queen is very restless, and instead of depositing her eggs in the cells, roams over the combs, and communicates her agitation to the whole colony.[8] The emigrating bees usually fill themselves with honey, just before their departure; but in one instance, I saw them lay in their supplies more than two hours before they left. A short time before the swarm rises, a few bees may generally be seen sporting in the air, with their heads turned always to the hive; and they occasionally fly in and out, as though impatient for the important event to take place. At length, a violent agitation commences in the hive; the bees appear almost frantic, whirling around in circles continually enlarging, like those made by a stone thrown into still water, until at last, the whole hive is in a state of the greatest ferment, and the bees, rushing impetuously to the entrance, pour forth in one steady stream. Not a bee looks behind, but each pushes straight ahead, as though flying "for dear life" or urged on by some invisible power, in its headlong career.

---

[7] *Again, this is a window of opportunity. If the swarm issues early enough during the day they can send the scouts out to find a new home and maybe leave the same day for the new home. Otherwise the bees have to maintain their cluster throughout a night and use up important reserves needed to start the new home.*

[8] *Langstroth attributes more control to the queen than modern biologists. Current theory is that the workers prevent the queen from laying eggs in order to reduce her weight so that she may be able to fly with the swarm.*

Often, the queen does not come out until many have left; and she is frequently so heavy, from the number of eggs in her ovaries, that she falls to the ground incapable of rising with her colony into the air. The bees soon miss her, and a very interesting scene may now be witnessed. Diligent search is at once made for their lost mother; the swarm scattering in all directions so that the leaves of the adjoining trees and bushes are often covered almost as thickly with anxious explorers, as with drops of rain after a copious shower. If she can not be found, they commonly return to the old hive, in five to fifteen minutes, though they occasionally attempt to enter a strange one, or to unite with another swarm.

The ringing of bells, and beating of kettles and frying pans, is probably not a whit more efficacious, than the hideous noises of some savage tribes, who, imagining that the sun, in an eclipse, has been swallowed by an enormous dragon, resort to such means to compel his snakeship to disgorge their favorite luminary.

Many who have never practiced "tanging," have never had a swarm leave without settling. Still, as one of the "country sounds," and as a relic of the olden times, even the most matter-of-fact beeman can readily excuse the enthusiasm of that pleasant writer in the London Quarterly Review, who discourses as follows:

"Some fine, warm, morning in May or June, the whole atmosphere seems alive with thousands of bees, whirling and buzzing, passing and re-passing, wheeling about in rapid circles, like a group of maddened bacchanals. Out runs the good housewife, with the frying-pan and key—the orthodox instruments for *ringing*—and never ceases her rough music, till the bees have settled. This custom, as old as the birth of Jupiter, is one of the most pleasing and exciting of the countryman's life; and there is an old colored print of bee-ringing still occasionally met with on the walls of a country inn, that has charms for us, and makes us think of bright sunny weather in the dreariest November day. Whether, as Aristotle says, it affects them through pleasure or fear, or whether, indeed, they hear* it at all is still as uncertain as that

---

* The piping of the queen has a shrill, metallic sound, which possibly may be overpowered by the ringing, so as to distract bees which intend to decamp, and cause them to alight.

philosopher left it; but we can wish no better luck to every bee-master that neglects the tradition, than that he may lose every swarm for which he omits to raise this time-honored concert."

If before its issue, a swarm has selected a new home, no amount of *noise* will compel them to alight, but as soon as the emigrating colony have all left the hive, they fly in a "bee-line" to the chosen spot. I have noticed, that such unceremonious leave-taking, though quite common when bees are neglected, seldom occurs when they are properly cared for.

When the Apiarian perceives that a swarm, instead of clustering, rises higher and higher in the air and means to depart, not, a moment must be lost: instead of empty noises, he should resort to means, much more effective to stay their vagrant propensities. Water or dirt thrown among them, will often so disorganize them as to compel them to alight. The most original of all devices for stopping them, is to flash the sun's rays among them, by a looking-glass! I never had occasion to try it, but an anonymous writer says he never knew it fail. If forcibly prevented from eloping, they will be almost sure to leave, soon after hiving, for their selected home, unless the queen is confined. If there is reason to expect desertion, and the queen cannot be confined, the bees may be carried into the cellar, and kept in total darkness, until towards sunset of the third day, being supplied in the mean time, with water and honey to build their combs. The same precautions must be used when fugitive swarms are re-hived.

It is always very easy to prevent a new colony from abandoning the movable-comb hive, by regulating the entrance so that, while a loaded worker-bee can just pass, the queen will be unable to leave; or a piece

---

- *(First, the queen can not "pipe" in the air. She needs a surface in order to make the sound. Secondly, while the bees may be able to hear the "tanging", a European honey bee swarm does not go very far and would likely land just about when the noise started.*

    Most of this procedure probably extends back to when the beekeeper had to follow a swarm to "claim" it from many others who would also like the swarm for their skeps. By banging on a metal pot you laid claim to the swarm. It seems that in post-medieval years the reason for this tanging has become corrupted into a totally different reason for the noise.)

of comb, with *unsealed* worker-brood, may be transferred to the new hive, when a swarm will seldom forsake it.[9]

It may generally be ascertained, soon after hiving a swarm, whether or not it intends to remain. If on plying the ear to the side of the hive, a sound be heard, as of gnawing or rubbing, the bees are getting ready for comb-building, and will rarely decamp.

If a colony decides to go, they look upon the hive in which they are put as only a temporary stopping-place, and seldom trouble themselves to build any comb. If the hive permits inspection, we may tell at a glance when bees are disgusted with their new residence, and mean to forsake it. They not only refuse to work with the characteristic energy of a new swarm, but their very attitude, hanging, as they do, with a sort of dogged or supercilious air, as though they hated even so much as to touch their detested abode, proclaims to the experienced eye, that they are unwilling tenants, and mean to be off as soon as they can. Numerous experiments to compel bees to work in observing-hives exposed to the full light of day, from the moment they were hived, instead of keeping them, as I now do, in darkness for several days, have made me quite familiar with all such do-nothing proceedings before their departure.

Bees sometimes abandon their hives very early in Spring, or late in Summer or Fall. Although exhibiting the appearance of natural swarming, they leave, not because the population is so crowded that they wish to form new colonies, but because it is either so small, or the hive so destitute of supplies, that they are driven to desperation.[10]

---

[9] *Honeycomb and honey usually work quite well for keeping a swarm in a "new" hive, though Langstroth's method of* unsealed *brood works almost without fail. Bees will generally not leave brood. Most often even when absconding the colony stop brood rearing before the colony abandons the hive. I have found, over the years, that if you have a good hive with comb or foundation, very few swarms will leave.*

I don't think I would rely on my hearing for the detection of the scratching noise made by bees that are going to stay in a new home, the method that Langstroth mentions next.

[10] *This behavior is called* **absconding**. *It may seem self defeating to leave a hive at some late date, but if the colony is starving or there is some other problem, the colony sees that change may possibly be better, for if they remain they will surely die as well. I have only seen absconding a few times with European honey bees, and in almost every case the condition causing the bees to leave was life threatening.*

Seeming to have a presentiment that they must perish if they stay, instead of awaiting the sure approach of famine, they sally out to see if they cannot better their condition. I have known a starving colony to leave their hive on a Spring-like day in December.

It may seem strange that the instincts of so provident an insect should not always impel it to select a suitable domicile before venturing to abandon the old home; since often, before they are housed again, they are exposed to powerful winds and drenching rains, which beat down and destroy many of their number.

I solve this bee-problem, like many others, by considering how the present arrangement conduces to the advantage of man.

Bees would have been of little service to him, if, instead of tarrying till he had time to hive them, their instincts had impelled them to decamp, without delay, from the restraints of domestication. In this, as in many other things, we see that what on a superficial view seemed an obvious imperfection, proves, on closer examination, to be a special contrivance to answer important ends.

To return to our new swam. The queen sometimes alights first, and sometimes joins the cluster after it has begun to form. The bees do not usually settle, unless she is with them; and when they do, and then disperse, it is frequently the case that, after first rising with them, she has fallen, from weakness, into some spot where she is unnoticed by the bees.

Perceiving a hive in the act of swarming, I, on two occasions, contracted the entrance, to secure the queen when she should make her appearance. In each ease, at least one-third of the bees came out before she joined them. As soon as the swarm ceased searching for her, and were returning to the parent-hive, being placed, with her wings clipped, on a limb of a small evergreen tree, she crawled to the very top of the limb, as if for the express purpose of making herself as conspicuous possible. The few bees which first noticed her, instead of alighting, darted rapidly to their companions; in a few seconds, the

---

The Africanized honey bee is much more inclined to abscond, and where these bees are found they often have absconding swarms. One of the reasons for this behavior is that the Africanized bees tend not to spend as much effort storing surplus honey and thus these colonies tend to exhaust their supplies and are thus forced into absconding more often.

whole colony was apprised of her presence, and flying in a dense cloud, began quietly to cluster around her. Bees when on the wing intercommunicate with such surprising rapidity, that telegraphic signals are scarcely more instantaneous.[11]

That bees send out *scouts* to seek a suitable abode, admits of no serious question. Swarms have been traced directly to their new home, in an air-line flight, either from their hive, or from the place where they clustered after alighting. Now this precision of flight to an unknown home, would plainly be impossible, if some of their number, by previous explorations, were not competent to act as guides to the rest.[12] The sight of bees for distant objects is so wonderfully acute, that, after rising to a sufficient elevation, they can see, at the distance of several miles, any prominent objects in the vicinity of their intended abode.

Whether bees send out scouts *before* or *after* swarming, may admit of more question. When a colony flies to its new home without alighting, the scouts must have been dispatched before swarming. If this were the usual course, we should expect every colony to take the same speedy departure; or if they should cluster for the convenience of the queen, or any bees over-fatigued by the excitement of swarming, we should look for only a transient tarrying. Instead of this, they often remain until the next day, and instances are not infrequent of a much more protracted delay. The stopping of bees in their flight to cluster again, is not inconsistent with these views; for if the weather is hot when they first cluster, and the sun shines directly upon them,

---

[11] *Langstroth did not know about queen pheromones, or Nasonov pheromone, that allow for this communication within a swarm. Many bees will have their Nasonov (scent) gland open and via the pheromone will communicate within the swarm. The odor will even attract bees from other hives, particularly drones. The Nasonov and queen pheromones are the glue that keeps the swarm together, particularly the Nasonov pheromone. We can now purchase synthetic pheromones in order to bait hive boxes to attract swarms. These odors, added to the right size box, a cavity of about 40 liter size—a standard hive body along with some comb, would be very effective in capturing swarms.*

[12] *We now know that there is a dance communication about the distance and compass angle to the site as well as size of the cavity. The many scouts may dance to indicate several different possible nests. When the scout-dancers come to an agreement (with about two-thirds of the scouts dancing for the same place), then the swarm departs to the new home.*

they will often leave before, they have found a suitable habitation. Sometimes the queen of an emigrating swarm, being heavy with eggs, and unaccustomed to fly, is compelled to alight, before she can reach their intended home. Queens, under such circumstances, are occasionally unwilling to take wing again, and the poor bees sometimes attempt to lay the foundations of their colony on fence-rails, hay-stacks or other unsuitable places.

Mr. Wagner, says, that he once knew a swarm of bees to lodge under the lowermost limb of an isolated oak tree, in a corn-field. It was not discovered until the corn was harvested, in September.[13] Those who found it, mistook it for a recent swarm, and in brushing it down to hive it, broke off three pieces of comb, each about eight inches square. Mr. Henry M. Zollickoffer of Philadelphia, informed me that he knew a swarm to settle on a willow tree in that city, in a lot, owned by the Pennsylvania Hospital; it remained there for some time, and the boys pelted it with stones, to get possession of its comb and honey.

The necessity for scouts or explorers seems to be unquestionable, unless we can admit that bees have the faculty of flying in an *"air line"* to a hollow tree which they have never seen, and which may be the only one among thousands where they can find a suitable abode.

These views are confirmed by the repeated instances in which a few bees have been noticed inquisitively prying into a hole in a hollow tree, or the cornice of a building, and have, before long, been followed by a whole colony.

Having described the method commonly pursued by a new swarm, when left to their natural instincts, we return to the parent stock from which they emigrated.

From the immense number which have abandoned it, we should naturally infer that it must be nearly depopulated. As bees swarm in the pleasantest part of the day, some suppose that the population is

---

[13] *These exposed wild (feral) colonies are not very common, and most occur when a swarm cannot find a suitable cavity such as a hollow tree. Possibly, they occur when by the time the scouts have found a nest site (or have come to an agreement on the best site) the bees have started enough comb that the queen began laying eggs. Once this happens the swarm will stay.*

Exposed nests rarely persist in the northern states as the honeycomb is robbed of much of its honey and the bees starve over winter. They would also have great difficulty with any snow that falls upon the nest as well.

replenished by the return of large numbers from the fields; this, however, cannot often be the ease, as it is seldom that many are absent from the hive at the time of swarming. To those who limit the fertility of the queen to four hundred eggs a day, the rapid replenishing of a hive, after swarming, must be inexplicable; but to those who have seen her lay from one to three thousand eggs a day, it is no mystery at all. Enough bees remain to carry or the domestic operations of the hive; and as the old queen departs only when there is a teeming population and when thousands of young are daily hatching, and tens of thousands rapidly maturing, the hive, in a short true, is almost as populous as it was before swarming.

Those who suppose that the new colony consists wholly of young bees, forced to emigrate by the older ones, if they closely examine a new swarm, will find that while some have the ragged wings of age, others are so young as to be barely able to fly.[14]

After the tumult of swarming is over, not a bee that did not participate in it, attempts to join the new colony, and not one that did, seeks to return. What determines some to go, and others to stay, we have no certain means of knowing. How wonderful must be the impression made upon an insect, to cause it in a few minutes so completely to lose its strong affection for the old home that when established in a hive only a few feet distant, it pays not the slightest attention to its former abode! When their new domicile is removed—after some have gone to the field—from the place where the bees were hived, on their return, they often fly for hours in ceaseless circles about the spot where the missing hive stood; and sometimes continue the vain search

---

[14] *In studies of the age of the workers in a swarm, it has been found that the bees within a swarm are mostly very young—average age of 5 days old. It is important for the swarm to be made up primarily of young bees since there will be no replacement bees for three weeks. This is true even if the queen starts laying the very first day the swarm finds a new home.*

This problem of getting started on the production of new bees is the same reason that I like to use the spray with syrup and direct release of queens in the installing of package bees. These direct-released queens will be laying in 24 hours even when installed on comb foundation. I have never found that there was any appreciable queen loss at introduction because of this direct release method. The quickness that these queens are able to lay eggs far outweighs any other concerns.

for their companions, until dropping from exhaustion, they perish in close proximity to their old home.[15]

It has been already been stated that, if the weather is favorable, the old queen usually leaves near the time that the young queens are sealed over to be changed into nymphs. In about a week, one of them hatches; and the question must be decided whether or not, any more colonies shall be formed that season. If the hive is well filled with bees, and the season is in all respects promising, it is generally decided in the affirmative; although, under such circumstances, strong colonies refuse to swarm more than once; while the repeated swarming of weaker ones often ruins both the parent-stock and its after-swarms.

If the bees decide to swarm but once, the first hatched queen, being allowed to have her own way, rushes immediately to the cells of her sisters, and stings them to death. The other bees probably aid her in this murderous transaction; they certainly tear open the cradles of the slaughtered innocents (Pl. XIV., Fig. 47, d), and remove them from the

---

[15] *I agree with Langstroth that this is quite a mystery. However, since most of the swarm is made up with very young bees they probably would have no recognition of the parent hive since they would not have yet taken orientation flights. It is these orientation flights that give the young bee cues to the location of the hive, though there is other evidence that the act of swarming itself causes the bees to forget the old home location.*

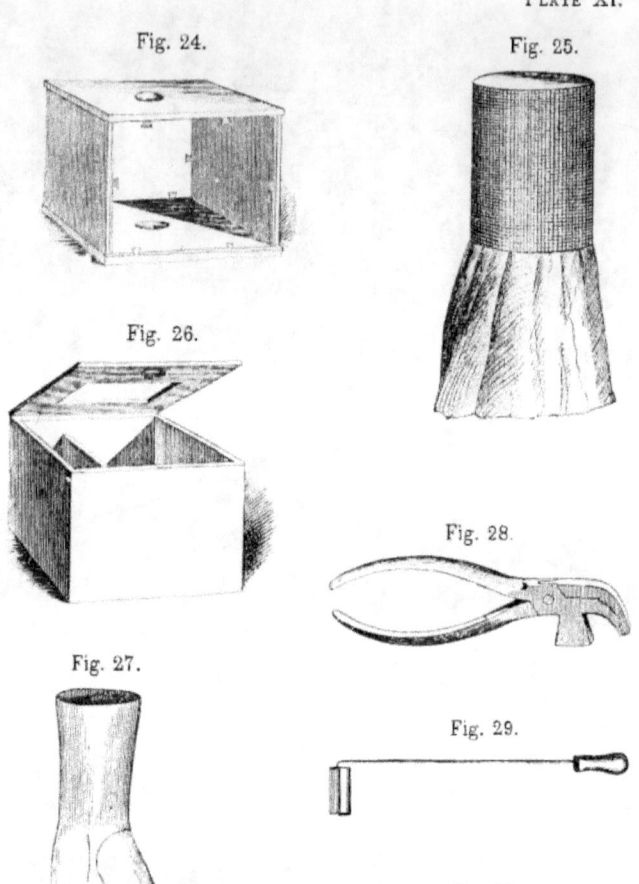

Plate XI.

cells. Their dead bodies may often be found on the ground in front of the hive.

When a queen has emerged from her cell in the natural way, the bees cut it down (Pl. XIV., Fig. 47, c), till only a small acorn cup remains; but if she met with a violent end, they usually remove the whole cell. By counting these acorn-cups, we can ascertain how many queens have hatched in a hive.[16]

If the bees of the parent-stock decide to send out a second colony, the first hatched queen is prevented from killing the others. A strong guard is kept over their cells, and as often as she approaches them with murderous intent, she is bitten, or given to understand by other most uncourtier-like demonstrations, that even a queen cannot, in all things, do just as she pleases.

Like some human beings who cannot have their own way, she is highly offended when thus repulsed, and utters, in a quick succession of notes, a shrill, angry sound, not unlike the rapid utterance of the words, "peep, peep." If held in the closed hand, she will make a similar noise. To this angry note, one or more of the unhatched queens will respond, in a somewhat hoarser key, just as, a cock, by crowing; bids defiance to its rivals.[17] These sounds, so entirely unlike the usual steady hum of the bees, or the fluttering noises of unhatched queens, are almost infallible indications that a second swarm will soon issue. They are occasionally so loud as to be heard at some distance from the hive. About a week after first swarming, the Apiarian should place his ear against the hive, in the morning or evening, when the bees are still, and if the queens are "piping," he will readily recognize their peculiar sounds. The young queens are all mature, at the latest, in sixteen days from the departure of the first swarm, even if it left as soon as the royal cells were begun. If, during this period, these notes are not heard, it is an infallible indication that the first hatched queen has no rivals; and that swarming, in that stock, is over for the season.

The second swarm usually issues on the second or third day after piping is heard; though they sometimes delay coming out until the

---

[16] *I suspect that this is not a very reliable method as I have found these queen cups present in older combs even when the colony had queen cells started nearby.*
[17] *This piping of virgin queens is very distinctive and once you have heard it you will always remember the sound.*

fifth day, in consequence of an unfavorable state of the weather. Occasionally, the weather is so extremely unfavorable, that the bees permit the oldest queen to kill the others, and refuse to swarm again. This is a rare occurrence, as young queens are not so particular about the weather as old ones, and sometimes venture out, not merely when it is cloudy, but when rain is falling. On this account, if a very close watch is not kept, they are often lost. As piping ordinarily commences about a week after first-swarming, the second swarm usually issues nine days after the first; although it has been known to issue as early as the third, and as late as the seventeenth; but such cases are very rare.

It frequently happens in the agitation of swarming, that the usual guard over the queen-cells is withdrawn, and several hatch at the same time, and accompany the colony; in which case, the bees often alight in two or more separate clusters. In my observing-hives I have repeatedly seen young queens thrust out their tongues from a hole in their cell, to be fed by the bees.[18] If allowed to issue at will they are pale and weak, like other young bees, and for some time unable to fly; but if confined the usual time, they come forth fully colored, and ready for all emergencies. I have seen them issue in this state, while the excitement caused by removing the combs from a hive, has driven the guard from their cells.

The following remarkable instance came under my observation, in Matamoras, Mexico. A second swarm deserting its abode the *second* day after being hived, settled upon a tree. On examining the abandoned hive five young queens were found lying dead on its bottom board. The swarm was returned, and, the next morning, two more dead queens were found. As the colony afterwards prospered, *eight* queens, at least, must have left the parent-stock in a single swarm!

Young queens, whose ovaries are not burdened with eggs, are much quicker on the wing than old ones, and frequently fly much far-

---

[18] *This is a very interesting observation by Langstroth. I have often seen queens cutting open their cells, but not being fed. A young bee is certainly hungry and will often get fed upon emergence, so this type of feeding of queens within their partly opened cells is probably not all that uncommon.*

ther from the parent-stock before they alight.[19] After the departure of the second swarm, the oldest remaining queen leaves her cell; and if another swarm is to come forth, piping will still be heard; and so before the issue of each swarm after the first. It will sometimes be heard for a short time after the issue of the second swarm, even when the bees do not intend to swarm again. The third swarm usually leaves the hive on the second or third day after the second swarm and the others, at intervals of about a day. I once had five swarms from one stock, in less than two weeks. In warm latitudes, more than twice this number of swarms have been known to issue, in one season, from a single stock.

In after-swarming, the queen sometimes re-enters the hive, after having appeared on the alighting-board. If she does this once, she will be apt to do it repeatedly, and the swarm, in each instance, will return to the mother hive.

In the Apiary of a friend in Matamoras, when his first swarm issued, there was no tree for it to alight on. The wind was so strong, that the bees did not leave the vicinity of their hives, but began to settle on a hive near their own. Although the queen was secured, with a portion of her colony, a large part of the swarm entered the adjoining stocks. When these stocks swarmed, although a tree had been set out for them to cluster on, the bees which had returned on the first occasion, did the same thing again, drawing with them the rest of their companions. The only way in which we could obtain a single swarm, was by *covering* with *sheets* all the hives in the Apiary as soon as one swarmed, and thus the bees, being unable to enter them, were compelled to

---

[19] *There are beekeepers that tell if the swarm has an old, young or virgin queen just by the height that the swarm settles in a tree. Certainly if the swarm is near the ground, and also near the parent hive, then the queen is almost assuredly an old queen, or one that had been laying eggs only a short time before swarming, and thus was not capable of flying very well.*

European-bee swarms usually do not go very far from the parent colony. Most are found within a few hundred yards, or less. The Africanized bees will sometime go miles during swarming. This is true even of reproductive swarms, as opposed to absconding swarms. Absconding swarms have a "reason" to go a long distance as they are trying to find new or better resources regardless of the consequences of leaving a hive. If they stay in the old location without food they will die. Thus, even though leaving is very risky it may be beneficial to the swarm in the end.

alight! It would he difficult to find a better illustration of the folly of neglecting the old adage, "A stitch in time saves nine."

After-swarms or casts—these names are given to all swarms after the first—seriously reduce the strength of the parent-stock; since by the time they issue, nearly all the brood left by the old queen has hatched, and no more eggs can be laid until all swarming is over. It is a wise arrangement, that the second swarm does not ordinarily issue until all the eggs left by the first queen are hatched, and the young mostly sealed over, so as to require no further feeding. Its departure earlier than this, would leave too few laborers to attend to the wants of the young bees. If after swarming, the weather suddenly becomes chilly, and the hive is thin, or the Apiarian continues the ventilation which was needed only for a crowded colony, the old stock being unable to maintain the requisite heat, great numbers of the brood often perish.

The effect on the profits of the Apiary, of too frequent swarming, is discussed in the next chapter. If the beekeeper wants no casts, he can easily prevent their issue from my hives. About five days after the first swarm comes out, the parent-stock may he opened, and all the queen-cells removed, except one. If done earlier than this, the bees may start others, in the place of those removed. Those only who have thoroughly tried both plans, can appreciate how much better this is, than to attempt to return the after-swarms to the parent hive. The Apiarian who desires by natural swarming to multiply his colonies as rapidly as possible, will find full directions in the sequel, [next chapter] for building up all after-swarms however small, so as to make vigorous stocks.[20]

It will be remembered, that both the parent-stock from which the swarm issues, and all the colonies, except the first, have a young queen. These queens never leave the hive for impregnation, until they

---

[20] *Modern beekeepers like to prevent all swarms since the reduction of the population, caused by swarming, reduces the surplus honey yield. Most management techniques are directed at producing the maximum population just short of producing a swarm.*

If after-swarms are found they almost always need extra care since they will have a delayed egg laying by the queen, as she needs a few days to mate and begin laying eggs. The bees meanwhile have begun dying and the colony will seldom become strong enough to survive the winter. In addition, most after-swarms have fewer bees than prime swarms.

are established as heads of independent families. They generally go out for this purpose, early in the afternoon of the first pleasant day, after being thus acknowledged, at which time the drones are flying most numerously. On leaving their hive, they fly with their heads turned towards it, often entering and departing several times, before they finally soar into the air. Such precautions on the part of a young queen are highly necessary, that she may not, on her return, lose her life, by attempting, through mistake, to enter a strange hive. More queens are thus lost than in any other way.[21]

When a young queen leaves for impregnation, the bees, on missing her, are often filled with such alarm that they rush from the hive, as if intending to swarm. Their agitation is soon quieted, if she returns in safety.

The drone perishes in the act of impregnating the queen. Although, when cut into two pieces, each piece will retain its vitality for a long time, I accidentally ascertained, in the Summer of 1852, that if his abdomen is gently pressed, and sometimes if several are closely held in the warm hand, the male organ will often be permanently extruded, with a motion very like the popping of roasted pop-corn; and the insect, with a shiver, will curl up and die, as quickly as if blasted with the lightning's stroke. This singular provision is unquestionably intended to give additional security to the queen, when she leaves her hive to have intercourse with the drone. Huber first discovered that she returned with the male organ torn from the drone and still adhering to her body.[22] If it were not for this arrangement, her spermatheca would not be filled, unless she remained so long in the air with the drone as to incur a very great risk of being devoured by birds. In one instance, some days after the impregnation of a queen, I found the

---

[21] *This reason for the loss of young queens is probably as true now as in Langstroth's time, with our crowded apiaries that have the same-color colonies all in a row.*

[22] *This adhering part of the drone sexual organ has been called the "mating sign." It was this presence that lead early apiculturists to think that a queen only mated once, since how would another drone be able to copulate. With the aid of video and movie cameras it is now known that the next drone removes this "plug" before he mates with the queen. This process of mating, removing the plug by the next drone, is repeated for several matings before the queen returns to the hive. It is only during the last mating of the mating flight, that the piece of the drone's penis is not removed, and thus leaves the "mating sign."*

male organ,* in a dried state, adhering so firmly to her body, that it could not be removed without tearing her to pieces.

---

* On page 40 of the English translation of Prof. Siebold's work on "Parthenogenesis" (that is, production without intercourse with the male) of Moths and Bees" may be found the following extract of a letter to Prof. Siebold, dated 21st July, 1853, from the celebrated German Apiarian, the Baron von Berlepsch.

"I succeeded, today, in impaling upon a pin, a queen which had flown out to copulate, just as she was about to re-enter the hive. The signs of copulation stand far out.*** Will you have the kindness to settle, by dissection: 1, if any, and what parts of the drone occur in the royal vulva; and 2, what is the condition of the seminal receptacle. If there be parts of the drone in the vulva, people will, at last admit that the drones are the males, and that the copulation takes place outside of the hive. *** Moreover, if you find the seminal receptacle filled with semen, Dzierzon's hypothesis—according to which the ovary is not fertilized, but the seminal receptacle filled with male drone-semen, by copulation—is raised into evidence."

Prof. Siebold says, that "he was able to establish, that those definitely formed parts in the vagina of the queen were nothing but the torn copulative organs of a male bee (drone). With this condition of the external sexual organs of the queen, the state of the internal generative organs also agreed exactly, for the seminal receptacle which is empty in all virgin female insects, was, in this queen, filled to overflowing with seminal filaments (spermatozoids)."

I give as interesting, in this connection, the following extract from my journal: "August 25th, 1852—Found the male organ protruding from a young queen; could not remove it without exerting so much force that I feared it would kill her. Dr. Joseph Leidy examined this queen-bee with the microscope, so as to demonstrate that—to use his words—'it was the penis and its appendages of a male, corresponding, in all its anatomical peculiarities, with the same organs examined, at the same time, in other drones. The testicles and *vasa deferentia* of these drones were found to be full of the spermatic fluid. The *spermatheca* of the queen was distended with the same semi-fluid, spermatic matter.' This one examination *demonstrates* that the drones are males and that they impregnate the queen by actual coition."

Prof. Siebold further says: "As in the act of copulation of the bees, the penis of a drone is completely protruded outwards, and as no particular muscular apparatus exists for the extrusion of the penis, the circumstance that the drones copulate in flight, has an important signification.*** During the movement of the wings, the different air-sacs of the tracheal system of the drone are filled with air, by which means these can act by pressure, in the interior of the body of the bee, upon the neighboring penis which is to be protruded."

"The following interesting experiment" (Parthenogenesis, p. 54) "was made by Berlepsch, in order to confirm the drone-productiveness of a virgin queen. He contrived the exclusion of queens at the end of September, 1854, and, therefore, at a time when there was no longer any males; be was lucky enough to keep one of them through the Winter, and this produced drone-offspring on the 2d of March, in the following year, furnishing fifteen hundred cells with brood. That this drone-bearing queen remained a virgin, was proved by the dissection which Leuckart undertook, at the request of Berlepsch. He

SWARMING AND HIVING. 127

The following facts will show that the impregnation of the queen by the drone, in the open air, may be made a matter of ocular demonstration: Lewis Shrimplin, of Wellsboro, Brook County, Virginia, purchased a movable-comb hive, in the Spring of 1857, into which he put a second swarm. Finding, after a few days, that the bees had built a number of very straight combs, he called some of his neighbors together, to witness the ease with which he could take out, and replace their combs. While standing in front of the hive, he saw the queen coming out, and the idea occurred to him to catch her, and tie a very fine silk thread to one of her thighs. This he accomplished successfully; and as she began to ascend,[*] the drones collected around her in very large numbers. After remaining in the air a short time, she returned to the entrance of her hive, exhibiting to the spectators the organs of the drone still protruding from her body.

The queen usually begins laying about two days after impregnation, and for the first season, lays almost entirely the eggs of workers; no males[*] being needed in colonies which will throw no swarm till

---

found the state and contents of the seminal pouch of this queen to be exactly of the same nature as those found in virgin queens. The seminal receptacle in such females never contains semen-masses, with their characteristic spermatozoids, but only a limpid fluid, destitute of cells and granules, which is produced from the two appendicular glands of the seminal capsule; and, as I suppose, serves the purpose of keeping the semen transferred into the seminal capsule In a fresh state, and the spermatozoids active, and, consequently, capable of impregnation."

By referring to pages 38, 39, the reader will see that Prof. Leidy dissected for me a drone-laying queen, nearly three years before this examination of Leuckart.

Prof. Siebold, in 1843, examined the spermatheca of the queen-bee, and found it after copulation, filled with the seminal fluid of the drone. At that time, Apiarians paid no attention to his views, but considered them, as he says, to be only *"theoretical stuff."* It seems, then, that Prof. Leidy's dissection (pp. 34, 35) was not, as I had hitherto supposed, the first, of an impregnated spermatheca.

[*] Dzierzon supposes that the sound of the queen's wings, when she is in the air, excites the drones. In the interior of the hive, they are never seen to notice her; so that she is not molested, even if thousands are members of the same colony with herself.

(We now know that it is the release of the queen pheromone while the queen is in flight that attracts the drones. While the queen pheromone is not very volatile it is enough so that within the drone congregation area the pheromone attracts the males for mating.)

[*] Huber supposed that male eggs were not developed in her ovaries until the second year; but as the sex depends upon the impregnation of the eggs, he was evidently mistaken. In warm climates, where after-swarms swarm again, drones are bred in large

another season. She is seldom treated much attention by the bees until after she has begun to replenish the cells with eggs; although if previously deprived of her, they show, by their despair, that they fully appreciated her importance to their welfare.

A first swarm will sometimes swarm again, about a month after it is hived; but in Northern climates this is a rare occurrence. In Southwestern Texas, I have known even second swarms to do the same thing, and colonies often swarm there in September and October, while in tropical climates, swarms issue at any season when forage is abundant. In our Northern and Middle States, swarming is usually over, three or four weeks after it begins. Inexperienced beekeepers, unaware of this, often watch their Apiaries, long after the swarming season has passed.

I shall now, while giving such directions for hiving swarms as may aid even some experienced Apiarians, attempt to make them sufficiently minute to guide those, who, having never seen a swarm hived, are apt to imagine that the process must he quite formidable. Experience in this, as in other things, will speedily give them the requisite skill and confidence; and the cry of "the bees are swarming" will often be hailed with even greater pleasure than an invitation to a sumptuous banquet.

The hives for the new swarms should be painted long enough beforehand to be thoroughly dry. The smell of fresh paint is well known to be very injurious to human beings, and is so detested by bees, that

---

numbers in hives having young queens. The bee is evidently a native of a hot climate, although it can live wherever there is a Summer long enough for it to prepare for Winter. Its complete development, however, can be witnessed only in tropical regions, and I am persuaded that many things which, in colder climates, have been regarded as fixed laws, are only exceptional adaptations to unfavorable circumstances.

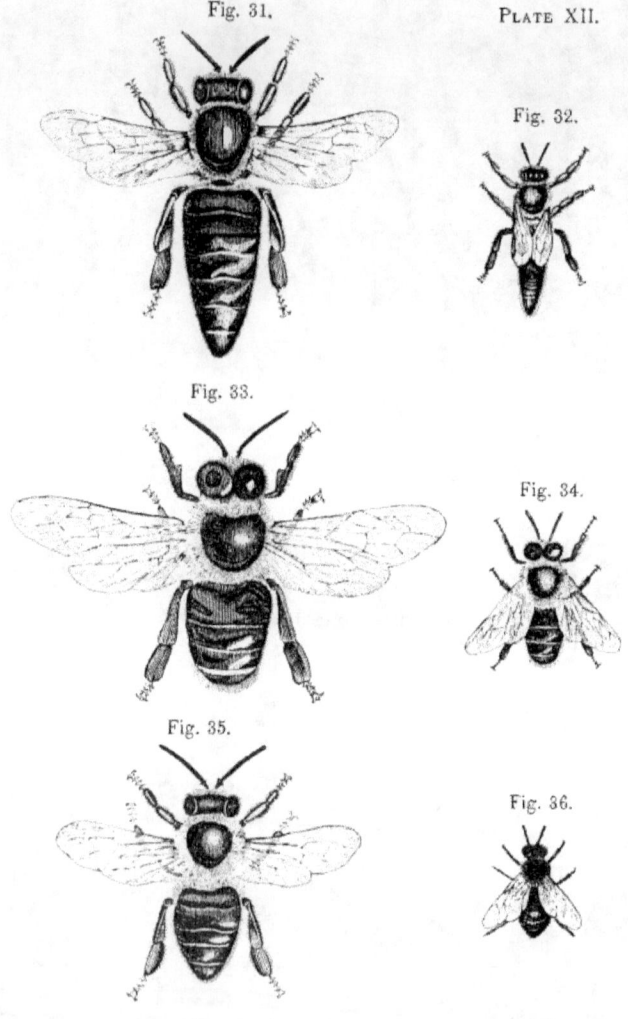

PLATE XII.

Fig. 31.
Fig. 32.
Fig. 33.
Fig. 34.
Fig. 35.
Fig. 36.

they will often desert a new hive sooner than endure it. If the hives cannot be seasonably painted, paints should be used which contain no white-lead, and which are mixed so as to dry as quickly as possible.[23]

The following recipe, taken from the Bienenzeitung, for a cheap and durable paint, for rough hives, is said to be preferable to oil paint: "Two parts, by measure, of fine sand, well sifted; one of best English cement*, one of curd, from which the whey has been well expressed; one of buttermilk. These are to be thoroughly mixed. The paint is to be applied, amid repeated stirring, to the hives, by means of a common paint-brush. A second coat is to be given after the lapse of half an hour. when this has become thoroughly dry, which will be in two or three days, it is to be brushed over lightly with a thin coat of boiled linseed oil, to which any desirable color may be given. The boards to which the paint is to be applied should not be planed, but remain rough as the saw leaves them. No more of the paint should be prepared at any one time, than can be used in the course of half an hour, as it quickly hardens. The hive may be used as soon as the paint stiffens."

Hives that have stood in the sun, ought never to be used for new swarms. Bees, when they swarm, being unnaturally excited and heated, often refuse to enter such hives, and at best, are slow in taking possession of them. The temperature of the parent-stock, at the moment of swarming, rises very suddenly, and many bees are often so drenched with perspiration, that they are unable to take wing and join the emigrating colony.[24] To attempt to make swarming bees enter a heated hive in a blazing sun, is, therefore, as irrational as it would be to force a panting crowd of human beings into the suffocating atmos-

---

[23] *Beekeepers still have the dilemma about paint, though most just use white paint as it is now considered "traditional". In the northern states a darker color would probably be more beneficial to the colony since the darker colors would attract more solar radiation. This extra heat would help the colonies in the Spring and the Fall. There may be a little period during mid summer when the colony may benefit from white paint. However, most of the time the hive would benefit with darker colors.*

* Roman, or common Hydraulic cement is probably meant, or would answer.

[24] *Bees do not perspire the way humans do, but they do use water to cool their body, but then generally only in flight. They spread water over the thorax and abdomen and the evaporating water will cool them. When they get excessively hot they will regurgitate nectar or water. If you find bees that have died from overheating they will generally be wet.*

phere of a close garret. If the process of hiving cannot be conducted in the shade, the hive should be covered with a sheet, or with leafy boughs.

In the movable-comb hive, the Apiarian can use all his good worker-comb, by fastening it in the frames. Such, however, is the shape of the artificial guide-combs in these frames, that the bees, even in an empty hive, will almost always build their combs with great regularity, if they are not furnished with too much empty room. I have, in a *few instances*, known them to build their combs directly across, from frame to frame, so that they could not be removed without cutting them to pieces. This may easily be prevented, by attaching a piece of guide comb to a single frame (see p. 72). While the hive should be set so as to incline from rear to front, to shed the rain, there ought not to be the least pitch from *side to side* or it will prevent the frames from hanging plumb, and compel the bees to build crooked combs. Drone-combs should never be put in the frames, or the bees will follow the pattern, and build comb suitable only for breeding a horde of useless consumers. Such comb, if white, may be used to great advantage in the surplus honey-boxes; if old, it should be melted for wax.

Every piece of good worker-comb, if large enough to be attached to a frame, should be used, both for its intrinsic value, and because bees are so pleased when they find such unexpected treasures in a hive, that they will seldom forsake it. A new swarm often takes possession of a deserted hive, well stored with comb; whilst, if dozens of empty ones stand in the Apiary, they very seldom enter them of their own accord. It once seemed to me that an instinct impelling them to do so, would have been much better for us than the present arrangement; but further reflection has shown me that, on the contrary, it would have been the fruitful origin of interminable broils among neighboring bee-keepers; and that in this, as in so many other things, the instincts of the honey-bee have been devised with special reference to the welfare of man.

When the frames[*] are first used for a new swarm, the rabbets on which they rest should be smeared with flour-paste; this will keep the

---

[*] For their proper adjustment, see Explanation of Plates.

frames firm, till they are fastened with propolis by the bees.[25] If hives are sweet and clean, the rubbing of them with various kinds of herbs or washes, is always useless, and often positively injurious.[26]

If there are no small trees or bushes, near the Apiary, from which the swarms, when clustered, can be easily gathered, limbs of evergreen or other trees may be fastened into the ground, a few rods in front of the hives, which will answer a very good temporary purpose. If there are high trees near his stocks, the bee-master, unless some special precautions are used will lose much time in hiving his swarms.

Having noticed that a new swarm will almost always alight wherever they see a mass of clustering bees, I find that they can be determined to some selected spot by an old black hit, or even a mullen stalk, which, when colored black, can hardly be distinguished, at a distance from a clustering swarm. A black woolen stocking or piece of cloth, fastened to a shady limb, in plain sight of the hives, and where the bees can be most conveniently hived, would probably answer as good a purpose. Swarms are not only attracted by the bee-like color of such objects, but are more readily induced to alight upon them, if they furnish something to which they can easily cling, the better to support their grape-like clusters. By proper precautions, before the first swarms issue, the beekeeper may so educate his favorites that they will seldom alight anywhere but on the spot which he has previously selected.[27]

---

[25] *The self-spacing or Hoffman top-bar frame was not invented until a later date, and thus Langstroth had to adjust his frames carefully. The addition of the wider end bars that self-spaced the frame was a good addition to the concept of the hanging frame of Langstroth. Modern beekeepers usually have no concept of how much time some of these improvements have made to beekeeping, since they started beekeeping with all of these improvements already in place. Most of these changes have probably allowed large-scale beekeeping to exist.*

[26] *I don't know which herbs might have been in general practice for rubbing into hives at the time of Langstroth. It is now known that some herbs, such as lemon balm, have oils and chemicals that are similar to those found in the Nasonov gland, or sent gland located the last abdominal tergite of the bee. This, Nasonov scent, is used by the bees to help keep all the bees aggregated into the swarm. Thus, by using a rubbing of lemon balm you will help attract swarms to the empty boxes.*

[27] *Probably because of chemical odors, swarms often alight in the same spot in the same tree year after year. There may be other reasons, as well, such as favorable height, the tree species, limb branching, etc.*

The Rev. Thomas P. Hunt, of Wyoming, Penn., has devised an amusing plan, by which he says that he can, at all times, prevent a swarm of bees from leaving his premises. Before his stocks swarm, he collects a number of dead bees, and, stringing them with a needle and thread, as worms are strung for catching eels, he makes of them a ball about the size of an egg, leaving a few strands loose. By carrying—fastened to a pole—this "bee-bob," about his Apiary, when the bees are swarming, or by placing it in some central position, he invariably secures every swarm!

It will inspire the inexperienced Apiarian with more confidence, to remember that almost all the bees in a swarm, are as a very peaceable mood, having filled themselves with honey before leaving the parent-stock. If he is timid, or suffers severely from the sting of a bee, he should, by all means, furnish himself with the protection of a bee-dress.

A new swarm should be hived as soon as they have quietly clustered around their queen; although there is no necessity for the headlong haste practiced by some, which, by exciting profuse perspiration, increases their liability to be stung. Those who show so little self-possession, must not be surprised, if they are stung by the bees of other hives, which, instead of being gorged with honey, are on the alert, and very naturally mistake the object of such excited demonstrations. The fact that the swarm has clustered, makes it almost certain, that, unless the weather is very hot, or they are exposed to the burning heat of the sun, they will not leave for at least one or two hours. All convenient dispatch, however, should be used in hiving a swarm, lest it send out scouts, which may entice it from the new hive, or lest other colonies issue, and attempt to add themselves to it.

If my hives are used, the whole entrance should be opened, that the bees may get in as soon as possible; and a sheet should be securely fastened to the alighting-board, to keep them from becoming separated, or soiled by dirt; for, if separated, they are a long time in entering and a bee covered with dust or dirt is very apt to perish. The common

---

If you want to entice swarms the "flower-pot" swarm traps along with a Nasonov pheromone lure and some old comb work very well. These traps should be placed about 15 feet high in a tree, and be in at least partial shade. If you place these traps in likely locations you can attract, and then use, many swarms each season.

hives should he propped up on the sheet, in such a way as to give the bees the readiest admission.

When the limb on which the bees have clustered can be easily reached, it should be shaken, with one hand, so that they may gently fall into a basket held under them, with the other. The basket should he open sufficiently to admit the air freely, but not enough to allow the bees to get through its sides. They should now be gently shaken or poured out on the sheet, in front of their new home. If they seem at all reluctant to enter it, gently scoop up a few of them with a large spoon, and shake them close to its entrance. As they go in with fanning wings, they will raise a peculiar note, which communicates to their companions the joyful news that they have found a home;[28] and in a short time, the whole swarm will enter, without injury to a single bee.

When bees are once shaken down on the sheet, they are quite unwilling to take wing again; for, being loaded with honey, they desire, like heavily-armed troops, to march slowly and sedately to their place of encampment. Bees are much obstructed in their travel, by any *corner*, or great inequality of surface; and if the sheet is not smoothly stretched, they are often so confused, that they take a long time to find the entrance to the hive. If they are too dilatory in entering the new hive, they may be gently separated, with a spoon, or leafy twig, where they gather in bunches on the sheet; or, they may he carefully "spooned up," and emptied before the entrance of the hive. If they cluster in the portico of my hive, they should be treated in the same way; or else the queen, mistaking this open place for her intended abode, may decamp with the bees.

---

[28] *The bees when fanning are also exposing the Nasonov gland at the tip of the abdomen that gives off this 'aggregation' pheromone. The odor is then blown toward the rest of the swarm that then moves in the direction of the odor.*

It is a real treat to see how quickly the bees, from a captured swarm that are poured out in front of a hive, orient toward the hive and start walking into it. This orientation takes only a minute or two to be seen. On these occasions you can usually smell the Nasonov pheromone yourself. At first the bees will be seen milling about in all directions. Then the Nasonov pheromone "odor trail" orients them toward the empty hive, and they all seem to turn at once and start walking into the hive. If you do this experiment you can see the little white gland (Nasonov) opened and exposed at the tip of the abdomen. When the bees expose this gland and fan their wings at the same time the odor is directed and thus the orientation of the bees toward the empty hive.

On first shaking them down into the hiving-basket, some will take wing, and others will remain on the tree; but if the queen has been secured, they will quickly form a line of communication with those on the sheet. If the queen has not been secured, the bees will either refuse to enter the hive, or will speedily come out, and take wing, to join her again. This happens oftenest with after-swarms, whose young queens, instead of exhibiting the gravity of an old matron, are apt to be frisking in the air. When the bees cluster again on the tree, the process of hiving must be repeated.[29]

If the Apiarian has a pair of sharp pruning-shears, and the limb on which the bees have clustered is so small, that it can be cut without jarring them off, they may be gently carried on it to the hiving-sheet.

If the bees settle too high to he easily reached, the basket may he fastened to a pole, and raised directly under them; when a quick upward push will secure most of the swarm. When the basket cannot be easily elevated to them, it may be carried to the cluster, and the beekeeper, after shaking the bees into it, may gently lower it, by a string, to an assistant below.

When a colony alights on the trunk of a tree, or on anything from which they cannot easily be gathered in a basket, fasten a leafy bough over them, without jarring, by a gimlet, and with a little smoke compel them to ascend it. If the place is inaccessible, they will enter a well shaded basket, inverted, and elevated just above the mass of the bees. I once hived a neighbor's swarm which settled in a thicket, on the inaccessible body of a tree, by throwing water upon them, so as to compel them gradually to ascend the tree, and enter an elevated box. If proper alighting places are not furnished, the trouble of hiving a swarm will often be greater than its value.

If two swarms cluster together, they may be advantageously kept together, if abundant room for storing surplus honey can be given

---

[29] *If you use a queen excluder to prevent a new swarm from leaving a hive, it is best to check them in about a day to see if there are eggs. If not, remove the queen excluder as most likely the swarm has a virgin queen that still needs to be mated, and will not be able to leave the hive with the excluder.*

The queen excluder can be placed either over the entrance or under the bottom hive body to prevent the swarm (queen) from leaving.

them, as in my hives.[30] Large quantities of honey are generally obtained from such stocks, if they issue early, and the season is favorable. If it is desired to separate them, take two hives, and give a portion of the bees to each, sprinkling them, both before and after they are shaken from the basket, sufficiently to keep them from taking wing to unite again. If possible, secure a queen for each hive. If both queens enter the same hive, one will quickly dispose of the other. The bees in the queenless hive will begin to leave as soon as they ascertain their condition. Prevent this, by shutting them up; and give them a queen, if you have one at your disposal; or supply them with a settled queen, nearly mature, taken from another hive. For reasons assigned in the next chapter, it will not do to compel them to raise a queen from worker-brood. If the Apiarian who uses the common hives does not succeed in getting a mature queen for each hive, the queenless one will go back to the old stock.

If, while hiving a swarm, the Apiarian wishes to secure the queen, the bees should be shaken from the hiving-basket, a foot or more from the hive, when a quick eye will generally see her as she passes over the sheet. If the bees are reluctant to go in, a few must be directed to the entrance, and care be taken to brush them back, when they press forward in such dense masses that the queen is likely to enter unobserved. An experienced eye readily detects her peculiar color and form. She may be taken up without danger, as she never stings,[31] unless engaged in combat with another queen.

It is interesting to witness how speedily a queen passes into the hive, as soon as she recognizes the joyful note announcing that her

---

[30] *Beekeepers often use swarms that are caught to strengthen weak hives. If the weak hive is very small I put a single sheet of newspaper (with a couple of small holes punched into the paper) over the hive body; then a queen excluder over the paper. I put an empty hive body on top and pour the swarm onto the excluder. I then add frames to the empty super and let the bees gradually join together by eating through the paper. The colony will have two queens for a short while, and you can either kill one or let the bees determine which one will survive. The queens cannot kill each other through an excluder but only one queen will be alive after a few weeks.*

[31] *I have been stung by a virgin queen when I tried to catch her as she was trying to fly away while I was preparing to instrumentally inseminate her, but generally they will not sting. They may act like they will sting when you hold them to either clip their wings or to mark them. It may be that the longer, curved stinger is just designed to slip through between the segmental plates of another queen.*

colony has found a home. She quickly follows in the direction of the moving mass, and her long legs enable her easily to outstrip, in the race for possession, all who attempt to follow her. Other bees linger around the entrance, or fly into the air or collect in listless knots on the sheet; but a fertile mother, with an air of conscious importance, marches straight forward, and looking neither to the right hand nor to the left, glides into the hive, with the same dispatchful haste that characterizes a bee returning fully laden from the nectar-bearing fields.

Persons unaccustomed to bees, may think that I speak about "scooping them up," and "shaking them out," almost as coolly as though giving directions to measure so many bushels of wheat; experience will soon convince them, that the ease with which they may be managed is not at all exaggerated.

The old-fashioned way of hiving swarms, by mounting trees, and cutting off valuable limbs, should be entirely abandoned; nor should the hive ever be put over the bees, so as to crush any of them, or endanger the life of the queen. A skillful bee-keeper with his hiving-basket, will often hive six or more swarms in the time required, by the old plan, for hiving one; and in large Apiaries managed on the swarming plan,[32] where a number of swarms come out on the same day, and there is constant danger of their mixing, this is an object of great importance.

Dr. Scudamore, an English physician, who has written a tract on the Formation of Artificial Swarms, says that he once knew as "many as ten swarms go forth at once, and settle and mingle together forming, literally, a monster meeting." There are instances recorded of a still larger number having clustered together. A venerable clergyman in Western Massachusetts, told me, that in the Apiary of one of his parishioners, five swarms once clustered together. As he had no hive which would hold them, they were put into a large box, roughly nailed together. When taken up in the Fall, it was evident that the five swarms had lived together as independent colonies. Four had begun their works, each near a corner of the box, and the fifth in the middle; and there was a distinct interval separating the works of the different

---

[32] *Langstroth is still speaking to the practice of killing half of the colonies each year and harvesting the honey and wax. This is the practice that the movable-comb hive almost totally replaced.*

colonies. In Cotton's "My Bee Book," is a cut illustrating a similar separation of two colonies in one hive. By hiving, in a large box, swarms which have settled together, and leaving them undisturbed till the following morning, they would probably be found in separate clusters, and might easily be put into different hives.[33]

Swarming bees make a singular hissing or whispering sound, which often causes other hives in the Apiary to swarm. This is a frequent occurrence with discouraged or dissatisfied stocks, and I have occasionally had swarms which had only immature queens in their hive issue, on hearing this sound. This peculiar swarming sound *may* be produced merely by the great numbers of bees flying idly, at such times, to and fro in the air; but it seems to me to differ in its character, as it certainly does in its effect upon the bees, from the noise produced by the ordinary flight of busy workers, however numerous. My observations on this point, have satisfied me that those Apiarians are mistaken who deny to the bee, the sense of hearing. This sense, on the contrary, seems to be acute.

If the Apiarian fears that another swarm will issue, to unite with the one he is hiving, he may confine its queen with my movable-blocks; or he may quickly envelope the swarming hive with a sheet. If his new colony has been shaken upon the swarming-sheet, he may cover it from the sight [34] of other swarms, with another sheet.

---

[33] *Dr. O. R. Taylor, Jr., in a somewhat artificial situation with Africanized bees in Mexico, collected 220 ± queens (swarms) into a single large mass of bees that measured 16-18 feet in one direction. I suspect that, as the "swarm" grew in size, the amount of pheromone that was being sent downwind was enough to bring in more and more swarms to the single location. The pheromone plume going downwind must have been very strong.*

[34] *The concept of the Nasonov and queen pheromone was not in their thinking in the 1850's.*

The hive, with the new swarm, should be removed to its permanent stand as soon as the bees have entered; or the scouts, on their return, will find them, and will often entice them to flee to the woods.[35] There is the more danger of this, if the bees remained long on the tree before they were hived. I have almost invariably found that swarms which abandon a suitable hive for the woods, were hived near the spot where they clustered, the bee-keeper intending to remove them in the evening, or early next morning. Bees which swarm early in the day, will generally begin to range the fields in a few hours after they are hived, or even in a few minutes, if they have empty comb; and the fewest bees will be lost, when the hive is removed to its permanent stand, as soon as the bees have entered it. If it is desirable, for any reason, to remove the hive before all the bees have gone in, the sheet, on which the bees are lying, may be so folded that the colony can be easily carried to their new stand, where the bees may enter at their leisure.

Swarms sometimes come off when no suitable hives are in readiness to receive them. In such an emergency, hive them in any old cask, or measure, and place them, with suitable protection against the sun, where their new hive is to stand; when this is ready, they may, by a quick jerking motion, be easily shaken out before it, on a hiving-sheet.

---

[35] *While the scouts might have some influence if another home site was better, I think, that when most swarms leave a hive after entering it is because of some unfavorable condition within the hive, e.g., moldy combs. This absconding by a swarm may happen when a beekeeper collects the swarm and puts it into a new hive. This hive was not on the "list" of possible sites. In a natural swarming situation the scouts have essentially "voted" in favor of the site where the swarm will go and thus are not likely to leave.*

This "voting" is done by a large majority of the scout bees all dancing to indicate the same nest site. They perform these dances on the surface of the swarm. Thus, if you see a swarm and can watch the dances you can possibly determine where the swarm is going. When most of the bees are dancing all the same dance then the swarm is ready to leave. It is really quite fun, and instructive, to watch a swarm work through this process of selecting a new home site. However, most often the swarm is too high to watch this whole process as the scout bees dance on the surface of the cluster. First, there are many different dances, and then fewer until only one or two locations, that are the most favorable, are signaled as possible nest sites.

SWARMING AND HIVING. 139

I have endeavored, even at the risk of being thought too minute, to give such directions as will qualify the novice to hive a swarm of bees, under almost any circumstances; knowing that however necessary, suitable information is seldom found even in the best treatises on bee-keeping. Vague or incomplete directions fail, at the very moment that the inexperienced attempt to put them into practice.

Natural swarming may, unquestionably, be made highly profitable; and as it is most obvious way of multiplying colonies, and requires the least knowledge or skill, it will undoubtedly be the favorite method with most bee-keepers, for many years, at least. I shall, therefore, show how it may be conducted more profitably than ever, by the use of my hives; many of its most embarrassing difficulties being effectively obviated.

1. A serious objection to reliance on natural swarming, is the vexatious fact, that most swarming-hives are so constructed, that, although bees often refuse to swarm at all, they cannot furnish to their crowded occupants the proper accommodations for storing honey. Under such circumstances, hordes of useless consumers often blacken, for months, the outside of the hives,[36] to the great loss of their disappointed owners. In the movable-comb hives, an abundance of store-room can always be given to the bees; so that, if indisposed to swarm, they have receptacles easily accessible, and made doubly attractive by empty comb, in which to store up any quantity of honey they can gather.

---

[36] *He is speaking here of the heat-crowded conditions that force bees to "hang out". When these bees cluster outside of the hive they cannot be useful within the hive.*

2. Another objection to natural swarming arises from the disheartening fact, that bees are liable to swarm so often, as to destroy the value of both the parent-stock, and its after-swarms. Experienced bee-keepers obviate this difficulty, by making one good colony out of two second swarms, and returning to the parent-stock all swarms after the second, and even this if the season is far advanced. Such operations often consume more time than they are worth. By removing all the queen-cells but one, after the first swarm has left, second swarming may be prevented in my hives; and by removing all but two, provision may be made for the issue of second swarms, and yet all further swarming be prevented. After-swarms, in many instances, have to be returned again and again, before one queen is allowed by the bees to destroy the others. In this way, a large part of the gathering season is wasted; as bees often seem unwilling to work with their wonted energy, so long as the pretensions of several rival queens are unsettled.[37] *

3. Another very serious objection to natural swarming, as practiced with the common hives, is, that it furnishes no facilities for making vigorous stocks of late and small swarms. The time and money devoted to feeding small colonies are usually wasted; as the larger portion of them never survive the Winter, and most of those that do, are so enfeebled as to be of little value. If they escape being robbed by

---

[37] *The problem here is that bees are governed almost totally by instinct. Thus, when they are involved in one behavior (swarming) they usually do not engage in another behavior such as foraging for nectar. This is why it is important to prevent the swarming instinct from starting as well as the swarming itself. Generally it is relatively easy to prevent the swarm from actually leaving, for example cutting out the queen cells as they are made, or by using a queen excluder to prevent the queen from leaving. It is more difficult to prevent the instinct to swarm from starting.*

* Before inventing the movable-comb hive, I obviated, as far as possible, the evils of after-swarming, by the following plan: the second swarm, as soon as hived, was placed on the top of the parent-stock, or so, that the entrances to the old and new colonies would be near together, and face the same way. If a third swarm issued, it was added, at sunset, to the second swarm, by placing the hive or box containing that swarm, on a sheet, and shaking out the third swarm before its entrance. In three or four days—sufficient time being given for the young queen to become impregnated—the bees in the after-swarms were added, in the same way, to the parent-stock. One queen would quickly kill the other, and then next morning, the conjoined swarms being on a familiar spot, would work as well as though they had never been separated. The comb which they had built in the new hive was used in the spare honey-boxes.

stronger stocks, or destroyed by the moth, they seldom recruit in season to swarm, and often, unless the feeding is repeated a second season, they perish at last. Doubtless, many of my readers, from their own experience, can endorse every word of these remarks; having found the attempt to multiply colonies, by nursing and feeding small swarms in the common hives, usually attended with nothing but loss and vexation. The more of such stocks a man has, the poorer he is; for by their weakness, they constantly tempt his strong swarms to evil courses; until at last, they prefer, as far as they can, to live by stealing, rather than by habits of honest industry; and even if the feeble colonies escape being plundered, they often become nurseries for raising a supply of moths, to infest his Apiary. [38]

Suitable directions are furnished, in the chapter upon Feeding Bees, for building up the smallest after-swarms into vigorous stocks, and for strengthening such colonies as are feeble in the Spring.[39]

---

[38] Langstroth may be venting a little frustration over poor beekeeping neighbors. However his contention that anyone using a box hive or straw skep would not be able to adequately take care of the colony and thus spred diseases and harbor wax moths was correct.

[39] One of the best rules for feeding a package-bee colony, a division (split) or a weak colony, is to keep a feeder of sugar syrup on the colony until they no longer take the syrup into the colony. If there is sufficient nectar in the field the bees will prefer to use it for making and storing honey. Feeding syrrp in the spring is especially important because of cold or rainy weather that prevents the foragers from leaving the colony.

4. As both the parent-stocks and the after-swarms very frequently lose their young queens after swarming, a hive by which this misfortune can be easily remedied, will be of great service to those who practice natural swarming. An intelligent bee-keeper once assured me that he should use one movable-comb hive is his Apiary, for this purpose, at least, even if it had no merit in other respects.

5. In the common hives, but little can be done to dislodge the bee-moth, when it has gained the ascendancy; whereas, in mine, it can be easily extirpated. (See remarks on the Bee-Moth.)[40]

6. In the common hives, it is difficult to remove an old queen where her fertility is impaired; whereas, in mine, it can easily be done;[41] and an Apiarian may always have queens in the full vigor of their reproductive powers.

Intelligent Apiarians will see, from these remarks, that with movable-comb hives, natural swarming can be carried on with greater certainty than ever before, many of the perplexing discouragements under which they have hitherto prosecuted it, being effectually remedied.

---

[40] *If for no other reason, wax worms and diseases, the movable-comb hive must have jumped out at beekeepers as a very important aspect of these hives. They would see these conditions before they became webbed mess, or a dead colony.*

[41] *It would have been nearly impossible to find a queen in a straw skep or box hive. In fact, I suspect that beekeepers looked at a skep as a unit that either lived, died, produced honey, or not. They just did not have any control over the hive.*

CHAPTER X.

ARTIFICIAL SWARMING.

The numerous efforts made for more than fifty years to dispense with natural swarming, show the anxiety of Apiarians to find some better mode of increasing their colonies.

Although, by the control of the combs bees may be propagated by natural swarming, with a rapidity and certainty hitherto unattainable, still, there are difficulties inherent to this mode of increase, and therefore incapable of being removed by any kind of hive. Before describing the various methods which have been contrived for increasing colonies by artificial means, these difficulties will be briefly enumerated, so that every bee-keeper may decide intelligently which is *his* best way to multiply his stocks.

1. The numerous swarms lost every year is a strong argument against natural swarming. An eminent Apiarian has estimated, that taking into account all who keep bees, one fourth of the best swarms are lost every season. While some beekeepers seldom lose a swarm, the majority suffer serious losses by the flight of their bees to the woods;[1] and it is next to impossible, even for the most careful, to prevent such occurrences, if their bees are allowed to swarm.

2. Natural swarming is objectionable, on account of the time and labor which it requires.

The Apiary must be closely watched during the whole swarming-season; and if this business is entrusted to thoughtless children, or careless adults, many swarms will be lost. If many colonies are kept, a

---

[1] *Langstroth at this point acknowledges that the woods must have many colonies of bees, a fact that he forgot when he was worried about mating in an earlier chapter. Wild (feral) colonies made up at least 50% of the bees before the parasitic mites invaded the United States and essentially removed, or greatly reduced, this source of bees. The lack of this feral population is of great concern to population biologists because they provided a much needed genetic diversity that is now lacking. A large proportion of all the commercially raised queens come from a very small number of stock lines.*

competent person should always be on hand, in the height of the season, to attend to the bees. Even the Sabbath cannot be observed as a day of rest; as the bee-keeper is often compelled to spend it in hard work among his bees. Although it is as proper for him to hive his bees on that day, as it is to take care of his other stock, still, the liability to such labor deters many from Apiarian pursuits.

Many merchants, mechanics, and professional men, who wish to keep bees, cannot superintend them during the swarming-season; and are thus often kept from a pursuit intensely fascinating to an inquiring mind.* No man who spends some of his leisure in studying the wonderful instincts of bees, will ever complain that he can find nothing to fill up his time, out of the range of his business or the gratification of his appetites. Bees may be kept with great advantage, even in large cities,[2] and those who are debarred from rural pursuits may still listen to their soothing hum and harvest annually their delicious nectar.

If the Apiarian could always be at home during the swarming-season it would still be oftentimes very inconvenient for him to attend to his bees. The farmer, for instance, may be interrupted in the business of hay-making, by the cry that his bees are swarming; and by the time he has hived them, perhaps a shower comes up, and his hay is injured more than the swarm is worth. Thus, the keeping of a few bees, instead of being a source of profit, may prove an expensive luxury; while in a large Apiary, the embarrassments are often seriously increased. If, after a succession of days unfavorable for swarming, the weather becomes pleasant, it often happens that several swarms rise at once, and cluster together; and not infrequently, in the noise and confusion fly off, and are lost. I have seen the bee-master, under such circumstances, so perplexed and exhausted as to be almost ready to wish he had never seen a bee.

---

* "Bee-life," says Prof. Siebold, "does not merely serve to furnish man with wax, honey and mead, but constitutes an extremely important link to the great and most multifariously-composed chain of animal existence."

[2] *I started keeping bees, as a teenager, in the city of Detroit. The houses were on about 40-foot lots. The first colony was kept in the garage and it had its flight entrance out the garage window. Only one or two neighbors ever knew that I had a colony of bees. Even under these city conditions the bees made about a 100 lbs. of surplus honey most years. The major nectar source was white clover* (Trifolium repens) *found in the lawns of the homes. Urban beekeeping is a large and growing population of beekeepers.*

PLATE XIII.

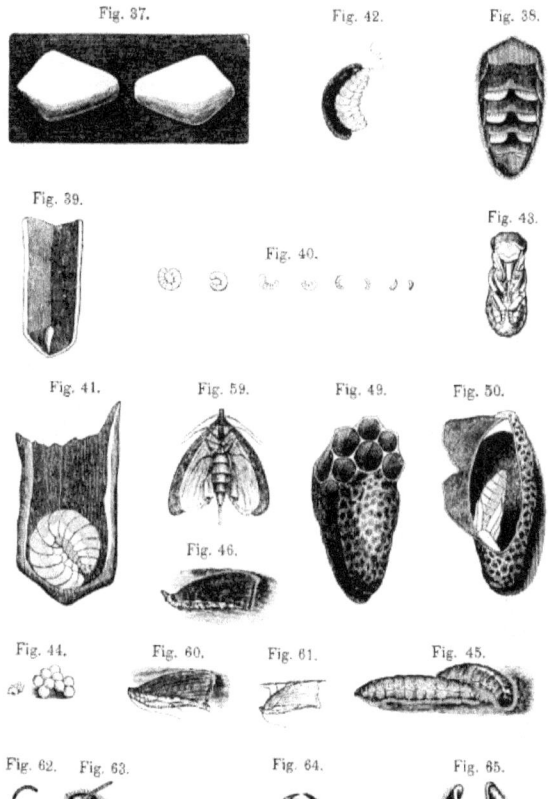

3. The multiplying of bees by natural swarming, must, in our country, almost entirely prevent the establishment of large Apiaries.

The swarming season is, with most bee-keepers, the busiest part of the year, and if they keep a large number of swarming-hives, they must devote nearly all their time, for a number of weeks, to their supervision; and at a season when labor commands the highest price, they may also be obliged to hire additional assistance.

To keep a few colonies in swarming-hives, often costs more than they are worth, while the supervision of a large number can he made profitable, only by those who can devote nearly all the Summer months to their bees. The number of such persons, in this country, must be very small;[3] and hence there are few who have succeeded in making bee-keeping anything more than a subordinate pursuit.

4. A serious objection to natural swarming, is the discouraging fact that bees often refuse to swarm at all; thus the Apiarian finds it impossible to multiply his colonies with any certainty or rapidity, even although he may be favorably situated for conducting bee-culture on an extensive scale.

Many of the most careful bee-keepers have fewer stocks than they had years ago, although they have sought to increase them to the extent of their power. Few intelligent Apiarians believe that there are half as many colonies in our Northern and Middle States, as there were twenty years ago; and most of them would abandon bee-keeping, if they did not regard it as a source of pleasant recreation, rather than of pecuniary profit; while others do not hesitate to say that much more money has, of late years, been spent upon patent hives, than those who have used them have realized from their bees.

---

[3] *Commercial beekeeping, as it is known today, did not exist in the 1850's. It took the invention of the movable-frame hive, extractors, and many such other inventions to make it possible to keep bees on a large, full-time basis. Most of the inventions that helped allow modern beekeeping were invented in the 20 or 30 years immediately after the invention of the movable-comb hive. Almost all of the inventions of modern beekeeping were made in the 19th century except for instrumental insemination of queen bees. It was the invention of the auto (and trucks) that really allowed the expansion into extensive out-yards that make up the large commercial beekeeping operations.*

Commercial beekeeping in Langstroth's time was most likely a beekeeper that had 200 colonies in one or two locations and sold all of his honey from his door.

It is an easy matter to make calculations on paper* almost as flattering as an imaginary tour to the gold mines of Australia or California. Only purchase a patent bee-hive, and if it fulfills the promises of its sanguine inventor, a fortune must be realized in a few years; but such are the disappointments resulting from bees refusing to swarm, that if the hive could remedy all other difficulties, it would still fail to answer the reasonable wishes of the experienced Apiarian. If every swarm of bees could be made to yield a profit of twenty dollars a year, the bee-keeper could not multiply his stocks, by natural swarming, so as to meet the demand for them; but would be entirely dependent upon the caprices of his bees, or rather upon the natural laws which control their swarming.

Every practical bee-keeper is aware of the uncertainty of natural swarming. Under no circumstances, can it be confidently relied on. While some stocks swarm regularly, and repeatedly, others, equally strong in numbers, and rich in stores, refuse to swarm, even in seasons in all respects highly propitious. Such colonies, on examination, will

---

* The following calculation of possible profits from bee-culture, taken from "Sydserff's Treaties [sic] on Bees," published in England, in 1792, is a perfect gem of its kind:

"Suppose a swarm of bees at the first to cost 10s; 6d., and neither them nor the swarms to be taken, but to do well, and swarm once every year"—bees must be naughty, indeed, if they dare to do otherwise!—"What will be the product for fourteen years, and what the profit, if each hive is sold at 10s. 6d.?

Years	Hives	Profits £ s. d.
1	1	0 0 0
2	2	1 1 0
3	4	2 2 0
4	8	4 4 0
**	**	* * *
14	8192	4300 16 0

"N.B.—Deduct 10s. 6d., what the first hive cost, and the remainder will be clear profit; supposing the second swarms to pay for hives, labor, &c." The modesty with which this writer, who seems to have had as much faith in his bees as in the doctrine that "figures cannot lie," closes his calculation at the end of fourteen years, is truly refreshing. No bee-keeper, on such a royal road to wealth, could ever find it in his heart to stop under twenty-one years, by which time his stocks would have increased to more than a million, when, *probably*, he would be willing to close his bee-business, by selling them for over two and three-quarter millions of dollars! The attention of all venders of humbug bee-hives, is respectfully invited to this antique specimen of the art of puffing.

often he found to have taken no steps for raising young queens. In some cases, the wings of the old mother are defective, while in others, she seems to prefer the riches of the old hive, to the risks attending the formation of a new colony. It frequently happens that, when all the preparations have been made for swarming, the weather proves so unpropitious that the young queens approach maturity before the old ones can leave, and are all destroyed. Under such circumstances, swarming, for that season, is almost certain to be prevented. The young queens are also sometimes destroyed, because of some sudden, and perhaps only temporary, suspension of the honey-harvest; for bees seldom colonize, even if all their preparations are completed, unless the blossoms are yielding an abundant supply of honey.[4] From these and other causes, which my limits will not permit me to notice, it has hitherto been found impossible, in the uncertain climate of our Northern States, for any but the most experienced and energetic Apiarians, to multiply colonies very rapidly by natural swarming. The numerous perplexities pertaining to natural swarming, have, for ages, directed the attention of cultivators to the importance of devising some more reliable method. for increasing their colonies.[*]

The ancient methods of artificial increase appear to have met with little success; but towards the close of the last century, a new interest was awakened on the subject, by the discovery of Schirach, a German clergyman, of the fact, previously known to a few, that bees are able to rear a queen from worker-brood. For want, however, of an acquaintance with some important principles in the economy of bees, his efforts met with but slender encouragement.

---

[4] *A swarm will not leave its store of honey in the parent hive if there is not some nectar available in the field. If a colony leaves when there is no replacement of its food it would be an absconding, and the whole colony will leave, hoping to find some nectar somewhere because it will starve where they are presently located.*

[*] Dr. Scudamore quotes Columella, who, about the middle of the first, century of the Christian Era, wrote twelve books on husbandry – "De re rustica"– as giving directions for making artificial swarms. Although he taught how to furnish a queen to a destitute colony, and how to transfer brood-comb, with maturing bees, from a strong stock to a weak one, he does not appear to have formed entirely new colonies by any artificial process. His treatise on bee-keeping shows not only that he was well acquainted with previous writers on the subject, but that he was also a successful practical Apiarian. Its precepts, with but few exceptions, are truly admirable, and prove that in his time bee-keeping, with the masses, must have been far in advance of what it now is.

Huber, after his splendid discoveries in the physiology of the bee, felt the need of some way of multiplying colonies, more reliable than that of natural swarming. His hive consisted of twelve frames, each an inch and a quarter in width, which were connected together by hinges, so that any one could be opened or shut at pleasure, like the leaves of a book. He recommends forming artificial swarms, by dividing one of these hives, and adding six empty frames to each half. After using his hive for years, I found that it could be made serviceable only by an adroit and fearless Apiarian. The bees fasten the frames with their propolis, so that they cannot easily be opened, without jarring the combs, and exciting their anger; or shut, without constant danger of crushing them. Huber nowhere speaks of having multiplied colonies extensively by such hives, and although they have been in use more than sixty years, they have never been successfully employed for such a purpose. If he had contrived a plan for giving his frames the requisite play, by suspending them on rabbets, instead of folding them together like the leaves of a book, he would have left much less room for subsequent improvements.[5]

"Dividing-hives," of various kinds, have been used in this country. The principle seems to have all the elements of success; and it was only after protracted experiment that I was able to ascertain that, however modified, such hives are all practically worthless for purposes of artificial swarming.

It is one of the laws of the hive, that bees which have no mature queen, seldom build any cells except such as are designed merely for storing honey, and are too *large* for the rearing of *workers*. Until my

---

[5] *Langstroth, here, implies how close others were to the invention of a movable-comb hive. The ability of the beekeeper (observer) to see what was happening within the hive was of great importance. Almost everyone who studied the problem at any length realized that fact. Thus, many minds were at work trying to solve the problem. While Dzierzon and Huber (or the servant, Burnens) were able to get into their hives and return them to their former state after they were through with their observations, they could not do this easily and/or exchange parts of one hive with another. Or take some combs from several hives and add them together to make up a new hive. I am not aware what people like Dzierzon thought of Langstroth's invention after the fact. I have to assume that they were delighted in the result, but maybe wondered why they did not think of the design themselves. There were other hives, it was just that they did not embrace the bee space concept and so the frames had to be cut apart for each examination. It was the concept of the movable-frame and the bee space that make Langstroth's hive different.*

perusal of Mr. Wagner's manuscript translation of Dzierzon, I thought that I was the only observer who had noticed the bearing of this remarkable fact on artificial swarming. It may, at first, seem unaccountable that bees should build only comb unfit for breeding, when their young queen will so soon require worker-cells for her eggs; but it must be borne in mind, that at such times they are in an *"abnormal,"* or unnatural condition. In a state of nature, they seldom swarm until their hive is full of comb; or if they do, their numbers are so reduced, that they are rarely able to resume comb building, until the young queen has hatched.

The determination of bees having no mature queen, to build comb designed only for storing honey, and unfit for rearing workers, shows very clearly the folly of attempting to multiply colonies by dividing-hives. Even if the Apiarian succeeds in dividing a colony, so that the queenless part proceeds to supply her loss, if it has bees enough to build sufficient new comb to make it of any value, it will build such as is designed only for storing honey; [6] using, chiefly for breeding purposes, the half of the hive containing the old comb. The next year, if this hive is divided, one half will contain nearly all the brood, while the other, having most of its combs fit only for storing honey, or raising drones, will be a complete failure.

Even with a Huber-hive, the plan of multiplying colonies by dividing a full hive into two parts, and adding an empty half to each, will be found to require a degree of skill and knowledge, far in advance of what can be expected of ordinary bee-keepers. The same remarks are substantially true of all frame or bar-hives which do not allow sufficient play between the parts to which the combs are attached; for, as

---

[6] *With comb foundation and drawn combs the problem of having the wrong size cells is not a concern as it was before the foundation press was invented. Some beekeepers use the different size combs (foundation now) to put into their honey supers. Since the queen is very reluctant to move into these combs the beekeeper does not have to use queen excluders to confine the queen to the broodnest. The number of cells per square decimeter has designated comb foundation sizes. Thus, 805 would be one of these odd sized foundations that do not allow brood rearing. The normal comb size would be about 825 cells per square decimeter.*

the bees usually build their combs slightly waving,[7] and some thicker than others, nearly insuperable practical difficulties will be found in making the necessary interchanges of comb, in such hives.

The attempt to multiply colonies by the common dividing-hives, will be found far more laborious and uncertain than by natural swarming. Every practical bee-keeper who has given it a fair trial, has been glad to abandon it, and return to the old-fashioned way.

Some Apiarians have attempted to multiply their colonies, by removing, when thousands of its inmates are ranging the fields, a strong stock to a new stand, and setting in its place an empty hive, with a piece of brood-comb, suitable for raising a queen. This method is still worse than the one just described. One half of the dividing hive was filled with breeding comb, while this empty having next to none, all that is built before the queen hatches, will be of a size unsuitable for rearing workers. The queenless part of the dividing-hive might also have contained a young queen almost mature, so that the building of large combs would have quickly ceased; for as soon as the young queen hatches, the bees commence building worker-combs.[*] When a new colony is formed by dividing the old hive, the queenless part has thousands of cells filled with brood and eggs, and young bees will be hatching for at least three weeks: by this time, the young queen will ordinarily be laying eggs, so that there will be an interval of not more than three weeks, during which the colony will receive no accessions. But when a new swarm is formed, in the way above described, not an egg will be laid for nearly three weeks, and not a bee hatched for

---

[7] *It is quite fascinating, to me, that the European bees build comb that is often waving, whereas the Africanized bees build them perfectly straight. It is one of the ways that scientists could tell whether a new colony was made up of European or African bees.*

[*] In attempting to rear artificial swarms by moving a full stock, my bees have built combs nearly four inches thick; and have afterwards pieced their lower edge with worker-cells, for the accommodation of the young queen. So uniformly do bees with an un-hatched queen build coarse, or drone-comb, that often a glance at the combs of a new colony, will show either that it is queenless, or that, having been so, it has just reared a new queen. It is not necessary that a queen should have commenced laying eggs to induce her colony to build worker-cells; I have know a strong swarm with a virgin queen, almost o fill their hive with beautiful worker-comb, before a single egg was deposited in the cells.

nearly six. During all this time, the colony will rapidly decrease,[*] and by the time the progeny of the young queen begins to mature, the new hive will have so few bees, that it would seldom be of any value, even if its combs were of the best construction.

After thoroughly testing this last plan of artificial swarming, I have found that it has not the least practical value; and as this is the method which Apiarians have usually tried, it is not strange that hitherto, they have almost unanimously condemned artificial swarming.

Another method of artificial swarming has been zealously advocated, which, seeming to require the smallest amount of labor or skill, would be everywhere practiced, if it could only be made effectual. A number of hives are to be connected by holes, so as to allow the bees to travel from any one to all the others. The bees, on this plan are to *colonize themselves*, and it is asserted that in due time, a single swarm, of its own accord, will form a large number of independent families, each possessing its own queen, and all living in perfect harmony.

This method, so fascinating in theory, though repeatedly tried with various ingenious modifications, has in every instance proved an entire failure. If the bees are allowed to pass from one hive to another, they will confine their breeding operations mostly to a single apartment, if it is of the ordinary size, and will use the others chiefly for storing honey. This is almost invariably the case, if the additional room is given by *collateral* or side boxes, as the queen seldom enters such apartments for the purpose of breeding; if, however, the new hive is directly *below* that in which the swarm was first lodged, and

---

[*] Every observing bee-keeper must have noticed how rapidly even a large swarm diminishes in number, for the first three weeks after it has been hived. So great is the mortality of bees during the height of the worker-season, that often, in less than that time, it does not contain one half its original number.

(This fairly rapid die-off of the bees should have been a clue to Langstroth as to the length of life of a honey bee. The average total life of a worker bee is only about a month, and any new, replacement bees would take three weeks to develop. This is the compelling reason that the swarm usually starts with a mated (old) queen, so that she can start egg laying for replacement bees immediately. If a young virgin left with the swarm (e.g., most afterswarms) she would have to mate (this would take at least two or three days) and then the young queen would take a few days to get up to speed in egg laying. This is time that the swarm just does not have if it is going to develop a population to store a supply of winter food and have sufficient population to overwinter.)

the connections are suitable, she will be almost certain to descend and lay her eggs in the new combs, as soon as they are begun by the bees. The *upper* hive being now almost entirely abandoned by her, the bees fill the cells with honey, as fast as the brood is hatched, their instinct impelling them to keep their stores of honey, if possible, above the breeding-cells. So long as bees have an abundance of room *below* their main hive, they very seldom swarm; but if it is on the *sides* of their hive, or *above* them, they often swarm rather than take possession of it.[8] In none of these cases, however, do they ever form independent colonies, *if left to themselves.*

The skillful Apiarian may, doubtless, *compel* his bees to rear an artificial colony, by separating from the main hive, by a slide, an apartment that happens to contain brood; but unless his hives admit of thorough inspection, as he can never know their exact condition, he will be far more likely to fail than to succeed. This plausible theory, therefore, to be reduced to even an empirical and precarious practice, requires more skill care, labor, and time, than are necessary to manage the ordinary swarming-hives.

The failure on the part of experienced, as well as inexperienced Apiarians, of so many attempts to increase colonies by artificial means, has led many to advocate the general use of non-swarming hives. In such hives, very large harvests of honey are often obtained from strong stocks of bees; but it is evident that if the formation of new colonies were generally discouraged, the insect would soon be exterminated.[9]

---

[8] *This theory of having expansion room below the brood area is in sharp contrast to modern thinking. Though, in defense of the idea, space within the broodnest is regarded as critical if swarming is to be prevented. I do not know of a particular set of experiments where this idea, set forth by Langstroth, has been tested..*

We now normally reverse brood chambers to give the room <u>above</u>, since the natural tendency is for the colony to move upwards, and we are trying to provide the most available space within the broodnest. These reversals allow the bees to better use the space at a time when the colony is expanding its population at a very rapid rate. The storage of honey is always at the top of the hive since this is the area that is the warmest during the winter and the cluster can move onto to replenish their food. Langstroth used only a single chamber for a broodnest and this would also probably affect his thinking.

[9] *Modern beekeepers still try to produce non-swarming colonies because they **do** produce more honey, and new colonies can be secured from other sources.*

Although the movable-comb hive may be made more effectual to prevent swarming than any with which I am acquainted, still there are some objections to the non-swarming plan which cannot be removed. To say nothing of its preventing the increase of stocks, bees usually work with diminished vigor, after they have been kept in a non-swarming hive for several seasons. This will be obvious to any one who will compare the super-abounding energy of a new swarm, with the more sluggish working of even a much stronger non-swarming stock.[10]

An old queen, whose fertility has become impaired, can be easily caught and removed, in the movable-comb hive; but when hives are used in which this cannot he done, the Apiary will contain queens that have passed their prime, and some which may die when there are no eggs from which others can be reared.

On no subject has the author of this work experimented more fully than on that of Artificial Swarming; and those bee-keepers to whom this chapter may, at first, seem needlessly diffuse, will find that it contains many important principles, which, in any other connection, would probably have required even more fullness of detail.

Before detailing the various methods of Artificial Swarming which may be practiced in the movable-comb hives, I shall describe one which may be used with almost any hive, by those who have sufficient confidence to manage bees.

About the season of natural swarming, what I shall call a *forced swarm*, may be obtained from a populous stock,* by the following process. Choose that part of a pleasant day, when many bees are abroad, and if any are clustered on the bottom-board or outside of the hive, puff among them a few whiffs of smoke—that from spunk is best—so as to drive them up among the combs. The bees will go up more readily if the hive is tipped back, or elevated by small wedges,

---

[10] *There is an unexplained vigor that seems to go with a swarm. There have been various theories for this "swarm vigor" though I do not know of any one idea that has prevailed. The swarm is made up of mostly young bees, but that would not account for all of the vigor. There are beekeeping management schemes that try to produce artificial swarms just to try and capture that swarm vigor.*

* "Driving succeeds best in warm weather, and with populous stocks; for if the combs be not worked down to the floor-board, the bees are apt to collect in the open space instead of ascending into the upper box."—BEVAN.

about one-quarter of an inch above the bottom-board. Have in readiness a box—which I shall call the *forcing box*—whose diameter is about the same with that of the hive from which you intend to drive the swarm. Lift the hive from its bottom-board without the slightest jar, turn it over, and carefully carry it off about a rod, as bees, if disturbed, are much more inclined to be peaceable, when removed a short distance from their stand. If the hive is gently placed upside down on the ground, scarcely a bee will fly out, and there will be little danger of being stung. The timid and inexperienced should protect themselves with a bee-dress, and may gently sprinkle the bees with sugar-water or blow more smoke among them, as soon as the hive is inverted. After placing it on the ground, the forcing-box must be put over it, and every opening between it and the hive, from which a bee might escape, should be stopped with paper, or any convenient material. The forcing-box, if smooth inside,† should have slats fastened one-third of the distance from the top, to aid the bees in clustering.

As Soon as the Apiarian has confined the bees, he should place an empty hive—which I shall call *the decoy-hive*—upon their old stand, which those returning from the fields may enter, instead of dispersing to other hives, to meet, perhaps, with a most ungracious reception. As a general rule, however, a bee with a load of honey or bee-bread, after the extent of his resources is ascertained, is pretty sure to be welcomed by any hive to which he may carry his treasure; while a poverty-stricken unfortunate that presumes to claim their hospitality is, usually, at once destroyed. The one meets with as flattering a reception as a wealthy gentleman proposing to take up his abode in a country village, while the other is as much an object of dislike as a poor

---

† In my own practice, I use a box, the inside edges of which are beveled, to facilitate the ascent of the bees, and the back hinged, so that it can be opened for seeing the queen as she goes up with them. The few bees that may escape, even if not full of honey, are too bewildered by their change of position, to make any attack.

(This confusion technique is very useful even inspecting a normally placed hive. Just tip it backwards and lay it on the ground [works best with two people]. The propolis holds the supers together. Then proceed to examine the combs from the bottom looking for queen cells, etc.. Then, return each hive body, to the original stand. You can remove frames at this time if you think you need to see more than you could see from the bottom. The bees are usually very confused while the hive is lying on the ground as they seemed to have lost their orientation, and they just do not know what to defend.)

man, who bids fair to become a public charge.

To return to our imprisoned bees: their hive should be beaten smartly with the palms of the hands, or two small rods, on the sides to which the combs are attached, so as to run no risk of loosening[*] them. These "rappings," although not of a very "spiritual" character, produce, nevertheless, a decided effect upon the bees. Their first impulse, if no smoke were used, would be to sally out, and wreak their vengeance on those who thus rudely assail their honied home; but as soon as they inhale its fumes, and feel the terrible concussions of their once stable abode, a sudden fear that they are to be driven from their treasures, takes possession of them. Determined to prepare for this unceremonious writ of ejection, by carrying off what they can, each bee begins to lay in a supply, and in about five minutes, all are filled to their utmost capacity. A prodigious humming is now heard, as they begin to mount into the upper box; and in about fifteen minutes from the time the rapping began—if it has been continued with but slight intermission— the mass of the bees, with their queen, will hang clustered in the forcing-box, like any natural swarm, and may, at the proper time, be readily shaken out, on a sheet, in front of their intended hive.

If the forced swarm could now be put on the old stand, and the parent-hive removed to a new place in the Apiary; or if the latter could be returned to its usual position, and the former be put somewhere else, it would simplify very much the making of artificial swarms. Neither method, however, can be pursued without serious loss; for if the position of a colony has been changed by the bee-keeper, the bees will not adhere to the new place, as they do when they swarm of their *own accord*.[11]

---

[*] There is little danger of loosening the combs of an old stock, but the greatest caution is necessary when the combs of a hive are new. If, in inverting such a hive, the *broad sides* of the combs, instead of their *edges*, are inclined downwards, the heat, and weight of the bees, may loosen the combs, and ruin the stock.

(Since Langstroth's time we have wired frames and/or wired foundation that strengthen the combs such that it isn't often the combs would loosen or fall out, however a new comb is still much more fragile than a comb that has had brood raised within, or even a comb that has had honey stored in it over several years.

[11] *This is related to the dilemma mentioned in the last chapter regarding the bees in a natural swarm not returning to the parent hive. That is, since this is not a swarm of*

In every case when the position of its hive has been changed, each bee, as it sallies out, flies with its head turned towards it, that by marking the surrounding objects, it may find its way back. If, however, the bees did not emigrate of their *own free will*, most of them appearing to forget that their location has been changed, return to the familiar spot; for it would seem that,

> "A 'bee removed' against its will
> Is of the same opinion still."

Should the Apiarian, ignorant of this fact, place the forced swarm on the old stand, and remove the *parent-stock* to a *new* place, the latter would lose so many of the bees which ought to be retained in it, that most of its unsealed brood would perish from neglect. If, on the contrary, he should remove the *forced swarm* to a *new* position, it would be so depopulated as to be of little value.

These difficulties may be obviated by removing either colony about half a mile from its former home, in which case, if forage is abundant, nearly all will remain in their proper hive. Some recommend that they should be carried off at least three miles; but I have found that this is unnecessary, unless there is a deficiency of blossoms in the immediate vicinity of their new home.[12] If the colonies are carried off the precautions given elsewhere * for moving bees must he carefully followed also the directions for retaining a sufficient number

---

young bees, many of these bees **do** *know the location of the parent-stock. When you force bees into the box you get all ages of bees rather than mostly new ones that go out with a natural swarm. The new, young bees found in a swarm do not know the location of the hive since they have never taken orientation flights.*

[12] *Since bees use visual cues when possible, the distance needed for moving a colony is relative to how far the bees would have foraged. Thus, Langstroth is right, but modified as to how good the foraging was near their <u>old</u> home. If the bees had not foraged very far they would not have learned the distant visual cues, and thus they would not be able to find their way back to the old location. Most beekeepers move a division, or split, two or three miles from the parent apiary. Almost any distance will be sufficient for the majority of the bees. Even if you leave a new, division colony within the apiary it is only the older bees that return to the parent colony. A beekeepers needs to remember this when making a division that remains within the apiary and add an additional number of bees to the division since some will return to the parent colony.*

* The copious alphabetical index at the end makes it easy to refer to any subject discussed in this book.

of bees in the parent-stock. Those not carried off must be put on their old stands.

As the transportation of colonies is laborious, and often times expensive, I shall describe the methods which, after years of experimenting, I have devised for dispensing with it. I have ascertained that, if a hive is removed most of the bees returning from abroad and alighting upon a neighboring hive, if kindly received, will not go back to their former stand. Even the *temporary* loss of their old home is followed by a distraction which makes on them such a permanent impression, that they mark their new location as carefully as a new swarm. Now I find that, on the same principle, nearly all the bees which have returned from the fields, while a swarm is being forced from the parent-hive, will enter this hive if it is put upon its old stand, and adhere to it afterwards wherever it may be placed.

As soon, therefore, as the bee-keeper has forced a swarm, the forcing-box must be gently lifted off, and set in a shady place where

entrance* closed until sunset. Unless this precaution is adopted, the bees in other hives, ascertaining its weak and queenless condition, may attempt to rob it.

If the stock from which the artificial colony was driven, were intending to swarm, it will contain maturing queens, one of which will soon take the place of the old one, as in natural swarming. If, no royal cells were in progress, the bees will proceed to construct them.

Artificial colonies should not be formed until drones have made their appearance, or the young queen may fail to be impregnated, and the parent-stock may perish.[13]

We return now to our forced swarm. The bees should be shaken out of the forcing-box, and hived like a new swarm, when, if placed on their old stand, they will work as vigorously as a natural swarm. If, they were driven, at first, into a hive which will suit the Apiarian, it may be returned to their old location, without disturbing the bees.

If, in driving the swarm, or in transferring it from the forcing-box the queen was not *seen*, it may be certainly known, in from five to fifteen minutes after the bees have entered their new hive, whether or not she is with them.

As soon as the bees are clustered in the hive, if they do not find her, a few will come out and run about, as if anxiously searching for something they have lost. The alarm is rapidly communicated to the whole colony; the explorers are rapidly reinforced, the ventilators suspend their operations, and soon the air is filled with bees. If they cannot find the queen, they return to their old stand, and if no hive is there, will soon enter one of the adjoining colonies. If their queen is restored to them soon after they miss her, those running out of the

---

* In closing the entrance, the bee-keeper will see that sufficient air is admitted, but not enough to chill the brood. If the weather should suddenly become very cool, and the hive is quite thin, it will be advisable to cover it with something that will aid in preserving its internal heat. The same precautions are often important in hives which have swarmed naturally.

[13] *It is the lack of good flight weather that limits mating of queens early in the year, rather than the lack of drones. We can raise queens in the middle of April, and the colonies have drones at that time. We do not have good, consistent mating weather until after the first of May. If a queen only has one mating day and then several cold days she may begin to lay eggs and will not mate again. By only mating with a few drones on the one day she would not have as full a spermatheca as if she had mated over several days. If you do not have good flight weather the presence of drones doesn't matter.*

hive will make a half-circle, and return; the joyful news is quickly communicated to those on the wing, who forthwith alight and enter the hive; all appearances of agitated running about on the outside of the hive, ceases, and ventilation, with its joyful hum, is again resumed.[*] If the bees remain quiet in the new hive, for about fifteen minutes, the queen is almost certainly with them.

If the Apiarian, in making his artificial swarm, does not see the queen, he must wait until the bees show, by their conduct, whether she is with them or not. If they begin to leave the hive, the entrance must be closed, to confine them until the parent-stock can be drummed again, and the queen, if possible, secured. If she cannot be induced to leave the parent-stock, and another cannot be had to supply her place, the bees must be returned, and the driving resumed at another time. A queen, however, which does not go up the first time, is very apt to persist in her refusal.

In forcing a swarm, I have directed that it be done when many workers are abroad, in order that they may be induced to adhere to the parent-stock. Many bee-keepers, however, may prefer to make their swarms early in the morning, or late in the afternoon, when few bees are at work. In this case, a proper number of adherents may be obtained for the parent-stock by shaking out the bees from the forcing-box on a sheet, that as they enter the hive in which they are permanently to reside, many may take wing, and return to the decoy-hive. If

---

[*] To witness these interesting proceedings, it is only necessary to catch the queen, and keep her until she is missed by her colony. For greater security, I usually confine her, when shaken from the bees, in a small paper-funnel, with twisted end, from which she may be easily taken.

It is a mistake to suppose that a swarm will not enter a hive unless the queen is with them. If some start for it, the others will speedily follow, all seeming to take for granted that the queen is somewhere among them. Even after they begin to disperse in search of her, they may often be induced to return, by pouring out a fresh lot of bees, which, by entering the hive with fanning wings, cause the others to believe that the queen is coming at last.

(This is a case where the Nasanov pheromone becomes dominant as the bees are fanning and exposing this scent gland. The queen pheromone is mostly transported by food exchange [trophalaxis] between worker bees. Thus, she is not missed until after they cluster, or take up residence in a hive.)

Bees which miss their queen, under such circumstances, will accept of any one that may be offered them, and may often be pacified with worker-comb.

the number is still too small, after most of the bees have entered the new hive, the sheet with some adhering to it may be carried to the decoy-hive. After these bees show that they miss their queen, by running in great confusion in and out and over the hive, the parent-hive must be presented to them, and when they have entered it, removed to a new position in the Apiary, and the forced swarm returned to the old stand. If one-quarter of the bees are left in the parent-stock, the supply will be ample; larger, indeed, than is usually left in natural swarming.[14]

If there are in the Apiary several old stocks standing close together, it is highly desirable in performing these various operations, that the decoy-hive, and that for the forced swarm, should be of the same shape and even color with that of the parent-stock. If they are very unlike, and the returning bees attempt to enter a neighboring hive, because it resembles their old home, the adjoining hives should have sheets thrown over them, to hide them from the bees, until the operation is completed.

I have sometimes obtained a supply of adhering bees for the parent-stock, by placing it on the old stand, and removing the forced swarm to a new location. The larger part of the bees will of course return to their former home; some, however, will remain with their queen, and begin to labor in the new hive. In two or three days, exchange the position of the two hives, when enough bees which have become accustomed to the new place, will return to it, to carry on their operations in the parent-stock. This plan has the advantage of retaining most of the bees in the parent-stock, until the cells for rearing young queens are begun;[15] it will also suit bee-keepers who are pressed for time, and are obliged to force their stocks, early in the

---

[14] *I think Langstroth is either counting the bees left after some afterswarms have also gone from the hive, or he overestimated the number of bees going with a swarm. Most apiculturists figure that only about half the bees leave the parent hive. Sometimes the contrast within the hive may make the number of bees left in the hive seem smaller than it actually is.*

[15] *This technique is also now used in making divisions or "splits." The same problem exists in that the older bees will return to the parent hive, however, and thus it is important to put more bees into the division than may seem necessary at the time of the split. Exchanging positions of the hives later is also a good way of equalizing the two colonies, much like Langstroth explains as follows in the text.*

morning or late in the afternoon, when but few bees are abroad in the fields.

If the parent-stock stands at some distance from others, and resembles in shape, size, and color, that intended for the forced swarm, a proper division of the bees may be effected as follows: Place the parent-stock about six inches to the right of the old stand, and the forced swarm as far to the left; so that the position of the old entrance shall be about equally distant from each. If either colony contains too few bees, it may be moved a little nearer to the old entrance; or it may be reinforced, after the bees have gone to work, by closing the entrance of the stronger hive until dark.

If the old stocks stand close together, some prefer another mode of forming the artificial swarm. After the bees have been driven from the parent-stock, the forced swarm is at once placed on the old stand, while the parent stock in which the proper number of bees has been left, is set in a cool place, and shut up—care being taken to give them air—until late in the afternoon of the third day. It may now be put on its permanent stand, and opened an hour or two before sunset, when the bees will take wing almost as if intending to swarm. Some will join the forced swarm on the old stand, but most, after hovering a short time in the air, will re-enter their hive. While the entrance was closed, thousands of young bees were hatched, and these, knowing no other home, will all unite in the labors of the hive. The imprisoned bees ought to be supplied with water, to enable them to prepare food for the larvae. In the common hive this may be injected with a straw through a gimlet-hole.[16] Where artificial swarming is practiced on a large scale, I have devised a plan which I very much prefer to any previously described. Let the Apiarian obtain a forced swarm [*] from some bee-keeper, a mile or two off or from one of his own stocks, carried that distance before the bees began to work in the Spring. Bringing it home, according to the directions subsequently given for transporting bees, let it be confined in a cool place, so as to have plenty of air. Late in the afternoon, or early next morning, let him force four or

---

[16] *A gimlet is now usually called an awl –a thin, sharp metal rod that is used to punch holes in wood or leather.*

[*] *If he delays artificial swarming until natural swarms begin to issue, he may use them in the same way.*

five[*] swarms, placing them, at once, on the stands of the parent-stocks, and these latter where it is intended they shall permanently remain. The forced swarm, brought from a distance, should now be shaken out on a sheet, a foot or more from a hive, and gently sprinkled, so as to prevent any bees from taking wing. With a saucer, scoop up, without hurting any of them, as many bees as you can, and carry them to the mouth of one of the old stocks, from which you have driven a swarm. Continue to do this, until you have about equally apportioned the bees, and if any remain on the sheet, carry it to the mouth of the hive which has received the least.[†] These bees, having no previous home in your Apiary, will adhere to the different hives in which they are placed, and thus, without any farther trouble, your parent-stocks and forced swarms will alike prosper.

One great advantage which this method has over all others, is, that it secures, so simply and effectually, the necessary number of bees for the parent-stocks. Inexperienced persons, instead of being perplexed to know how many bees they shall leave in the forced stocks, may drive from them, if they can, every bee. If the bee-keeper can not conveniently obtain a swarm from a distance, he may use, for this purpose, the first natural swarm which comes off in his own Apiary; and by delaying to make artificial colonies until natural swarms begin to issue, every such swarm may be used for forming at least four artificial swarms. Or, by the method recommended by Dr. Dönhoff, of Germany, he may secure a colony, which, when divided in the way

---

[*] An expert will force them all in the time usually taken by a novice to force one. As soon as a forcing-box is placed over one hive, he will remove another from its stand, and then the rest, and in drumming them will pass from one to another, so as to lose not a moment's time in the whole operation. Ten artificial swarms, or even more, may be made, in this way, in less than an hour after sunrise or before sunset.

(I don't think that modern beekeepers would go to this much work to create a new hive. Maybe this is the reason that the package bee industry has become so successful. But even in this specialized part of modern beekeeping, the movable-comb hive makes the production of package bees quite possible.

An easy way to equalize colonies of bees to just move a weak hive to the location of a strong hive and the strong hive to the weak hive location. This move accomplishes two things. The first is to possibly prevent the strong colony from swarming, and also to make the weak colony a productive one.)

[†] The queen should be looked for, and the hive noted to which she is given. If she has entered the empty hive, she may be easily secured.

above mentioned, will adhere to their new locations: "On an evening, when the next day promises to be clear and warm, drive out a swarm, and set it in the place of the parent-stock. Next day, when it is warm, pour some honey among the bees in the box, and in a few hours they will swarm." *

The directions given for the formation of artificial colonies, differ, in some important respects, from any furnished by other writers, and are so simple that any one accustomed to handle bees can easily follow them. They enable the Apiarian, let him use what hive he will, to be entirely independent of natural swarming.

It will be obvious, however, that artificial swarming, to be successful, requires a knowledge of the laws which control the breeding of bees. Those, therefore, who are ignorant of the economy of the beehive, cannot safely depart from the old-fashioned mode of management; as emergencies which they are unprepared to meet, may at any

---

* A forced swarm may be made to adhere to its new location as follows: Secure their queen, when they are shaken out of the hive; and when they show that they miss her, confine them to their hive, until their agitation has reached its height. Then open the hive, and as the bees begin to take wing, present to them their queen (see p. 159). When they have clustered around her, they may be treated like a natural swarm. To do this with every forced swarm would take too much time; but it would answer well when the forced swarm is to be divided, as above, into four or five parts.

Mr. P. J. Mahan, of Philadelphia, informs me that he has several times succeeded in making an old colony adhere to a new place in the Apiary, by beating the hive, after the bees have been shut in, even at the risk of slightly injuring some of its combs. When it is opened, the bees will fly out in great numbers, but nearly all will return to their hive on the new stand.

moment occur. An Apiarian may use the common hives * a whole life-time, and, unless he gains his information from other sources, may yet remain ignorant of some of the most important principles in the physiology of the honey-bee: while any intelligent cultivator may, with movable-combs, in a single season, verify for himself the discoveries which have been made only by the accumulated toll of many observers, for more than two thousand years.

By the aid of movable-comb hives, artificial swarming may be easily and quickly performed. An empty hive, with its frames properly arranged, must be in readiness to receive the new swarm; and before carrying the parent-stock from its stand, a little smoke should be puffed into the entrance, which should then be closed with the movable-blocks. Remove, now, one or two of the tins that cover the holes on the spare honey-board (Pl. VIII., Fig. 21), and blow smoke into the hive, until the bees begin to make a loud humming, then the honey-board may be loosened with a knife, and safely removed, care being taken to set it on its edge, so as not to crush the bees with which its under surface is usually covered. No danger need be apprehended from these bees, as they are completely bewildered by their sudden

---

* "An opportunity of beholding the proceedings of the queen, in hives of the usual form, is so very rarely afforded, that many Apiarians have passed their lives without enjoying it; and Reaumur himself, even with the assistance of a glass-hive, acknowledges that he was many years before he had that pleasure." —BEVAN

Swammerdam, who wrote his wonderful treatise on bees, before the invention of glass hives, was obliged to tear hives to pieces in making his investigations! When we see what important results these great geniuses obtained, with means so imperfect, it compared with the facilities which the veriest tyro may now possess, it ought to teach us a becoming lesson of humility.

The sentiments of the following extract from Swammerdam, ought to be engraved upon the hearts of all engaged in investigating the works of God: "I would not have any one think that I say this from a love of fault-finding" – He had been criticizing some incorrect drawings and descriptions – "my sole design is to have the true face and disposition of Nature exposed to sight. I wish other may pass censure, when due, on my works; for I doubt not that I have made many mistakes, although I can, from the heart, say that I have not, in this treatise designed to mislead. *** The desire of writing is so prevalent, that men publish books filled only with the fancies of their brain, and thus misrepresent God and his works. God forbid that I should ever do this. Truth, and a religious scrupulousness of mind, ought everywhere to prevail in describing natural things; for they are the Bibles of the divine miracles. If he who writes aims to deceive himself and others, let him know that in due time all things will be revealed."

ARTIFICIAL SWARMING.

exposure to the light, and removal from the hive. Any of the large "*supers*"* used in my hives, or any other box of suitable dimensions, may now be set over the bees, into which they may be driven, in the way described on page 155. A little more smoke blown into the entrance of the hive, will obviate the necessity of much rapping, and materially quicken the ascent of the bees.† After they have been driven from the parent-stock, the directions must be followed which have already been so minutely described.

Whenever the bee-keeper learns how to handle safely the movable-frames—full directions for doing which will soon be given—he may dispense with the forcing-box, and make his swarms by lifting out the frames from the parent-stock, and shaking the bees from them, by a quick jerking motion, upon a sheet, directly in front of the new hive.[17] As soon as a comb is deprived of its bees, it should he returned to the parent-stock. If one or two combs containing brood, eggs, and stores, are given to the forced swarm, it will be much encouraged, and will need no feeding, if the weather should be unfavorable. In removing the frames, the bee-keeper should look for the queen, and give the comb on which she is, to the forced swarm, without shaking off the bees.[18] If he does not see her on the combs, he will seldom fail to notice her, after a little practice, as she is shaken on the sheet, and crawls towards the new hive. The queen is seldom left on a frame after it has

---

* This term is used by Apiarians to designate any upper box placed over the main lower-hive. An empty hive, like that in Pl. Fig. 1., or a hive like that in Pl. III., Fig. 2. — if inverted — will answer for a forcing-box.

(*Modern beekeepers often talk of all boxes as supers regardless if they are for brood or for honey. Though mostly, call any <u>empty</u> box a super. This description is not very accurate but coveys to other beekeepers the sense of a box with combs.*)

† Time will be saved by arranging (p. 162) to force several swarms at once.

[17] *It would seem to me that if anyone, in those days, had any doubts about the usefulness of the movable-comb hive prior to this explanation of handling of individual frames, the doubts would have ceased. Today we make these kinds of manipulations all the time, either in removing honey or exchanging brood frames. However, before the movable-frame hive such easy manipulations were not possible.*

[18] *The main reason to give the queen to the forced-swarm (or division) is that more bees will remain in a hive when there is a queen present. Since many of the older bees will return to the original location it is therefore always better to tip the balance in favor of the new hive. If you are making a division with a purchased queen you would want to put more bees with the new queen as the bees will still associate with the odor of the old queen, and more bees will remain, or return, to the old queen.*

been shaken so that most of the bees fall off. As soon as the necessary number of bees have been transferred to the new hive, the precautions previously given must be used to obtain adhering bees for the parent-stock. If the proper allowance of bees is secured for the parent-stock by the method described on page 162, the hive for the forced-swarm may be placed at once on the old stand, and the bees from the parent-stock shaken from the frames upon a sheet, so placed that they can easily run into their new hive.

If the forced swarms were made a short time before natural swarming would have taken place, some of the parent-stocks will contain a number of maturing queens, which may be removed, a few days before hatching, and given to such as have started none. By making a few forced swarms, about a week or ten days before the time in which the most are to be made, there will be an abundance of sealed queens, almost mature, so that every parent-stock may have one. If an unhatched queen can be given, on her frame, to each stock that needs it, so much the better; but if there are not enough frames with sealed queens, while some contain two or more, the bee-keeper must proceed as follows:

With a sharp pen-knife, carefully remove a piece of comb, an inch or more square, that contains a queen-cell; and in one of the combs of the hive to which this cell is to be given, cut out a place just large enough to receive and hold it m a natural position. If it is not secure, apply, with a feather, a little melted wax, where the edges meet, and the bees will soon fasten it to suit themselves.

Unless *very* great care is used in transferring a royal cell, its inmate will be destroyed, as her body, until she is nearly mature, is so exceedingly soft, that a slight compression of her cell—especially near the base, where there is no cocoon—generally proves fatal. For this reason, it is best to defer removing them, until they are within three or four lays of hatching. A queen-cell, nearly mature, may be known by its having the wax removed from the lid, by the bees, so as to give it b*rown* appearance.[19]

---

[19] *If you raise your own queens by the grafting method you* **will** *know the age fairly closely and can move the cell when it is about to emerge –about 10 days after grafting, or the 13-14th day after the egg was laid. Otherwise you will have to use the method*

ARTIFICIAL SWARMING.

The forcing of a swarm ought not to be attempted when the weather is so cool as to chill the brood; and never unless there is sufficient light not only to enable the Apiarian to see distinctly, but for the bees that take wing to direct their flight to the entrance of their hive. Bees are always much more irascible when their hives are disturbed after it is dark, and as they cannot see where to fly, they will alight on the person of the bee-keeper, who will be almost sure to be stung. It is seldom that night work is attempted upon bees, without the operator having occasion to repent his folly. If the weather is not too cool, early in the morning, before the bees are stirring, is the best time for most operations, as there will then he the least danger of annoyance from robber-bees.

To some of my readers, it may appear almost incredible that bees can be dealt with in the summary ways that have been described, without becoming greatly enraged; so far, however, is this from being the case, that in my operations, I often use neither smoke, sugar-water, nor bee-dress, although I by no means advise the neglect of such precautions. While the timid, if unprotected, are almost sure to be stung, there is something in the determined aspect and movements of a courageous and skillful operator, that seems often to strike bees with instant terror, so that they become perfectly submissive to his will.[20]

Artificial swarms may be created with perfect safety, even at mid-

---

described by Langstroth. If you miss knowing the age, or graft too old larvae, a young virgin will emerge and kill all other queen cells that you are trying to raise.

[20] *I have heard this told many times and have seen no proof that bees can smell 'terror'. Many persons frightened of bees have a tendency to swing their hands at bees of any that come too close. This movement often brings on a sting or two. Bees can smell different things and recognize different colors. Usually it is the odor of the first sting that elicits additional stings, as bees do recognize the alarm pheromone from the first sting. If you want to impress a new beekeeper or observer, get them stung first and they will then be bothered, or even stung again, by the bees and you will look like the 'master' beekeeper that doesn't get stung. New beekeepers also tend to make more unsure or erratic movements. These actions will also result in more stings. Bee gloves more than likely have stings and alarm odors all over them, and beginning beekeepers tend to wear these as a protection against being stung, yet may attract more stings than they prevent. Gloves have their place, however, and if they prevent too many stings, or provide some security, then they should be used. I find gloves useful when removing honey. This is usually when bees are robbing and defensive.*

day, as the thousands of bees returning their loads, never make an attack, while those at home can be easily pacified.

The arrangement which permits the top of the movable-comb hive to be easily removed, and the sugar-water to be sprinkled upon the bees, before they attempt to take wing, has great advantages. If the hive opened on the side, like Dzierzon's, it would be impossible to make the sweetened water run down between all the ranges of comb, and it would be necessary to use smoke[*] in every operation. The use of smoke frequently causes the queen to leave the combs, for greater security. This often causes great delay in the formation of artificial swarms by removing the frames, and in operations where it is desirable to catch the queen, or to examine her upon the comb.

Huber thus speaks of the pacific effect produced upon the bees by the use of his leaf-hive: "On opening the hive, no stings are to be dreaded, for one of the most singular and valuable properties attending my construction, is its rendering the bees tractable. I ascribe their tranquillity to the manner in which they are affected by the sudden admission of light; they appear rather to testify fear than anger. Many retire, and entering the cells, seem to conceal themselves." Huber has here fallen into an error which he probably would not have made, had he been able to use his own eyes. The bees are, indeed, bewildered by the sudden admission of light, and will enter the cells,

---

[*] After using smoke sometime two or three times a day, to open a hive upon which I was experimenting, I found that, at last, the cunning creatures, instead of filling themselves with honey, rushed out to attack me! A colony will never refuse the sweetened water, however often it may be presented to them.

Fig. 47.   Plate XIV.

unless provoked by a sudden jar, or the breath of the operator; not, however, "to conceal themselves;" but imagining that their sweets, thus unceremoniously exposed, are to be taken from them, they gorge themselves almost to bursting, to save what they can. They will always appropriate the contents of the open cells, as soon as their frames are removed from the hive.

It is not merely the sudden admission of light, but its introduction from an *unexpected quarter*, that for the time disarms the hostility of the bees. They appear, for a few moments, almost as much confounded as a man would be, if without any warning, the roof and ceiling of his house should suddenly be torn from over his head. Before they recover from their amazement, the sweet libation[*] is poured upon them, and their surprise is quickly changed into pleasure; or they are saluted with a puff of smoke, which, by alarming them for the safety of their treasures, induces them to snatch whatever they can.[21] In the working season, the bees near the *top* are gorged with honey; and those coming from *below* are met in their threatening ascent, either by an avalanche of nectar, which, like "a soft answer," most effectually "turneth away wrath," or a harmless smoke, which excites their fears, but leaves no unpleasant smell behind. No genuine lover of bees ought ever to use the sickening fumes of tobacco.

The greatest care should be taken to repress, by the sweetened water or smoke, the *first* manifestations of anger; for as bees communicate their sensations to each other with almost magic celerity, while a whole colony will quickly catch the pleased or subdued notes uttered by a few, it will be roused to instant fury by the shrill note of anger from a single bee. When once they are thoroughly excited, it will be

---

[*] If, when the hive is first opened, honey-water is used, instead of sugar-water or smoke, in sprinkling the bees, its smell will be very apt to entice marauders from other hives. When the honey-harvest is abundant – and this is the best time for forcing swarms – bees are seldom inclined to rob, if proper precautions are used. It is sometimes difficult to induce them to notice honey-combs, even when put in an exposed situation..

[21] *Since the mechanical air-blast smokers that we know and use had not yet been invented, using sugar syrup would seem as easy as using smoke. Nowadays we can not be troubled by using syrup, though at times its use is very effective at calming bees and to stop bees from fighting. It is very helpful to have a small sprayer full of sugar syrup whenever you go to the apiary.*

found very difficult to subdue them, and the unfortunate operator, if inexperienced, will often abandon the attempt in despair.

It cannot be too deeply impressed upon the beginner, that nothing irritates bees more than breathing upon them or jarring their combs. Every motion should be deliberate, and no attempt whatever made to strike at them If inclined to be cross, they will often resent even a quick *pointing* at them with the finger, by darting upon and leaving their stings behind. A novice, or a person liable to be stung will, of course, protect his face and hands.

Directions have been given (p. 165), for removing the spare honey-board from the hive. As soon as it is disposed of, the Apiarian should sprinkle the bees with the sweet solution. This should descend from the watering pot in a fine stream, so as not to *drench* the bees, and should fill upon the *tops* of the frames, as well as between the ranges of comb. The bees, accepting the proffered treat, will begin to lap it up, as peaceably as so many chickens helping themselves to corn. While they are thus engaged, the frames which have been glued fast to the rabbets by the bees, must be very gently pried loose; this may be done without any serious jar, and without wounding or enraging a single bee; the rabbets being wide enough to allow the frames to be pried from the *rear* to the *front,* or *vice versa.*[22] If the rabbets were only just wide enough to receive the shoulders of the frames, it would be necessary, in loosening the frames, to pry them laterally, or towards each other, by which they might be brought so close together, as to crush the bees, injure the brood, disfigure the combs, or even kill the queen.[23]

---

[22] *Most modern frames do not have this moving or prying space at the ends of the top-bar. I suspect that the average beekeeper has not liked the shorter frames since the frame often will drop into the hive, especially if the super is moved or transported any distance, such as when bees are moved to pollinate a crop. These gaps, or spaces, at the ends of the frames did allow the frame to be broken free of the propolis that may be holding it fast. With self-spacing frames the frame can be moved against the frame next to it without fear of crushing bees.*

[23] *The self-spacing end bars of frames (Hoffman) were invented at a later time to avoid this problem. However, if I find a queen on a comb I take extra care when putting it back into a hive and slide it over next to an adjoining frame carefully so as not to injure a queen.*

ARTIFICIAL SWARMING. 171

The frames may be all loosened for removal in less than a minute:[*] by this time the sprinkled bees will have filled themselves, or if all have not, the intelligence that sweets have been furnished, will diffuse an unusual good nature through the honied realm. The Apiarian should now *gently* push the third frame from either end of the hive, a little nearer to the fourth frame; and then the second as near as he can to the third, to get ample room to lift out the end one, without crushing its comb, or injuring any of the bees. To remove it, he should take hold of its two shoulders which rest upon the rabbets, and carefully lift it, so as to brush no bees by letting it touch the sides of the hive, or the next frame. If it is desired to remove any particular frame, room must be gained by moving, in the same way, the adjoining ones on each side. As bees usually build their combs slightly waving, it will be found impossible to remove a frame safely, without making room for it in this way; and if the tops of the frames have not sufficient play on the rabbets, and between each other, the frames cannot be lifted out of the hive, without crushing the combs, and killing the bees. In handling the frames, be careful not to incline them from their *perpendicular*, or the combs will he liable to break from their own weight, and fall out of the frames.[24]

If more combs are to be examined, after lifting out the outside frame, set it carefully on end, near the hive,[†] when the second one may be easily moved towards the vacant space, and lifted out. After examination, put it in the place of the one first removed; in the same

---

[*] Without smoke or sweetened water, ten minutes may be spent in opening and shutting a single frame in a Huber-hive, and even then some of the bees will probably be crushed. The great caution recommended by Huber in opening his hives, shows that he did not know how to make himself independent of the anger of the bees.

[24] *Modern frames are often wired to prevent the sagging and falling of the foundation, or we use the wired or plastic base foundation. If the frame has been used for a year or more it is usually quite secure as well. It is the new comb that may be soft and may fall out if handled roughly. This is especially true if the frame is full of honey and the temperature is warm. Obviously, the larger (deeper) frames have more difficulty with sagging than the shallow frames. The embedded wired foundation, or the plastic based foundation, gives great strength to a comb. This is especially true for the shallow, or medium depth frames. This is just one of the reasons that I prefer the medium depth hive boxes.*

[†] If the frames, as they are removed, are put into an empty hive, they may be protected from the cold, and from robber-bees.

way, examine the third, and put it in place of the second, and so proceed until all have been examined. If the bees are to be removed, they must, of course, be shaken off on a sheet, as previously described. If the comb first taken out will fit, it may be put in the place of that last taken out; if it will not fit, and cannot be made to do so by a little trimming, the frames must be slid on the rabbets back to their former places, when this first comb may be returned to its old position.

The inexperienced operator, who sees that the bees have built some small pieces of comb between the outside of the frames, and the sides of the hive, or slightly fastened together some parts of their combs, may imagine that the frames cannot be removed at all. Such slight attachments, however, offer no practical difficulty to their removal.* The great point to be gained, is to secure a single comb on each frame; and this is effected by the use of the triangular comb-guides.

If bees were disposed to fly away from their combs, as soon as they are taken out, instead of adhering to them with such remarkable tenacity, it would be far more difficult to manage them; but even if their combs, when removed, are all arranged in a continued line, the bees, instead of leaving then, will stoutly defend them against the thieving propensities of other bees.

In *returning* the frames, care must be taken not to crush the bees between them and the rabbets on which they rest; they should be put in so *slowly*, that a bee, on feeling the slightest pressure, may have a chance to creep from under them before it is hurt. In shutting up the hive, the surplus honey-board should be carefully slid on, so that any bees which are in the way may be pushed before it, instead of being crushed. A beginner will find it to his advantage to practice—using an empty hive—the directions for opening and shutting hives, and lifting out the frames, until confident that he fully understands them. If any bees are where they would be imprisoned by closing the upper cover,

---

* If sufficient room for storing surplus honey is not given to a strong stock, in its anxiety to amass as much as possible, it will fill the smallest accessible places. If the bees build comb between the tops of the frames, and the under side of the spare honey-board, it can be easily cut off, and used for wax. If this shallow chamber were not used, they would fasten the honey-board to the frames so tightly, that it would be very difficult to remove it; and every time it was taken off, they would glue it still faster, so that, at last, it would be well nigh impossible, in getting it off, not to start the frames so as to crush the bees between the combs.

ARTIFICIAL SWARMING.

it should be propped up a little, until they have flown to the entrance of the hive: (Pl. VII., Fig. 20.)

An artificial colony may be made in five minutes from the time a hive is opened, if the queen is seen as quickly as she often is, by an expert. Fifteen minutes is, on an average, ample time to complete the whole work. In less than a week, if the weather is pleasant, an Apiarian with a hundred old stocks, by devoting to them a few hours every day, can, without any assistance, easily finish the business of swarming for the whole season.

But if the formation of artificial swarms is delayed, as it always should be, till near the time * for natural swarming, how can the beekeeper, unless constantly on hand, escape the risk of losing some of his best swarms? If he prefers to dispense entirely with natural swarming, he may deprive his *fertile* queens of their wings: (see chapter on Loss of the Queen.) As an old queen leaves the hive only with a new swarm, the loss of her wings †in no way interferes with her usefulness, or the attachment of the bees. If, in spite of her inability to fly, she is bent on emigrating, though she has a "will," she can find "no way," but helplessly falls to the ground, instead of gaily mounting into the air. If the bees find her, they cluster around her, and may be easily secured by the Apiarian; if she is not found, they return to the parent-stock, to await the maturity of the young queens. As soon as the piping of the first hatched queen is heard (p. 121), the Apiarian may force his swarm, unless—having fair warning of their intentions he prefers to allow them to swarm m the natural way. The number of queens nearly ready to hatch which are usually found in such a stock, may be profitably used in the swarming season.

---

* It will be easy — with movable-comb hives — to determine, by an occasional inspection, when the season for natural swarming is approaching.

(In modern thinking we do not want colonies to begin swarming 'behavior,' as it is very difficult to stop the behavior even if you stop the actual swarming. Honey bees are driven by instinctive behaviors. And many of these types of behaviors are only terminated when they are completed. Thus, when the colony has begun swarming the bees will stop foraging well and change many other patterns until the swarm has gone, or there is some other effective termination. Since swarming can only happen when there is at least some nectar available in the field, the beekeeper will lose some of the potential crop even if the swarm is prevented from leaving the hive. Thus, it is important to stop the swarming behavior from even beginning.)

As the queen can not get through an opening 5/32nds [*] of an inch high, which will just pass a loaded worker, if the entrance to the hive be contracted to this dimension, she will not be able to leave with a swarm: (see Pl. III, Figs. II, 12).

This method of preventing swarming,[†] requires great accuracy of measurement, for a very trifling deviation from the dimensions given, will either shut out the loaded workers, or let out the queen. It should be used only to imprison old queens; for young ones, if confined to the hive, cannot be impregnated. These blocks, if firmly fastened, will exclude mice from the hive in the Winter. When used to prevent all swarming, it will be necessary to adjust them a little after sunrise and before sunset, to allow the bees to carry out any drones that have died.

Some bee-keepers while reading these various processes for making artificial swarms, have probably thought that it would be much

---

[*] Huber does not give the size necessary for confining a queen; but he speaks of adjusting a glass tube, so as to pass out a worker, and not a queen. The smallest queen I ever saw, could not pass through my blocks. Although the workers are at first slightly annoyed by them, they soon become accustomed to them, as they do not confuse them, by presenting the entrance in a new place. The ventilation not depending on this contracted entrance, abundance of air can be given to the bees, when the blocks are adjusted to confine the queen.

(In modern beekeeping we would put a queen excluder over the entrance to prevent a queen for leaving. The space or gap of the excluder is about the same as given here.)

[†] Ill health, for the last two Summers, has prevented me from giving this method of swarming such a full trial that I can confidently endorse it, except for temporary purposes; though I have little doubt that it may be made entirely to prevent the issue of swarms. If so, it will be a great service to those who fear to open a hive to remove the royal cells, or cut off the wings of a queen. If as soon as piping is heard, the entrance is contracted for about a week, the bees may allow the young queens to engage in mortal combat. In this case, the blocks might be used to prevent the issue of second as well as first swarms. If the simple turning over of two blocks will prevent all swarming, and with any ulterior evil consequences to the colony, it will meet the wants of a large class of bee-keepers.

The difference between theoretical conjectures and practical results is often so great, that nothing in the bee-line, or indeed in any other line, should be considered as established, until by being submitted to rigorous demonstration, it has triumphantly passed from the mere regions of the brain, to those of actual fact. A theory which may seem so plausible as almost to amount t positive demonstration, when put to the working test, may be encumbered by some unforeseen difficulty, which speedily convinces even the most sanguine that it has no practical value. Nine things out of ten may work to a charm, and yet the tenth may be so connected with the other nine, that its failure renders their success of no account.

ARTIFICIAL SWARMING.

better to double the colonies by transferring half the combs and bees of a full stock to an empty hive; but for reasons already assigned (p. 156), such a course, though apparently more simple, would be injurious to the bees.

Having detailed the methods which can be most advantageously used for doubling stocks in one season, by artificial swarming, it seems proper to discuss the question whether it will be best to aim at a rate of increase more or less rapid than this.[*]

The Apiarian who aims at obtaining much surplus honey in any season, cannot, usually, at the farthest, more than double his stocks; nor even that, unless all are strong, and the season is favorable. If in any season that is not favorable, he attempts a more rapid increase, he must not only expect no surplus honey, but must even purchase food for his bees, to keep them from starving. The time, care, skill, and food required in our uncertain climate for the rapid increase of colonies, are so great, that not one bee-keeper in a hundred can make it *profitable*; while most who attempt it, will be almost sure, at the close of the season, to find themselves in possession of stocks which have been *managed to death*.

To make this matter plain, let us suppose a colony to swarm. Nearly forty pounds of honey will be ordinarily used by the new swarm an filling their hive with comb. If the season is favorable, and the swarm large and early, the bees may gather enough to build and store this comb, and a surplus besides. If the parent-stock does not swarm again, it will rapidly replenish its numbers, and having no new comb to build in the main hive, will be able besides to store up a generous allowance in the upper boxes. If, however, the season should be unfavorable, neither the first swarm nor the parent-stock can ordinarily gather more than enough for their own use; and if the honey-harvest is very defi-

---

[*] As soon as persons find that colonies can be multiplied at will, they are very apt to so overdo the matter, as to risk losing their bees. Notwithstanding repeated cautions to "make haste slowly," some have multiplied so rapidly, as to ruin their stocks, and bring great discredit on my hive, and system of management. Others will probably do the same thing; for it would seem that nothing but a sad experience of its folly, in bee-keeping, as well as in other pursuits, can ever convince men of the danger of "making haste to be rich." If, in spite of all that can be said, the inexperienced will persist in the rapid multiplication of stocks, it is hoped that they will at least have candor enough to attribute their losses to their own folly.

cient, both may require feeding. The bee-keepers profits in such an unfortunate season, will be the increase of his stocks.

If the parent-stock is weak in the Spring, the early honey harvest will pass away, and the bees be able to obtain very little from it. During all this time of meagre accumulations, the orchards may present

> "One boundless blush, one white empurpled shower
> Of mingled blossoms;"

and tens of thousands of bees from stronger stocks may be engaged all day in sipping the fragrant sweets, so that every gale which "fans its odoriferous wings" about their dwellings, dispenses

> "Native perfumes, and whispers whence they stole
> Those balmy spoils." *

By the time the feeble stock is prepared—if at all—to swarm, the honey-harvest is almost over, and the new colony, instead of gathering enough for its own use, may starve, unless fed. Bee-keeping, with colonies which are feeble in the Spring, except in extraordinary seasons and locations, is emphatically nothing but "folly and vexation of spirit."

I have shown how a handsome profit may, in a favorable season be realized from a strong stock, which has swarmed early, and but once. If the parent-stock throws a second swarm, unless it issues early, and the honey-season good, it will seldom prove of any value, if managed on the ordinary plan. It usually perishes in the Winter, unless previously destroyed, and the parent-stock will not only gather no surplus

---

* The scent of the hives, during the height of the gathering season, usually indicates from what sources the bees have gathered their supplies.
(I suspect that Langstroth had a very good nose, as most beekeepers would have difficulty in determining exactly what crop, or flowers, that the bees were foraging on. Some nectars have a very distinct smell, such as goldenrod, but most do not. Many honeys have a distinctive taste that characterizes them, but not an odor that most could tell within the bee hive This is especially true since usually the colony is collecting more than one type of nectar at a time.)

honey—unless it was secured before the first swarm issued—but will often perish also. Thus the novice who was so delighted with the rapid increase of his colonies begins the next season with no more than he had the previous year, and with the entire loss of all the time bestowed upon his bees.

With the movable-comb hives, the death, of the bees may be prevented, and all the feeble colonies made strong and powerful; but only by abandoning the idea of obtaining a single pound of surplus honey. From the parent stock, and first swarm, combs containing maturing brood must be taken to strengthen the weak swarms, and instead of being able to store their combs with honey, they will be constantly tasked in replacing those taken away, so that when the honey-harvest closes, they must be fed to save them from starving.

Any one intelligent enough to keep bees, can, from these remarks, understand exactly why colonies cannot be rapidly multiplied, in ordinary seasons, and yet be made to yield large supplies of surplus honey. Even the doubling of stocks will often be too rapid an increase for the greatest yield of spare honey.

I would strongly dissuade any but the most experienced Apiarians, from attempting, at the furthest, to do more than treble their stocks in one year. Another book would be needed, to furnish directions for rapid multiplication, sufficiently full and explicit for the inexperienced; and even then, most who should undertake it, would be sure, at first to fail. With ten strong stocks of bees, in movable-comb hives, in one propitious season, I could so increase them, in a favorable location, as to have, on the approach of Winter, one hundred good colonies; but I should expect to purchase hundreds of pounds of honey, devoting nearly all my time to their management, and bringing to the work the experience of many years, and the judgment acquired by numerous lamentable failures.[*]

---

[*] In one season, being called from home after my colonies had been greatly multiplied, the honey-harvest was suddenly cut short by a drought, and I found, on my return, that most of my stocks were ruined. The bees not having been fed, had gone into the groceries, and perished by hundreds of thousands.

(I think it is very interesting to see the nature of a truly social animal in such times as starvation. One spring, while teaching my apiculture class, I had a very large and thriving colony. Since it was so good it was not examined for a period of time. Then one morning, of a class day, I took the cover off the colony and it was full of bees so I de-

A *certain* rather than a *rapid* multiplication of stocks, is most needed. A single colony, doubling every year, would in ten years increase to 1,024 stocks, and in twenty years to over a million! At this rate, our whole country might, in a few years, be stocked with bees; an increase of one-third, annually, would soon give us enough. This latter rate of increase should be encouraged, even if, in the Fall, the stocks are reduced (see Union of Stocks), to the Spring number;[25] as, in the long run, it will both keep the colonies in the most prosperous condition, and secure the largest yield of honey.

I have never myself hesitated to sacrifice several colonies, in order to ascertain a single fact; and it would require a large volume, to detail my various experiments on the single subject of artificial swarming. The practical bee-keeper, however, should never lose sight of the important distinction between an Apiary managed principally for purposes of observation and discovery, and one conducted exclusively with reference to pecuniary profit.[*] Any beekeeper can easily experiment with my hives; but he should do it, at first, only on a small scale, and if pecuniary profit is his object, should follow my directions, until he is sure that he has discovered others which are better. These cautions are given to prevent serious losses in using hives which, by facilitating all manner of experiments, may tempt the inexperienced into rash and unprofitable courses. Beginners, especially, should follow my directions as closely as possible; for, although they may doubtless

---

termined that I would show the class the colony that afternoon. We examined the colony about 4-5 hours after I looked under the cover. The bees were almost all dead! During that time the colony had lost nearly 30,000 bees. All that was left was the queen and a few dozen attendants. The colony had shared its honey until they all died in a very short time.. Had I taken the time to look into the colony in the morning I could have saved them. The experience points up that honey bees have what I call a "corporate stomach". They share food equally, and also many pheromones, through constant exchange of food The queen will be the last bee to starve in these cases.

The food exchange where they pass on the chemical information within the hive is called throphalaxis. The bees communicate all kinds of information by this method; queen loss, nectar sources, brood conditions, etc.)

[25] *This is one of the best rules of beekeeping —to take your Winter losses in the Fall and then you can divide the surviving colonies the following Spring. If you unite two weak colonies, or one weak colony with a strong one, you will save the colony and then the following spring you can divide the colonies to regain your original number, if desired.*

[*] Prof. Siebold says that Berlepsch told him that some of his hives "had been very much prejudiced by the various scientific experiments."

be modified and improved, it can only be done by those experienced in managing bees.

Let me not be understood as wishing to intimate that perfection has been so nearly attained, that no more important discoveries remain to be made. On the contrary, I should be glad if those who have time and means would experiment on a large scale with the movable-comb hives; and I hope that every intelligent bee-keeper who uses them, will experiment at least on a small scale.[26] In this way, we may hope that those points in the natural history of the bee still involved in doubt, will, ere long, be satisfactorily explained.

The practical bee-keeper should remember that *the less he disturbs the stocks on which he relies for surplus honey, the better*. Their hives ought not to be needlessly opened, and the bees should never be so much interfered with, as to feel that they hold their possessions by an uncertain tenure; as such an impression will often impair their zeal for accumulation.* The object of giving the control over every comb in the hive, is not to enable the bee-keeper to be necessarily taking them in and out, and subjecting the bees to all sorts of annoyances. Unless he is conducting a course of experiment, such interference will be almost as silly as the conduct of children who dig up the seeds they have planted, to see how much they have grown.

Having described how forced swarms are made, both in common

---

[26] *Beekeepers are always 'experimenting' with their hives. I suspect there are at least two reasons for this behavior. First, beekeepers are probably fascinated by the biology of bees, much like Langstroth. And secondly, the use of one or two hives is not a great loss if the experiment is a failure. It is, of course, how change and new ideas get started. There are some things that would be difficult to change and that is the Langstroth size hive in the United States. This invention was so successful that you would be hard pressed to find a different kind of hive* in use *today. I have often thought that the dimensions of the Langstroth hive are not necessarily the best, but to change the hive size would require the loss of billions of dollars in inventory. Therefore, unless a person could show a substantial increase in the production of honey, the size of the supers will remain.*

* These remarks apply more particularly to stocks engaged in storing honey in receptacles *not in the main hive*. The experience of Dzierzon and myself, shows that opening the hives ordinarily interrupts their labors for only a few minutes.

 (Modern day research has shown that even minor disruption to the hive during a nectar flow will reduce the total honey produced by the colony, and a full examination of the broodnest reduces the yield even more. It will take a colony approximately one day to recover from a full inspection, as far as honey production.)

and movable-comb hives, when the Apiarian wishes in one season to *double* his colonies, I shall now show how he can secure the largest yield of honey, by forming only one new colony from two old ones.

When it is time to form artificial colonies, drum[27] a strong stock—which call *A*—so as to secure *all* its bees, and put the forced swarm on the old stand. If any bees are abroad when this is done, they will join this new colony. Remove to a new stand in the Apiary a second strong stock—which call *B*—and put *A* in its place. Thousands of the bees that belong to *B*, as they return from the fields,† will enter *A*, which thus secures enough to develop the brood, rear a new queen, and gather, if the season is favorable, large surplus stores.

If *B* had been first forced, and then removed, it would (p. 156) have been seriously injured; but as it loses fewer bees than if it had swarmed, and retains its queen, it will soon become almost as powerful as before it was removed.*

This method of forming colonies may be practiced, on any pleasant day, from sunrise until late in the afternoon; for if no bees are abroad to recruit the drummed hive, it may be shut up, until it can be put upon the stand of any strong stock which has already begun to fly with vigor. Of all the methods which I have devised for practicing artificial

---

[27] *As noted previously, (page 28) this drumming was a common practice for moving bees from a box hive, and thus well understood by beekeepers in the 1850's. It is so rarely practiced today that most beekeepers would not understand the reference.*

† It is quite amusing to observe the actions of these bees, when they return to their old stand. If the strange hive is like their own in size and outward appearance, they go in as though all was right, but soon rush out in violent agitation, imagining that by some unaccountable mistake, they have entered the wrong place. Taking wing to correct their blunder, they find, to their increasing surprise that they had directed their flight to the proper spot; again they enter, and again they tumble out, in bewildered crowds, until at length if they find a queen, or the means of raising one, they make up their minds that if the strange hive is not home, it looks like it, stands where it ought to be, and is, at all events, the only home they are likely to get. No doubt they often feel that a very hard bargain has been imposed upon them, but they are generally wise enough to make the best of it. They will be altogether too much disconcerted to quarrel with any bees that were left in the hive when it was forced, who on their part give them a welcome reception.

* Might not a forced swarm be made to adhere to a new location, by thoroughly shaking them in an empty box —see note on p. 163 — and then setting them on their new stand, and permitting them to fly? The queen might be confined, for safety, in a queen-cage.

ARTIFICIAL SWARMING. 181

swarming,[†] with almost any kind of hive, this appears to be one of the simplest, safest, and best. It not only secures a reasonable increase of colonies, but maintains them all in high vigor; and in ordinary seasons will yield, in good locations, more surplus honey, than. if all increase of colonies was discouraged.[28] If every bee-keeper would adopt this plan, our country might soon be like the ancient Palestine, "a land flowing with milk and honey."

In all the modes of artificial increase thus far given, the parent or *mother-stock*—as I shall call it in this connection—after parting with the forced swarm, was either supplied with a sealed royal cell, or left to raise a new queen from worker-brood. *By the use of movable-comb hives, it may be at once supplied with a fertile young queen.* Before showing how this is done, its extraordinary advantages will be described.

It sometimes happens that the mother-stock, when deprived of its queen, perishes, either because it takes no steps to supply her loss, or because it fails in the attempt. If it raises several queens, it may become reduced by after-swarming; and, at all events, its young queen must run the usual risks in meeting the drones. When all goes right, it will usually be from two to three weeks before any eggs are laid in the mother-stock; and when the brood left by the old queen has all matured, the number of the bees will so rapidly decrease, before any of the brood of the young queen hatches, that she will not have a fair chance, seasonably to replenish the hive.

[†] The Apiarian, by treating a natural swarm as he has been directed to treat a forced one, can secure an increase of one colony from two; and of all the methods of conducting natural swarming, in regions where rapid increase is not profitable, this is the best, provided the colonies do not stand too close together, and the hives used in the process are alike in shape and color.

[28] *The production of swarms to produce more honey is certainly against modern apiculture thinking. Yet, in defense of Langstroth, he makes these artificial swarms early so they would be in the position to produce honey. This would be much like our starting with package bees except that these artificial swarms would have brood with them so they should develop faster than a package. Maybe these artificial swarms would not produce as much as a well managed colony today, but certainly more than the average colony of the 1850's. In many ways these artificial swarms could be compared with our "splits" or divisions early in the spring. Remember the mindset of the beekeepers of the day. That is, up until this time beekeeping consisted of producing swarms to replace the colonies that you had to kill every year in order to harvest your honey from a box hive.*

Again; while the system that gives no hatched queen to the mother-stock, exposes it to be robbed if forage is scarce, the presence of a fertile mother emboldens it to a much more determined resistance.

If the mother-stock has not been supplied with a fertile queen, it cannot, for a long time, part with another colony, without being seriously weakened. Second swarming—as is well known—often very much injures the parent-stock, although its queens are rapidly maturing; but the forced mother-stock may have to start theirs almost from the egg. By giving it a fertile queen, end retaining enough adhering bees to develop the brood, a moderate swarm may be safely taken away in ten or twelve days, and the mother-stock left in a far better condition than if it had parted with two natural swarms. In favorable seasons and localities, this process may be repeated four or five times, at intervals of ten days, and if no combs are removed, the mother-stock will still be well supplied with brood and mature bees. Indeed, the judicious removal of bees, at proper · intervals, often leaves it, at the close of the Summer, better supplied than non-swarming stocks with maturing brood; the latter having—in the expressive language of an old writer—"waxed over fat."* I have had stocks which, after part-

---

· If a strong stock of bees, in a hive of moderate size, is examined, at the height of the honey-harvest, nearly all the cells will often be found full of brood, honey or bee-bread. The great laying of the queen is over — not as some imagine, because her fertility has decreased, but simply for want of room for more brood. A queen in such a colony, or in a hive having few bees, often appears almost as slender as one still infertile; but if she has plenty of bees and empty comb given to her, her proportions will soon become very much enlarged. (p.47.)

(It appears to me that Langstroth has here described what modern package-bee producers practice every Spring, i.e., taking a portion of the bees when the colony is in the log [rapid] phase of growth. The colony can give up many bees with this type of removal. Package bee shippers may take up to two pounds of bees [8,000 bees] every week. At this time in the growth of the colony the queen is laying a maximum number of eggs per day [≈ 3,000], and the package-bee shippers just try to keep the colony at this maximum growth phase, at least during the shipping season.)

* Columella had noticed that, in very productive seasons, strong stocks, if left to themselves, fill up their brood-combs with honey, instead of rearing young bees. He advises the unskillful, instead of being pleased with this apparent gain, to shut up their hives every third day, and thus compel the bees to attend to breeding! This gives the queen a chance to deposit eggs in the cells from which the young bees hatch, before they are filled with honey; and no better plan can be devised for the common hives.

ARTIFICIAL SWARMING. 183

ing with four swarms in the way above described, have stored their hives with buck-wheat honey, besides yielding a surplus in boxes.

This method of artificial increase, which resembles natural swarming, in not disturbing the combs of the mother-stock, is not only superior to it, in leaving a fertile queen, but obviates almost entirely all risk of after-swarming; for the old queen, when given to the forced swarm, very seldom attempts to lead forth a new colony (p. 128); and the young one, which is gives to the mother-stock, is equally content—except in very warm climate—to stay where she is put. Even if the old queen is allowed to remain in the mother-stock, she will seldom leave, if sufficient room is given for storing surplus honey; and it makes no difference—as far as liability of swarming is concerned—where the young one is put.†

In the movable-comb hives, a few of the combs nearest the ends may be taken out, and as many empty frames put between every two of the central combs; these will at once be supplied with combs, in which the queen will deposit eggs. It would seem that, while the instincts of the bees teach them to rear all the eggs deposited in cells, their avaricious propensities often — as in human beings — get the better of them, so that they give their queen no chance to lay, and thus incur the risk of perishing, in order to become over-rich.

† I have frequently noticed that after-swarms are much less inclined than first swarms to build drone-comb –their young queens seldom laying many drone-eggs the first season. If we can cause the new colonies to fill their hives almost entirely with worker-combs, merely by supplying them with young queens, bee-keeping will take another important step in advance.

(I am sure that Langstroth later welcomed the development of beeswax foundation in worker-size cells. He indicates late in the book that "foundation" has been developed and that he was going to try it.

This idea of young queens not producing many drones would also be an another incentive for beekeepers to keep young queens in their hives Though, as I have said elsewhere, the presence of a few, or even hundreds, of drones is not all that much loss to the colony. These drones supply heat and heat dissipation throughout the brood nest, especially in the cool part of the night and morning. Honey bees also desire a "balance" of individual castes, and so will produce drones when there is an absence within the hive. This is especially true early in the season when the biology says it is swarming time.

I do not like an excess of drones within a colony and with modern comb foundation this is seldom a problem since all of the cells are worker size. However, the colony will make at least some drone cells. In a natural comb situation, such as a hollow tree, the number of drone cells has been determined and the number approaches 25 percent. This number is too many as the drones use up honey to develop as well as honey during their development.)

The bee-keeper can *double* his stocks in one season, even better in this way, than by the method described on page 162; and in favorable seasons and locations, this rate of increase will yield a large surplus of honey.

For bee-keepers who may desire a more rapid increase of colonies, I shall give the methods, which—after years of experimenting—I have found to be the best; referring them to the cautions already given, lest, at the end of the season, they find that their fancied gains consist only of large investments in *dearly bought experience*. If they are *cautious* and *skillful* in good seasons and locations, they may safely increase their colonies threefold, and may, possibly, by liberal feeding, increase them five or six-fold, or even more.[29]

The plan of artificial swarming, described on page 180, when combined with the giving of a fertile young queen to the mother-stock, instead of stopping short with an increase of one from two, may be expanded to any rate of increase that can possibly be secured; while it has this admirable peculiarity, that each step in advance is entirely independent of any that are subsequently to be made; and the process may be stopped at any time when forage fails, or the bee-keeper chooses—from any cause—to carry it no farther.

If it is used for doubling the stocks, proceed as follows: Let a fertile young queen be given to *A* (p. 180) and soon as it is forced, and in ten days force a swarm from *B*, which I shall call *D*. Put *D* on the stand of *B*, and after removing *A* to a new place, set *B* where *A* stood, giving to *B* a fertile young queen. If another colony, *E*, is to be formed, make it in the same way, by forcing *A*, and transposing with *B*; and so continue, by the transposition of *A* and *B*—forcing the new colony alternately from each—to make successively, at intervals of about ten days, *F, G, H*, &c.; *A* and *B* being supplied with a fertile queen as often as they are forced.

---

[29] *With any such artificial increase the problem will usually end up that the colonies do not have enough honey stored to allow them to get through the following winter. If you feed bees heavily, you take some toll on the longevity of the bees, and this will also impact winter survival. It is wise to limit your increases to two or three per year.*

ARTIFICIAL SWARMING.

To make this process more intelligible, let *A* and *B* represent the first positions, in the Apiary, of the original stocks:

Original stocks,		***A, B***.
Position after 1st forcing,		***C, A, B***.
"	" 2nd "	***C, B, D, A***.
"	" 3rd "	***C, A, D, E, B***.
"	" 4th "	***C, B, D, E, F, A***.
"	" 5th "	***C, A, D, E, F, G, B***.
"	" 6th "	***C, B, D, E, F, G, H, A***.

By looking at this table,[*] it will be seen that the new colonies, *C, D, E*, &c., always remain *undisturbed* on the stands where they are first put.

Dzierzon has noticed the great number of bees which may, at intervals, be removed from a stock-hive, *if it only retains a fertile queen, and sufficient adhering bees*; and says that he has known as many bees to be lost, in a single day, from a strong stock, by high winds [†] or sudden storms, as would suffice to make a respectable swarm.

This able Apiarian, who unites to the sagacity of Huber, an immense amount of practical experience in managing bees, has for years

---

[*] This table is not intended to recommend setting hives in rows, close together, A and B may be *anywhere* in the Apiary, and C, D, E, & C, as far apart as is at all desirable. (See Chap. on Loss of Queen.)

[†] If forage is very abundant, bees are almost crazy to get it, however windy the weather, and some Apiarians, on such days, confine them to their hives.

(I suspect that they were overly concerned about losing bees to the wind. If the wind is too strong bees just do not fly. And when they do fly it is often close to the ground where the wind speed is the lowest. Honey bees can fly about 15 mph, and if the wind speed is greater than that the bee cannot make headway. This is when they will fly close to the ground as the wind is less there. When winds become greater than the 15 mph the bees just do not fly.

It is probably true that a new, young bee that firsts flies on a windy day may end up drifting to other hives, especially to those hives at the ends of rows. This fact may be the basis for the statement of bees being lost on windy days. Drifting of bees is a real phenomenon and can involve many bees. We have found that five percent, or more, of the bees foraging from a colony were raised in another colony. The real reason that young bees drift is that we often make our hives the same color and line them up in precise rows—something that would not happen in nature.)

formed his artificial colonies chiefly by removing the forced swarms to a distant Apiary. Though this plan has some decided merits, and might suit two persons—sufficiently far apart—who could agree to manage their bees as a joint concern, the expense of transporting the bees makes it objectionable to most bee-keepers. From the beginning, my plans for artificial increase were mainly with reference to a single Apiary; and it would seem from the recent discussion in the Annual Apiarian Convention (p. 20), that the German bee-keepers are fast adopting the same method.

By making holes on the inside of the bottom-board of my hives *—the glass ones excepted—artificial swarming may be practiced in a way approaching still nearer to natural swarming than any yet described. About a week or ten days before the artificial swarm is to be made, put an empty hive *C*, on the top of a strong stock *A*—making the entrance of *C* to face in the opposite way from that of *A*—and uncover the holes in the bottom-board of *C*, so that the bees may pass from *A* to *C*. A number of the young bees, as they go out to work, will use the upper entrance, so that when a colony is driven from *A*, and the mother-stock is put in place of *C*, it will have the requisite number of adhering bees: the forced swarm being put into *C*, and taking the stand of *A*, it will secure, as it ought, the most of the mature bees. In a few days, the upper hive may be set down close to the other, and *gradually* [30] removed to any convenient distance, and its entrance made to face in any direction. The same process may be repented, at intervals, with the mother-stock, until as many new colonies are

---

* These holes are similar to those in the spare honey-board (Pl. VIII., Fig. 21), and are closed in the same way, when not in use. They permit the bees to communicate, where the hives are piled one on the top of the other; and the upper hive may be used as a place for the storage of surplus honey in small boxes, or (Pl. X., Fig. 23) in large or small frames.

[30] *It is only possible to move a colony a few feet a day, and that is why Langstroth commented earlier that two beekeepers could work this scheme—each beekeeper putting his new hives in the other's apiary for a while and then moving them back to their original location. Once the new hives had been started they are moved to the second apiary where drifting back to the mother-hive would not be a problem. The distance between apiaries should be about two miles, or more. With modern vehicles, the establishment of a second apiary, if even for a temporary period, is relatively easy. In Langstroth's time out-apiaries were not common. The fact that they had too many colonies for the locations could account for some of the low colony yields that were reported.*

ARTIFICIAL SWARMING.   187

formed as may be desired.[*] If the Apiarian does not aim at a very rapid increase, he can take from the mother-stock, in forcing it, two or three of its combs which are best filled with *sealed* brood, so that the artificial swarm will have recruits before its new brood matures.

If the new colony is forced by removing the frames (p. 165), the bees may be shaken on a sheet directly in front of *A*, and allowed to enter it again; the combs being all transferred to *C*, unless the bee-keeper wishes to return a few to the parent-stock.

With a fertile queen, a new colony may be formed by simply reversing the positions of *A* and *C,* when the bees are in full flight; and after the lapse of a few days, if *C* is weaker than *A*, the position of the colonies may be again reversed: or *A* and *C* may be reversed, *end for end*, without *lifting* one from the other; or the comb containing the queen may be left in *A*, and the others transferred to *C*, when the bees are in full flight. Other methods still will suggest themselves to the expert.

To those who have learned to open the hives and remove the combs, and who use but one Apiary, this way of making artificial swarms—which I call the *piling mode*—will probably prove to be the best. It does not *confuse* the bees, by presenting to them a new entrance, or a hive having a *strange smell*, and retains in the mother-stock adult bees enough to gather water, and attend to all necessary out-door work. In the Apiarian Convention of 1857, which was largely attended, and where the question of artificial swarms with one Apiary, was fully discussed, Dzierzon recommended a method as much like this as the plan of his hives would permit.

I shall now show how, by means of movable-comb hives, fertile young queens may always be kept on hand, to supply the forced mother-stocks: About three weeks before *A* (p. 180) is to be forced, take from it, as late in the afternoon as there is light enough to do it, a comb containing worker-eggs, and bees just gnawing out of their cells, and put it, with the mature bees that are on it, into an empty hive. If there are not bees enough adhering to it to prevent the brood

---

[*] I find, by referring to my Journal, that I devised this method in the Summer of 1854, when using frames in hives which, like Dzierzon's, opened at both ends. I soon ascertained that such hives – even with my frames – did not give suitable facilities for managing bees.

from being chilled during the night, more must be shaken into the hive from another comb. If the transfer is made so late in the day that the bees are not disposed to leave the hive, enough will have hatched, by morning, to supply the place of those which may return to the parent-stock. A comb from which about one-quarter of the brood has hatched, will almost always have eggs in the empty cells, and if all things are favorable, the bees, in a few hours, will usually begin to raise a queen.*

If the comb used in forcing such a colony which I shall call a *nucleus*—was removed at a time of day when the bees upon it would be likely to return to the parent-stock, they should be confined to the hive, until it is too late for them to leave; and if the number of bees, just emerging from their cells, is not large, the entrance to the hive should be closed, until about an hour before sunset of the next day but one (see p. 161). The hive containing this small colony, should be properly ventilated and shaded—if thin—from the intense heat of the sun; it should always be well supplied with honey and water.† Suitable precautions should also be taken to guard against the loss of its young queen, when she leaves the hive to meet the drones. (See Chap. on Loss of Queen.)

The best way of forming a nucleus, with movable-comb hives, will be by setting an empty hive over a full stock, in the way already described (p. 186): when enough bees begin to make use of the upper

---

* I have known about a tea-cup of bees, confined in a dark place, to begin, within an hour, enlarging cells for raising a queen.

*(It has been shown that a small number of bees **can** raise a few cells, and mate a queen, if the nucleus can provide enough heat. Early in the spring having enough heat can be a major problem. If the nuc is put above another colony with a double, wire-mesh screen separating the colonies the heat from the lower colony will help keep the brood [and queen cells] warm.)*

† Whenever the position of a colony is so changed as to interrupt for a few days the flight of the bees, it will be advisable to supply them with water in their hive, as the want of it is often fatal to the brood.

(Water is important to brood rearing, but I think that Langstroth did not appreciate how much the bees respond to the need to collect water. The foraging bees would normally find water in most regions where they forage, and of course nectar itself has a great amount of water. In dry, or desert, areas water does need to be added to feeders on top of colonies, especially if little or no nectar is coming into the hive. Otherwise brood rearing will cease. Also, if the colony is confined, without being able to forage, then water would have to be added for the same reason.)

ARTIFICIAL SWARMING.

entrance, a brood comb, with adhering bees, may be transferred to it, and the connection between the two hives closed.[31] If the bees are reluctant to enter the upper hive, they may be encouraged to do so by placing honey there, in a feeder—keeping the outside entrance closed against robbers—and they may afterwards be allowed to pass out through the upper hive. In a few days this nucleus may be set down, and gradually removed, so that another hive may be put on the mother stock.

If all things are favorable, this nucleus, by the time A is forced, will have a fertile queen, which may be given to A, when the bees that return from the fields show that they realize (page 158), their queenless condition. The comb belonging to the nucleus, with all the bees that are on it, may then be given to the artificial colony, C. Or, if the beekeeper prefer, he may give to A its own queen, and give the young one —with the precautions subsequently described—to C.

If the stocks are to be *doubled* a second nucleus must be formed, by taking about ten days later, a brood-comb from B, and giving the second queen to the second artificial colony, D.*

If the colonies are to be multiplied more rapidly still, then from the first nucleus only its queen must be taken, after she has begun to lay, and her colony will at once begin to raise another. If she is removed before she has laid any eggs, the comb of the nucleus—after all the

---

[31] *Langstroth often points out the advantages of the movable-comb hive, but does not tout its advantage here. He just says to put a frame of brood into this upper 'division'. It is this flexibility that made his hive so superior In our modern beekeeping we just take this movement of combs within, or between, hives as axiomatic. It is this flexibility of moving combs from one colony to another that has made the movable-comb hive so essential to modern beekeeping.*

* Those who rely entirely on natural swarming, may often secure fertile queens, by catching the supernumerary young queens of after-swarms (p. 122, and hiving them, with a few bees, in any small box containing a piece of worker-comb.

{Having extra laying queens in an apiary is very advantageous. Many beekeepers keep one or two nucleus hives within the apiary These are usually 2,3 or 4 frame boxes that contain standard frames, either deep or shallow. When they find a colony with a poor queen, or one that is queenless, they remove the old queen, pick up two frames from the nuc, with the queen in the center, and put them into the colony needing the queen. Since she is a laying queen, surrounded by her bees, the introduction is almost always successful. This would also be a good time to spray the colony with some sugar syrup to aid in the union of two colonies. The nucleus is then allowed to raise a new queen, which after mating, could be used in a subsequent re-queening)

bees are shaken from it—must be returned to *A* or *B*, and replaced with another that is well supplied with eggs: and if, at any time, the number of bees in the nucleus is too small, it may be reinforced by exchanging its comb for one that is as full of hatching brood as when it was first formed (p. 188). The same process must be adopted with the second nucleus, and thus—at regular intervals—enough queens may be obtained from the two, to multiply the colonies to any desired extent.

To make this matter perfectly plain, let us suppose that *C* is to be forced on the 1st of June and *D, E, F*, &c., at intervals of ten days.[*] Then, as before, *C, A*, and *B* (p. 185), represent the positions of the colonies on the 1st of June, and the other columns, their places on the 10th, 20th, &c. Now, let *I* and *II* represent the nuclei—I use this name when speaking of more than one nucleus and I^1, II^1 represent them when each has a queen; I^2, II^2, when each has raised its second queen; I^3, II^3, when each has its third, and so on, it being always understood that *I, II* without the small numbers above them, indicate that the nuclei are at that time rearing queens. The first nucleus will be formed May 10th, and the second May 20th.

May 10th, *I*,   June 20th, I^2, *II*,
" 20th, *I, II*,   " 30th, *I*, II^2,
June 1st, I^1, *II*,   July 10th, I^3, *II*,
" 10th, *I*, II^1,   " 20th, *I*, II^3, &c., &c.

As it may often be desirable to remove the queen of a nucleus, before she has begun to lay eggs, if her colony is supplied with, a sealed royal cell from another nucleus, no time will be lost, and much trouble saved.[32]

---

[*] Of course, no one will imagine, that operations which depend so much on season, climate, and weather, can always be conducted with the mathematical accuracy with which they are set forth in such an illustration.

[32] *The practice of taking up a queen before she has laid eggs is somewhat risky. Experts probably can determine that the young queen is just about to begin oviposition, but it may happen that she has not finished mating. In Langstroth's time they thought that a queen only mated with one drone, whereas we now know that they mate with a dozen or more drones, and over several days duration. So just seeing a mating sign would not mean that a queen is fully mated, only that she had mated at least once. If a queen is*

The following, from the pen of Rev. Mr. Kleine, one of the ablest German Apiarians, will be interesting in this connection: "Dzierzon recently intimated that, as Huber, by introducing some royal jelly into cells containing worker-brood, obtained queens, it may be possible to induce bees to construct royal cells where the Apiarian prefers to have them, by inserting a small portion of royal jelly in cells containing worker-larvae! If left to themselves, the bees often so crowd their royal cells together"—See Pl. XV. "that it is difficult to remove one, without fatally injuring the others; as, when each a cell is cut into, the destruction and removal of the larva usually follows. To prevent such losses, I usually proceed as follows: When I have selected a comb with unsealed brood, for rearing queens, I shake or brush off the bees, and trim off, if necessary, the empty cells at its margin. I then take an unsealed royal cell—which usually contains an excess of royal jelly—and remove from it a portion of the jelly, on the point of a knife or, pen, and by placing it on the inner margin of any worker-cells feel confident that the larvae in them will be reared as queens; and as these royal cells are *separate*, and on the *margin* of the comb, they can be easily and safely removed. This is another important advance in practical bee-culture, for which we are indebted to the sagacity of Dzierzon." *Bienenzeitung,* 1858, p. 199. *Translated by Mr. Wagner.*[33]

---

very elongated and full of eggs, even though you find no eggs in cells, you probably can assume that she has finished mating and will begin oviposition very soon. As I wrote above, queens sometimes take several days to complete mating. The length of time needed for mating often depends on favorable weather. Queens, and especially drones, fo not fly out for mating until the day becomes warm and usually sunny. Thus, mating happens in the early afternoons. When the weather turns cold or rainy during a queens mating period is when they sometimes become poorly mated. After spending some time after the first mating flight confined to the hive a queen may start to develop her ovarian tubules (oviarioles) and commence oviposition.

[33] *While this is not grafting of queen larvae as we know it today, it was obviously the first step! By cutting the edge of the comb the method is very close to the method that we call in the United States the Miller queen rearing method. C. C. Miller developed this cut-edge queen rearing at a later time. Maybe we should call it the Dzierzon method as it would appear that he developed the method first.*

If the spare queen-cells are cut out (p. 166) from *I*, before the first queen matures, other nuclei may be formed by similar processes; indeed, with movable combs, any number of queens may be raised, and kept where, when wanted, they can be readily secured.*

Both the original nuclei, *I* and *II*, and those made from their sealed queens, may be formed by bringing from another Apiary, in a small box, the few adhering bees which are wanted (p. 162); and as many maybe returned in it, to be used for a similar purpose. The expert will also be able to catch up adhering bees, by *slightly* moving * the parent- stocks (p. 161), and in various other ways, which will readily suggest themselves.

---

* Dzierzon estimates a fertile queen to be worth, in the swarming season, one-half the price of a new swarm.

* If the adhering bees are thus obtained, and there is not a cluster of bees on the brood-comb, they may be so dissatisfied with its deserted appearance, as to refuse to stay. If they intend to submit to this system of forced colonization, they will, however much agitated at firs, soon join the cluster of bees on the comb; otherwise, they will quickly abandon the hive, carrying off with them all that were put in with the comb.

While it is admitted that bees can raise a queen from any worker-egg or young larva, it is certain the workers of any age are able or disposed to do it?

Huber speaks of two kinds of workers: "One of these is, in general, destined for the elaboration of wax, and its size is considerable enlarged with full of honey; the other immediately imparts what it has collected, to its companions; its abdomen undergoes no sensible change, or it retains only the honey necessary for its own subsistence. The particular function of the bees of this kind is to take care of the young, for they are not charged with provisioning the hive. In opposition to the wax-workers, we shall call them small bees, or nurses.

"Although the external difference be inconsiderable, this is not an distinction. Anatomical observations prove that the stomach is not the same: experiments have ascertained that one of the species cannot fulfill all the functions shared among the workers of a hive. We painted those of each class with different colors, in order to study their proceedings; and these were not interchanged. In another experiment, after supplying a hive, deprived of a queen, with brood and pollen, we saw the small bees quickly

Fig. 48.   PLATE XV

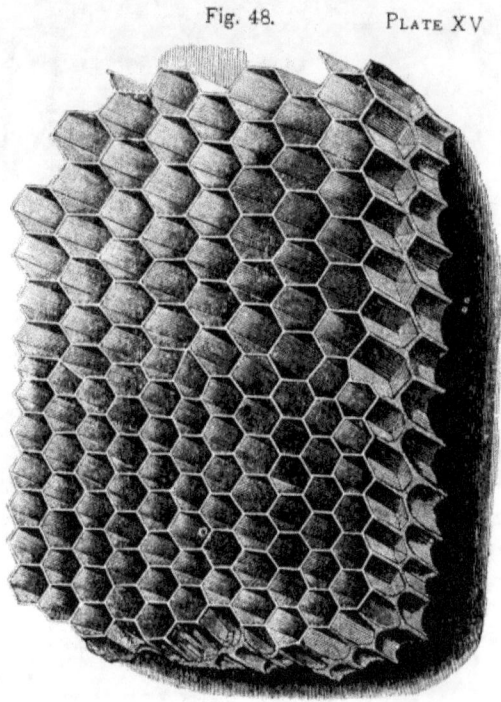

Fig. 58.

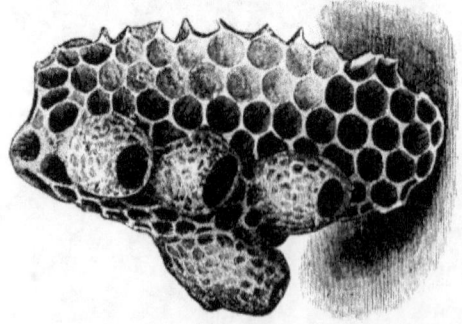

One queen can be made to supply several hives with brood, while they are constantly engaged in raising spare queens. Deprive two colonies, 1 and 2, at intervals of a week, each of its queen, using these queens for artificial swarms. As soon as the royal cells in 1 are old enough for use, remove them, and give 1 a queen from another hive, 3. When the royal calls in 2 are removed, this queen may be taken from 1 where she will have laid abundantly and given to 2. By this time, the queen cells in 3 being sealed over, may he removed, and the queen restored to her own stock. She has thus made one circuit, and supplied 1 and 2 with eggs; and after replenishing her own hive, she may be sent again on her perambulating mission. By this device, I can obtain, from a few stocks, a large number of queens.

A few days after a nucleus is formed, it should be examined, and if royal cells are not begun, or there are no larvae in them, the bees must be shaken from the comb, which should then be exchanged for another.

Bees sometimes commence queen-cells, which, in a few days, are found to be untenanted. At the second attempt they usually start a larger number, and seldom fail of success. Does practice make them more perfect? or were some of the necessary conditions wanting at first?

The following able communication, from the pen of Dr. Dönhoff, may throw some light on this subject:—"Dzierzon states it as a fact, that worker bees attend more exclusively to the domestic concerns of the colony in the early period of life; assuming the discharge of the more active outdoor duties only during the later periods of their existence. The Italian bees furnished me with suitable means to test the correctness of this opinion.

"On the 18th of April, 1855, I introduced an Italian queen into a colony of common bees; and on the 10h of May following, the first Italian workers emerged from the cells. On the ensuing day, they emerged in great numbers, as the colony had been kept in good condition by regular and plentiful feeding. I will arrange my observations under the following heads:

"1. On the 10th of May the first Italian workers emerged; and' on the 17th they made their first appearance outside of the hive. On the next day, and then daily till the 29th, they came forth about noon, disporting in front of the hive, in the rays of the sun. They, however, manifestly, did not issue for the purpose of gathering honey or pollen, for during that time none were noticed returning with pellets; none were seen alighting on any of the flowers in my garden; and I found no honey in the stomachs of such as I caught and killed for examination. The gathering was done exclusively by the old bees of the original stock, until the 29th of May, when the Italian bees began to labor in that vocation also—being then 19 days old.

"2. On the feeding troughs placed in my garden, and which were constantly crowded with common bees, I saw no Italian bees till the 27th of May, seventeen days after the first had emerged from the cells

"From the 10th of May on? I daily presented to Italian bees, in the hive, a stick dipped in honey. The younger ones never attempted to lick any of it; the older occasionally seemed to sip a little but immediately left it and moved away. The common bees always eagerly licked it up, never leaving it till they had filled their honey-bags. Not till the 25th of May did I see any Italian bee lick up honey eagerly, as the common bees did from the beginning.

"These repeated observations force me to conclude that, during the first two weeks of the worker-bee's life, the impulse for gathering honey and pollen does not exist, or at least is not developed; and that the development of this impulse proceeds slowly and gradually. At first the young bee will not even touch the honey presented to her; some days later she will simply taste it, and only after a further lapse of time will she consume it eagerly. Two weeks elapse before she readily eats honey, and nearly three weeks pass, before the *gathering* impulse is sufficiently developed to impel her to fly abroad, and seek for honey and pollen among the flowers.[34]

I made, further, the following observations respecting the domestic employment's of the young Italian bees.

---

[34] *Langstroth has the division of labor time interval just about right. It usually takes nearly three weeks before a bee becomes a forager. However, if conditions dictate she may start foraging as early as the first week, e.g., a bee in a new swarm. Once a bee starts foraging they do not live a long time. All the problems of nature come into play such as insect eating birds, rain, wind, etc. that cause the foragers to die. They also just wear out their wings when foraging which causes them to die in the field. Seeing worn wings is one way to tell the age of a forager bee in a hive.*

"1. On the 20th of May, I took out of the hive all the combs it contained, and replaced them after examination. On inspecting them half an hour later, I was surprised to see that the edges of the combs, which had been cut on removal, wore covered by Italian bees exclusively. On closer examination, I found that they were busily engaged in re-attaching the combs to the sides of the hive. When I brushed them away, they instantly returned, in eager haste, to resume their labors.

"2. After making the foregoing observations, I inserted in the hive a bar from which a comb had been cut, to ascertain whether the rebuilding of comb would be undertaken by the Italian bees. I took it out again a few hours subsequently, and found it covered almost exclusively by Italian workers, though the colony, at that time, still contained a large majority of common bees. I saw that they were sedulously engaged in building comb; and they prosecuted the work unremittingly, whilst I held the bar in my hand.* I repeated this experiment several days in succession, and satisfied myself that the bees engaged in this work were always almost exclusively of the Italian race. Many of them had scales of wax visibly protruding between their abdominal rings. These observations show that, in the early stage of their existence, the impulse for comb building is stronger than later in life.

---

* I have had a queen which continued to lay eggs in a comb, after it was removed from the hive.

(Such docile behavior is not all that uncommon as I have had a queen continue laying eggs many times when I happened to remove a frame with the queen. I suspect it is more common with certain strains of bees than with others. It also probably occurs more often with older queens than with a young, newly laying queen. The change to bright sunlight would seem a great deterrent to normal behavior, but sometimes another behavior (oviposition) conquers the change in the environment. She is more likely to hide from the light, and will quickly move to the shade side of the frame when you are examining it.)

"3. Whenever I examined the colony during the first three weeks after the Italian bees emerged, I found the brood-combs covered principally by bees of that race: and it is, hence, probable that the brood [†] is chiefly attended to and nursed by the younger bees. The evidence, however, is not so conclusive as in the case of comb building, inasmuch as they *may* have congregated on the brood-combs because these are warmer than the others.

"I may add another interesting observation. The feces in the intestines of the young Italian bees was viscid and yellow; that of the common or old bees was thin and limpid, like that of the queen bee. This is confirmatory of the opinion, that, for the production of wax and jelly, the bees require pollen; [35] but do not need any for their own sustenance."—*B. Z.* 1855, p 163. S. Wagner

If the colonies are to be multiplied rapidly, the nuclei must never be allowed to become too much reduced in numbers or to be destitute of brood or honey. With these precautions, the oftener their queen is taken from them, the more intent they usually become in supplying her loss.

[†] I once had a colony which, after it had been queenless for some time, not only refused to make royal cells, but even devoured the eggs which were given to them. Similar facts have been noticed by other observers. When a colony which refuses to rear a queen, has a comb given to it containing maturing bees, the motherless innocents will at once proceed to supply their loss. Dr. Dönhoff's observations account for these facts.

[35] *We now know that pollen is needed for the production of brood food, worker jelly and royal jelly, but not for wax. Honey bees use sugar (honey) to produce the beeswax. The young bees would have many remains, or shells, of the pollen grains within their hind-gut. Since young bees usually do not fly, and have therefore not yet defecated, it is easy to see how this idea would have been generated by the bee scientists of the time.*

There is one trait in the character of bees which worthy of profound respect. Such is their indomitable energy and perseverance, that under circumstances apparently hopeless, they labor to the utmost to retrieve their losses, and sustain the sinking State. So long as they have a queen, or any prospect of raising one, they struggle vigorously against impending ruin, and never give up until their condition is absolutely desperate. I once knew a colony of bees not large enough to cover a piece of comb four inches square, to attempt to raise a queen. For two whole weeks they adhered to their forlorn hope; until at last, when they had dwindled to less than one half of their original number, their new queen emerged, but with wings so imperfect that she could not fly. Crippled as she was, they treated her with almost as much respect as though she were fertile. In the course of a week more, scarce a dozen workers remained in the hive, and a few days later, the queen was gone, and only a few disconsolate wretches were left on the comb.[36]

Shame on the faint-hearted of our race, who, when overtaken by calamity, instead of nobly breasting the stormy waters of affliction, meanly resign themselves to an ignoble fate, and perish, where they ought to have lived and triumphed! and double shame upon those who living in a Christian land, thus "faint in the day of adversity," when if they would only believe the word of God, they might behold, with the eye of faith, his "bow of promise" spanning the still stormy clouds, and hear his voice of love bidding them trust in Him as a "Strong Deliverer!"

---

[36] *The description aptly describes a colony that has a failed queen, for whatever reason. It also would apply to those hopelessly queenless colonies that become laying worker colonies. Usually, colonies that have a un-fertile queen, or one that has laying workers, there are several, to many, small-sized drones that were produced in worker cells. In the case that Langstroth describes, it would appear that the colony became very weak and thus died out without any production of drones.*

In the previous editions of this work, with other methods of artificial swarming, very full directions were furnished for increasing colonies, by giving to the nuclei a second comb with maturing brood, as soon as their queens began to lay eggs, and then, at proper intervals, a third and a fourth, until they were strong enough to take care of themselves.[37] This mode of increase is laborious, and requires skill and judgment which few possess: it is also peculiarly liable to cause robbing among the bees, requiring the hives to be too frequently opened, to remove the combs needed in the various processes. As a number of nuclei are to be simultaneously strengthened, the Apiarian cannot complete his artificial process by a single operation and must always be on hand, or incur the risk of ending the season with a number of starving colonies. For these and other reasons, I much prefer the methods which I have devised, for dispensing with so much opening of hives and handling of combs. If, however, any of the new colonies are weak enough to need it, they may be helped to combs from stronger stocks.

---

[37] *In any method of forced increase, with the idea of making as many colonies as possible, the beekeeper is sacrificing **all** honey production.*

The spreading of brood between colonies, or by exchanging their stands, will equalize colonies as much as possible. I agree with Langstroth that care must be given so that enough honey stores are provided, or syrup given, so that the colonies are able to winter. Most modern beekeepers feel that a colony can by split two, or sometime three, times in a season. These divisions are for a rapid increase in colonies and not for producing a crop of honey in that year. If you make more than about three divisions the resulting colonies are usually too weak to survive a northern winter. This is true even if you add honey, they just don't produce enough bees to make a good winter cluster.

Whatever method of artificial increase is pursued by the Apiarian, he should never reduce the strength of his mother-stock, so as seriously to cripple the reproductive power of their queens. This principle should be to him as "the law of the Medes and Persians, which altereth not:" for while a queen, with an abundance of worker comb and bees, may, in a single season, become the parent a number of prosperous families, if her colony, at the beginning of the swarming season, is divided into three or four parts, not one of them will ordinarily acquire stores enough to survive the Winter.

If the Apiarian is in the vicinity of sugar-houses, confectioneries, or other tempting places of bee-resort, he will find his stocks, both old and new, so depopulated by their zeal for ill gotten gains, as to be in danger of perishing. *In such situations, all attempts at rapid increase are entirely futile.*

Artificial operations of all kinds are most successful when bee-forage is abundant; when it is scarce they are quite precarious, even if the colonies are well supplied with food.

When bees are not busy in honey-gathering, they have leisure to ascertain the condition of weak stocks, which are almost certain to be robbed, if they are incautiously opened. When forage is scarce, the hives should be opened before sunrise, or after sunset, or when very few bees are flying abroad; and if it is necessary to open them at other times, they must be removed out of the reach of annoyance from other colonies. The Apiarian who does not guard against robbing, will seriously impair the value of his stocks, and entail upon himself much useless and vexatious labor.[38] *Beware of demoralizing bees, by tempt-*

---

[38] *Maybe Langstroth was right in commenting about the poor seasons for bees in the 20 years prior to writing this book. He certainly had trouble with robbing. I never think that there will be much robbing as long as there is nectar coming in.. But it also could be that because of the transportation problem the average beekeeper had too many colonies in one location. If you have too many colonies there just are not enough nectar resources and the colonies end up robbing each other.*

I have found the use of the anti-robbing screens prevents most natural robbing. And as such the constant stinging that often occurs during robbing periods. The making and use of these screens I described in [Gleanings in] Bee Culture, Vol. 110:92, 109 (1982). These screens will cut down on stinging very much during those periods when little or no nectar is coming into the colony.

ing them to rob each other!

In an Apiary where hives very *unlike*, in *size, shape,* and *color,* are *crowded together*, artificial operations will often be exceedingly hazardous, as the bees will be continually liable to enter the wrong hives. If the stocks must be kept very close together, even if the hives are all of the same color and pattern, it will be best to carry to a second Apiary, either the forced swarms, or the mother-stock from which they were made.

The bee-keeper has already been reminded that *caution is needed in giving to bees a stranger-queen*. Huber thus describes the way in which a new queen is usually received by a hive:

"If another queen is introduced into the hive within *twelve* hours after the removal of the reigning one, they surround, seize, and keep her a very long time captive, in an impenetrable cluster, and she commonly dies either from hunger or want of air. If eighteen hours elapse before the substitution of a stranger-queen, she is treated, at first, in the same way, but the bees leave her sooner, nor is the surrounding cluster so close; they gradually disperse, and the queen is at last liberated; she moves languidly, and sometimes expires in a few minutes. Some, however, escape in good health, and afterwards reign in the hive. If *twenty-four* hours elapse before substituting the stranger-queen, she will be well received, from the moment of her introduction.[39]

"Reaumur affirms, that, should the original queen be removed, and another introduced, this new one will be perfectly well received from the beginning. He induced four or five hundred bees to leave their hive, and enter a glass box, containing a small piece of comb. At first, they were in great agitation, but from the moment that he presented a

---

[39] *The safest way to introduce a queen is by the use of the 'push-in' cage. I use approximately a three-inch square made of 8-mesh hardware cloth. The cage is cut and folded such as to make a shallow cage with sides about ¾ in. high. These sides are pushed into the comb surface, which allows about 3/8 to 1/2 inch for the queen to move about. The workers will feed her through the wire and she will begin to lay eggs. Once she is laying (usually 2-3 days) the bees will not kill her.* **However, do not remove the cage until she is laying eggs.** *Other queen cages also work for introduction, it is just that they are not as sure as the push-in cage.*

new queen the tumult ceased, and the stranger was received with all respect.

"I do not dispute the truth of this experiment, but Reaumur's bees were too much removed from their natural condition to allow him to judge of their instincts and dispositions. He has himself observed, that their industry and activity are affected by reducing their numbers too much. To render such an experiment truly conclusive, it mast be made in a populous hive; and on removing the native queen, the stranger must be immediately substituted in her place."

It would seem, from his use of the word *immediately*, that Huber must have been aware of the fact, that if a strange queen is given to a colony, before this agitation is calmed down (p. 158), and before royal cells are begun, she will usually be well received. If the bees of a colony are made to fill themselves with honey by drumming, smoking, or giving them liquid sweets, and often, if they are removed to a new stand, they will readily accept of any queen offered them, in place of their own.[40]

---

[40] *I found, because of a shortness of time, that a new queen could be introduced into a colony by using a sugar-syrup spray on the bees AND on the queen. Sometimes she comes out a little worse for wear, as later you will see that the bees did take most of the hair (setae) from her thorax. This type of queen introduction works best if the queen is already laying. There is something about a laying queen, even from another colony, that allows her not to be killed by a strange colony. A couple of time I have found two queens within a colony, and both were marked! How did this happen? A logical explanation is that I removed a frame from a colony with the queen on it, and to protect the queen from damage rested it against the colony next to it. When returning the comb to the original colony the queen was not checked for. The queen crawled off of the comb leaning against the colony and into the queen-right colony next door. Since she was a laying queen she was accepted by the strange colony.*

Bees, in possession of a fertile queen, are often quite reluctant to accept of an unimpregnated one in her stead; indeed, it requires much experience to be able to give a strange queen to a colony, and yet be sure of securing for her a good reception. In several instances, the workers have stung a strange queen to death, while I was holding her in my fingers, to be able to remove her if she was not kindly welcomed. To prevent accidents, it will be well to confine a queen when given to a strange colony—in what the Germans call a "queen-cage," which may be made by boring a hole into a block, and covering it with wire gauze, or any perforated cover.[41] The bees will cultivate an acquaintance with the imprisoned mother, by thrusting their antenna through the openings, and the next day she may be safety given to them. Queens bent on escaping to the woods, may be confined in the same way. A pasteboard box, pierced with holes, answers equally well, or even a match box, properly scalded.

If the cage is put with its small openings over one of holes on the spare honey board, or set inside of the hive, the bees will be as quiet as though the queen had her liberty. Such a cage will be very convenient for any temporary confinement of a queen.

In catching a queen, she should be gently taken, with the fingers, from among the bees, and if none are crushed; there is no risk of being stung. The queen, although she will not sting, even if roughly handled, will sometimes, when closely confined, bite the hand of the operator so as to cause a little uneasiness—her jaws, which are intended for gnawing into the base of the royal cells, being larger and stronger than those of a common bee. If she is allowed to fly, she may be lost, by attempting to enter a strange hive.

As a fertile queen can lay several thousand eggs a day, it is not strange that she should quickly become exhausted if taken from the bees. "*Ex nihilo nihil fit*"—from nothing, nothing comes—and the arduous duties of maternity compel her to be an enormous eater. After

---

[41] *Frank Benton is recognized as having made the first mailing cage to ship bees back from Europe. It would appear that he had read this section and was able to adapt it to his cage. Benton made it a 3-hole cage instead of one, but it would appear that the idea was Langstroth's. As you will see by what follows he even mailed queens in his cage. This is just another case of where Langstroth was years ahead of his fellow beekeepers.*

an absence from the bees of only fifteen minutes, she will solicit honey, when returned; and if kept away for an hour or upwards, she must either be fed by the Apiarian, or have a few bees, gorged with honey, given to her to supply her wants. One which I sent by express in a queen-cage, with a suite of well fed workers, arrived in safety, at the Apiary of a friend, on the next day.

Great caution is not only requisite in giving a hive a strange queen, but in all attempts *to mix bees belonging to different colonies.* Bees having a fertile queen will almost always quarrel with those having an unimpregnated one; and this is one reason why a furious contest, in which thousands perish, often ensues when new swarms attempt to mingle.

Members of different colonies appear to recognize their hive companions by the sense of smell and if there should be a thousand stocks in the Apiary, any one will readily detect a strange bee;[42] just as each mother in a large frock of sheep is able, by the same sense, in the darkest night, to distinguish her own lamb from all the others. It would seem, therefore, that colonies might always be safely mingled, by sprinkling them with sugar water, scented with peppermint or any other strong odor, which would make them all smell alike.

A few seasons ago, however, I discovered that bees often recognize strangers by their actions, even when they have the same scent; for a *frightened bee curls himself up with a cowed look*, which unmistakably proclaims that he is conscious of being an intruder. If, therefore, the bees of one colony are left on their *own stand* and the others are suddenly introduced, the latter, even when both colonies have the same smell, are often so frightened that they are discovered to be strangers and are instantly killed. If, however, *both* colonies are removed to a *new stand*, and shaken out together on a sheet they will

---

[42] *We have found in some of our recent studies that drifting between colonies is much greater than we previously thought. It may be very common even during active nectar and pollen gathering, and we have had bees drift more than a mile into other apiaries. When bees are moved for pollination the figures approach 25% of the foragers were reared in other colonies. This accounts for some of the rapid transfer of diseases and parasitic mites that we have seen. I am sure that bees did not expect all their homes to be square boxes all painted the same color!*

peaceably mingle, when scented alike.*

If, when two colonies are put together, the bees in the one on the old stand are not gorged with honey, they will often attack the others, which are loaded, and speedily sting them to death, in spite of all their attempts to purchase immunity, by offering their honey. Mr. Wm. W. Cary, of Coleraine, Massachusetts, who has long been an accurate observer of the habits of bees, unites colonies very successfully, by alarming those that are on the old stand; as soon as they show, by their notes that they are subdued, he gives them the new comers. The alarm which causes them to gorge themselves with honey (p. 27), puts them, doubtless, upon their good behavior, long enough to give the others a fair chance.[43]

It has been stated already, that a queen-bee cannot be induced to sting, by any kind of treatment, however severe. The reason of this strange unwillingness will be obvious, when we consider that the preservation of her life is indispensable to the existence of the colony, and that, although the loss of her sting would be fatal to herself it could avail no more for their defense, in case of an attack, than the single sword of a Washington or a Wellington could decide a great battle. While the common bees are ready to sally forth and sacrifice

---

* I find substantially the same thing recommended, in 1778, by Thomas Wildman (page 230 of the 3rd edition of his valuable work on Bees), who says, that bees will "unite while in fear and distress, without fighting, as they would be apt to do, if strange bees were added to a hive in possession of its honey."

Of all the old writers, Wildman appears to have made the nearest approaches to the modern methods of taming and handling bees. Twenty-five years before Huber's investigations on the origin of wax, this acute observer had noticed the scales of wax on the abdomen of the workers; and he was so thoroughly convinced that wax was secreted from honey, that he recommended feeding new swarms, when the weather is story, that they may sooner *build comb* for the eggs of the queen.

Mr. Wagner refers me to "Orerbeck's *Glossarium Melliturgium*" —BREMEN, 1765, p. 89 – in which the origin of wax is claimed, more than 20 years before the date of that work – say 1745 – for a Hanoverian Pastor, named Herman C. Horsbostel. He gave his discoveries to the world in the so-called "HAMBURGH LIBRARY," vol. 2, p. 45; and they are so particularly described as to leave no doubt of their correctness.

[43] *Another idea could be that the various pheromones have been mixed up, or the alarm pheromone covered up by the different pheromones. Usually bees will not attack strange bees that are full of honey. Otherwise, I cannot explain the great drifting that occurs and these drifting bees not being killed. If a forager returns to a hive laden with nectar and enters a strange colony she would almost always be accepted.*

their lives on the slightest provocation, a queen bee only buries herself more deeply among the clustering thousands, and will never use her sting, except when engaged in mortal combat with another queen. When two rivals meet, they clinch, at once, with every demonstration of the most vindictive hatred. Why, then, are not both often destroyed? We can never sufficiently admire the provision so simple, and yet so effectual, by which such a calamity is prevented. A queen never stings, unless she has such an advantage that she can curve her body under that of her rival, so as to inflict a deadly wound, without any risk to herself—the moment the position of the two combatants is such that neither has the advantage, but both are liable to perish, they not only refuse to sting, but disengage themselves, and suspend their conflict for a short time!

The following interesting statements were furnished to the *New England Farmer* (Oct. 1855), by Hon. Simon Brown, Lieutenant-Governor of the Commonwealth of Massachusetts, in 1855.

"On the 17th of July last, we placed in our dining. room window one of Mr. Langstroth's observing bee-hives, constructed of glass, so that all the operations of the bees could be plainly and conveniently seen. A comb about a foot square was placed in it, containing some brood, with plenty of workers and drones, but *without a queen*. The hive was then carefully observed by one of the ladies of the family, who has given us the following account of their doings.

"'The first business the bees attended to, was to commence cells for a queen, and they prosecuted it with energy for two days. At the end of that time, a queen was taken from another colony and placed with them, upon which they pulled down the cells they had made, in less than half the time it had required to construct them, and then began to piece out and repair the comb which needed a corner. The queen at once commenced laying, and soon filled the unoccupied cells, when she was again removed, and the bees once more began the construction of queen cells.

"'The young bees now began to hatch forth and in two weeks the family increased so fast as to make it necessary for them to prepare to emigrate. They had built six queen cells, and in about twelve days the first queen was hatched. As soon as she was fairly born, she marched rapidly, and in the most energetic manner, over the comb, and visited the other cells in which were the embryo queens, seeming at times furious to destroy them. The workers, however, surrounded her, and prevented such wholesale murder. But for two days she was intent upon her fell purpose, and kept in almost continuous motion to effect it. On the fourteenth day, the second queen was ready to come out, piping and making various noises to attract attention.

"A part of the colony then seemed to conclude that it was time to take the first queen and go, but by some mistake she remained in the hive after the swarm had left. The second queen came out as soon as possible after the others had gone, and then there were now *two* hatched queens in the hive! they ran about on the comb, which was now nearly empty, so that they could be distinctly seen. But they had not, apparently, noticed each other, while the workers were in a state of great uneasiness and commotion, seeming impatient for the destruction of one of them. The mode they adopted to accomplish it was of the most deliberate and cold-blooded kind. A circle of bees kept one queen stationary, while another party dragged the other up to her, so that their heads nearly touched, and then the bees stood back, leaving a fair field for the combatants, in which one was to gain her laurels, and the other to die! The battle was fierce and sanguinary. They grappled each other, and, like expert wrestlers, strove to inflict the fatal blow by some sudden or adroit movement. But for some moments the parties seemed equally matched; no advantage could be gained on either side. The bees stood looking calmly on the dreadful affray, as though they themselves had been the heroes of a hundred wars. But the battle, like all others, had its close; one fell upon the field, and was immediately taken by the workers and carried out of the hive. By this time, the bees which had swarmed made the discovery that their queen was missing, and although they had been hived without any trouble, came rushing back, but not in season to witness the fatal battle, and the fall of their poor slain queen, who should have gone forth with them to seek a future home." *

---

* "We introduced a queen into a hive," says Huber, "after painting her thorax, to distinguish her from the reigning queen. A circle of bees formed so closely around the stranger, that in scarcely a minute she lost her liberty. Other workers at the same time collect around the reigning queen, and restrained her motions. * * * They retained their prisoners only when they appeared to withdraw from each other; and if one, less re-

The Apiarian has already been reminded of the importance of securing straight worker-combs for his stocks. To a stock-hive, such combs are like, cash capital to a business man; and so long as they are fit for use they should never be destroyed (p. 60).* Those who have plenty of good worker-comb, will unquestionably find it to their advantage to use it in the place of the artificial guides (Pl. I, Fig. 2, w).† Those who use the guides, should examine a swarm two or three days after it is hived, when, by a little management, any irregularities in their combs may be easily corrected. Some combs may need a little compression, to bring them into their proper. positions, and others may even require to be cut out, and fastened as guides in other frames; but no pains should be spared to see that they are all right, before the work has gone so far as to make it laborious to remedy any defects.[44] If a colony is small it ought to be confined, by a movable partition, to such a space in the hive as it can occupy with comb—as well for its encouragement, as to economize its animal heat, and guard against

---

strained, seemed desirous of approaching her rival, all the bees forming the clusters gave way, to allow her full liberty of attack; then, if they showed a disposition to fly, they returned to enclose them."

* Mr. S. Wagner has a colony over 21 years old, whose young bees appear to be as large as any others in his Apiary.

(We have found in modern times that it is not the supposed reduction in cell size that calls for replacement of the combs. We have found that the wax has picked up many chemicals over time, and the replacement with new foundation helps the colony by allowing more larvae to mature (not killed by pesticides). Beeswax is a sink for all types of lipophilic chemicals such as pesticides. These can accumulate in the wax and reduce the number of eggs and larvae that mature. It has been found that a systematic replacement of the combs keeps this problem at a minimum. There seems to be no absolute guide for this replacement, however a common goal, or guide, is to replace 20 percent of the coms every year. This replacement may be a little work, but the colony will be stronger because of the comb replacement. One of the easiest ways to recycle combs is each spring, when the cluster of bees is at the top of the hive, examine the bottom box [now empty] and remove all dark and empty combs possible. These combs were the previous seasons brood coms, and as such may have been in the colony for some years.)

† See Explanation of Plates of Hives, for a description of the various styles of movable frames.

[44] *New comb may be moved and shaped almost too easily. But as Langstroth indicates, this is the time to make changes otherwise you have the misshapen comb for the rest of its life. This is one of the advantages of the plastic-base foundation as it is difficult for it to be moved out of shape.*

irregularities in comb-building. Varro, who flourished before the Christian Era, says (Liber III, Cap. xviii), that bees become dispirited, when placed in hives that are too large.

The possession of five frames of straight worker-comb, may be made to answer an admirable end, if given to a new swarm, so as to alternate with its empty frames. After the bees have had possession of them two or three days, they may be politely informed that these worker combs were only loaned to them as patterns, and their new combs may be alternated with empty frames. Five combs may thus be used for many successive swarms.

As the artificial guides increase the expense of the frames, and cannot *be invariably relied on*, the practical Apiarian will aim, as far as possible, to dispense with their use. I have devised a plan which will be elsewhere described for superseding them, and enabling the beginner to compel his bees, without any comb, to build in the frames with entire regularity. [45]

It must be obvious to every intelligent bee-keeper, that the perfect control of the combs of the hive is the soul of a system of practical management, which may be modified to suit the wants of all who cultivate bees. Even the old-fashioned beekeeper can, with movable combs, destroy his faithful laborers quite as speedily as by setting them over a sulphur-pit; thus preserving his honey from disgusting fumes, while he secures it on frames from which it may be conveniently cut, and preserves all empty comb for future use (p. 71).

As many who would like to keep bees are so much afraid of being stung, that they object entirely even to natural Swarming, how, it may be asked, can such persons open hives, lift out the combs, shake or brush off the bees, and practice other processes which seem like bearding a lion in its very den? The truth is, that some persons are so timid, or suffer so dreadfully when stung, that they are every way disqualified from having anything to do with bees, and, ought either to

---

[45] *Comb foundation has taken care of this problem. I suspect that one of the real advantages for using foundation is that the combs are usually straight. It must have been a real dilemma to have a movable-comb hive and then not be able to remove the combs because the combs were crossed such that they could not be removed. This is why Langstroth is advocating the alternation of new (empty) frames with ones full of comb. This would make the new ones reasonably straight. I find that it is not a bad idea, even with foundation, as then one frame does not become too thick.*

have none upon their premises, or to entrust the care of them to others. With the directions furnished in this treatise, almost any one, however, by using a bee-dress, can learn to superintend bees with very little risk. I find, in short, that the risk of being stung is really diminished by the use of my hives; although it is very difficult for those who have not seen them in use, to believe that this can be so.

The ignorance of most bee-keepers of the almost unlimited control which may be peaceably acquired over his bees, has ever been regarded by the author of this treatise as the greatest obstacle to the speedy introduction of movable-comb hives. He might easily have invented contrivances which, by adapting themselves to this ignorance, would, at first, have proved much more lucrative to him, had be thought it just, either to the community or to himself, to have taken such a course. Such ignorance has led to the invention of costly and complicated hives.* all the ingenuity and expense lavished upon which, are known by the better informed, to be as unnecessary as a costly machine for lifting up bread and butter, and gently pushing it

---

* I have before me a small pamphlet, published in London in 1851, describing the construction of the "Bar and Frame hive" of W. A. Munn, Esq. The object of this invention is to *elevate* frames, one at a time, into a case with glass sides, so that they may be examined without risk of annoyance from the bees. Great ingenuity is exhibited by the inventor of this very costly and very complicated hive, who seems to imagine that smoke "must be injurious both to the bees and their brood." Even if a little smoke is so injurious, the Apiarian, by sweetened water, or by drumming upon a hive, after closing its entrance, and cause the bees to fill themselves with honey (p. 27), when all their combs may be safely lifted out.

A Huber-hive, or one with movable bars, may be much more safely managed than any one which proposes to elevate the frames, without permitting them to be pushed apart (p. 150). A single hive, the arrangements of which are such as to maim and irritate bees, is more to be dreaded in an Apiary than a thousand of proper construction; as it educates bees to regard their keeper in the light of an *enemy.*

On p. 15, I have spoken of the bar-hive, as at least one hundred years old. From "A Journey into Greece, by George Wheeler, Esq.," made in 1675-6, it appears that it was, at that time, in common use there, and, probably, even then an old invention; he describes how it was used for forming artificial swarms, and removing spare honey. As the new swarms were made by dividing the combs between two hives, and no mention is made of giving the queenless one a royal cell — These old observers were probably acquainted with the fact that they could rear one from the worker-brood. Huber says; — Monticelli, an Neapolitan Professor, claims that the plan of artificial swarming was borrowed from Favignana, and that the practice is so ancient that even the Latin names are preserved by the inhabitants in their procedure.

into the mouth and down the throat of an active and healthy child.

The Rev. John Thorley, in his "Female Monarchy," published at London, in 1744, appears to have first introduced the practice of stupefying bees by the narcotic fumes, of the "puff ball" (*Fungus pulverulentus*), dried till it will hold fire like tinder. The same effect has been produced by pushing a rag, saturated with chloroform or ether, into the entrance of the hive, and closing all tight, to prevent the escape of the fumes. The bees soon drop motionless from their combs, and recover again after a short exposure to the air.

Some of my readers may suppose that such an easy mode of stupefying bees would very greatly facilitate the removal of combs; but, however valuable to those ignorant of the great law, that a gorged bee never volunteers an attack, to the better informed, narcotics of all kinds are, for general purposes, worse than useless. *Living* bees may be easily made to get out of the way; but *drunken* ones, like drunken men, are constantly liable to be maimed or killed. There is a large class of bee-keepers—not bee-masters—who desire a hive which will give them, however ignorant or careless, a large yield of honey from their bees. They are easily captivated by the shallowest devices, and spend their money and destroy their bees, to fill the purses of unprincipled men. There never will be a "royal road" to profitable bee-keeping. Like all other branches of rural economy, it demands care and experience; and those who are conscious of a strong disposition to procrastinate and neglect, will do well to let bees alone, unless they hope, by the study of their systematic industry, to reform evil habits which are well nigh incurable.

While I feel increasingly sanguine that the movable-comb hive [*] will be extensively used by skillful bee-keepers, I well know the difficulty of rapidly introducing any system of management which is much in advance of current knowledge; even a perfect hive (p. 116) would require years to win its way into general use. It is only of late years, that the splendid discoveries of Huber—like the writings of Bruce on the Sources of the Nile—have emerged from the clouds of

---

[*] The day on which I contrived the movable-frames, I wrote as follows, in my Bee-Journal: – "The use of these frames will, I am persuaded, give a new impulse to the easy and profitable management of bees; and will render the making of artificial swarms an easy operation."

ridicule and aspersion in which they were so long enveloped; and even now, to describe a tithe of the wonders of the bee-hive, however thoroughly they have been demonstrated, is, unfortunately, in the estimation of many of our oldest bee-keepers, to deserve the name of a fool, a liar, or a cheat.

A photo of L. L. Langstroth about the time he wrote this book. The photograph was not found in the original edition.

Preface to Chapter XI

When you read this chapter you again realize the many hours that Langstroth had to have put in observing honey bees, and in this case queen bees. Research over the last 150 years has given us an advantage in that we know more about the chemistry and physiology of the queen. However, most of what you read here is just good beekeeping knowledge. For example, his ideas on the causes and cures for drifting were probably at least 120 years ahead of his time. That is, the painting of the front of the hives different colors and painting on different shapes—both of which help bees and queens from getting lost on their initial flights.

CHAPTER XI.

LOSS OF THE QUEEN.

THAT the Queen-Bee is often lost, and that her colony will be ruined unless such a calamity is seasonably remedied, ought to be familiar facts to every bee-keeper.

Queens sometimes die of disease or old age, when there is no brood to supply their loss. Few, however, perish under such circumstances; for either the bees build royal cells, aware of their approaching end, or they die so suddenly as to leave young brood behind them. Queens are not only much longer lived (p. 58) than the workers, but are usually the last to perish in any fatal casualty.[1] As many die of old age, if their death did not ordinarily occur under favorable circumstances, it would cause, yearly, the loss of a very large number of colonies. As they seldom die when their strength is not severely taxed in breeding, drones are usually on hand to impregnate their successors.*

---

[1] *See my footnote on page 178 about a colony dying from lack of food and only the queen and a few bees surviving.*

* In preparing my stocks for Winter, I found—on the 21st of October, 1856—two which had sealed queens. As the drones were not killed, in some of the hives, until after the 1st of November, these queens might have been impregnated, if the weather had not become very cold. When examined on the 21st day of February, these stocks had each a few *sealed drones* and larvae, while weaker stocks had much brood. The following is an extract form Prof. Leidy's description of these queens:—"Their ovaries were filled with eggs, from a mere point to such as measured four-fifths of a line long, and one-eighth of a line broad. Their spermathecae were filled with muscoid, granular matter, and epithelial cells, *and did not contain a trace of spermatic filaments.*" While the intestines of these queens contained only a little limpid excrement, the rectum of a worker, examined at the same time, was filled with an enormous quantity of a dark, offensive substance.

These drone-laying colonies were supplied with queens from other stocks, which, when opened in April, were found to have raised queens in February. One queen was laying worker, and the other drone-eggs, and the former must have been impregnated in March, and probably by some of the brood of the drone-laying queens. Might not a few drone-laying queens be kept to advantage in large Apiaries?

(I think you will find some drones in colonies throughout the winter, certainly enough to mate with a virgin. The major problem is not the presence of drones but the lack of suitable mating (flying) weather. This is true in early Spring or late Fall. Drones and queens do not normally fly until the temperature is probably into the 70°'s

Young queens are sometimes born with wings so imperfect that they cannot fly (p. 39); and they are often so injured in their contests with each other, or by the rude treatment they receive when driven from the royal cells (p. 121), that they cannot leave the hive for impregnation.

We have yet, however, to describe under what circumstances the majority of hives become queenless. *More queens, whose loss cannot be supplied by the bees, perish when they leave the hive to meet the drones, than in all other ways.* After the departure of the first swarm, the mother-stock and all the after-swarms have young queens which must leave the hive for impregnation; *their larger size and slower flight* make them a more tempting prey to birds, while others are dashed, by sudden gusts of wind, against some hard object, or blown into water: for, with all their queenly dignity, they are not exempt from mishaps common to the humblest of their race.

In spite of their caution to mark the position and appearance of their habitation (p. 125), *the young queens frequently make a fatal mistake and are destroyed by attempting to enter the wrong hive.* This accounts for the notorious fact, that ignorant beekeepers, with forlorn and rickety hives, no two of them look just alike, are often more successful than those whose hives are of the best construction. The former—unless their hives are excessively crowded lose but few queens, while the latter lose them almost in exact proportion to the taste and skill which induced them to make their hives of *uniform size, shape, and color.*[2]

I first learned the full extent of the danger of crowded Apiaries, in the Summer of 1854. To protect my hives against extremes of heat and cold, they were ranged, side by side, over a trench, so that, through ventilators in their bottom-boards they night receive, in Summer, a cooler, and in Winter, a much warmer air, than the external atmosphere. By this arrangement—which failed entirely to answer

---

and usually with sunshine. If the weather is bad for a long period the queens become unmated drone layers.)

[2] *Here in the United States we have a strong tendency to paint our hives white and put them into perfect rows, either for esthetics or convenience. The bees (workers or queens) probably drift more because of this reason. As Langstroth says later, ancient bees did not see two hollow trees exactly alike so they did not have a drifting problem.*

its design—many of my colonies became queenless, and I soon ascertained under what circumstances young queens are ordinarily lost.

From the great uniformity of the hives in size, shape, color, and height, it was next to impossible for a young queen to be sure of returning to her hive. The difficulty was increased, from the fact that the ground before the trench was free from bushes or trees, and no hive—except the two end ones,[3] which did not lose their queen—could have its location more easily remembered, from its relative position to some external object. Most of the hives thus placed, which had young queens, became queenless, although supplied with other queens, again and again; and many, even of the workers were constantly entering hives adjoining their own.

If a traveler should be carried, in a dark night, to a hotel in a strange city, and on rising in the morning, should find the streets lined with buildings precisely like it, he would be able to return to his proper place, only by previously ascertaining its number, or by counting the houses between it and the corner. Such a numbering faculty, however, was not given to the queen-bee; for who, in a state of nature, ever saw a dozen or more hollow trees or other places frequented by bees, standing close together, precisely alike in size, shape, and color, with their entrances all facing the same way, and at exactly the same height from the ground!

On describing to a friend my observations on the loss of queens, he told me that the management of his hens, he had fallen into a somewhat similar mistake. To economize room, and to give easier access to his setting hens, he had partitioned a long box into a dozen or more separate apartments. The hens, in returning to their nests, were deceived by the similarity of the entrances, so that often one box contained two or three unamiable aspirants for the honors of maternity, while others were entirely forsaken. Many eggs were broken, more

---

[3] *This is one of the major drawbacks of drifting and that is the end-of-row colony ends up with more bees. I have seen the end-of-the-row drifting very serious in certain pollination placements. These colonies then want to swarm, as they are too full of bees.*

The other more serious problem is, of course, the loss of a queen on the return from a mating flight. The simple answer to the problem is to paint colors on the hives or a different color for each hive. See Langstroth's footnote on the next page, the recommendation is not only for a color but also for a pattern as well. Good advice if you want to reduce drifting of bees and especially queens.

were addled, and hardly enough hatched to establish one mother as the happy mistress of a flourishing family. Had he left his hens to their own instincts, they would have scattered their nests, and gladdened his eyes with a numerous offspring.

Through the length and breadth of our land, bee-keepers who suffer heavy losses, from the proximity and similarity of their hives, unsuspicious of the true cause of their misfortunes, impute them to the bee-moth, or some of the many enemies of the bee. Judge Fishback, of Batavia, Ohio, informed me, in the Fall of 1854, while on a visit to his large Apiary, that he had for many years guarded against the loss of young queens, by painting the fronts of his hives of different colors, and making their entrances face in various ways.* Every bee-keeper, whose hives are so arranged that the young queens are liable to make mistakes, must count upon heavy losses. If he puts a number of hives, under circumstances similar to those described, upon a bench, or the shelves of a bee-house, he can never keep their number good without constant renewal. The first swarms, and those stocks which do not swarm, as they retain their fertile queens, will do well enough; but many of those that swarm will be robbed by other bees, or fall a prey to the moth, or gradually dwindle away.

As the bee-keeper, from limited space or other reasons, may prefer to keep his colonies close together, I have devised a way of effecting it, without resulting the loss of the young queens:—

If he relies upon natural swarming, he should remove the *mother-stock*, as soon as it has swarmed, to a *new position*, giving it two or

---

* John Mills, in a work published at London, in 1766, gives (p. 98) the following directions:—"Forget not to paint the mouths of your colonies with different colors, as red, white, blue, yellow, &c., in form of half-moon, or square, that the bees may the better know their own home." Such precautions preserved the stocks from becoming queenless, although they were not adopted for that end.

(A way to limit drifting is to arrange your hives in a serpentine fashion with each colony facing in a slightly different direction. Additionally, painting the colonies different colors, or if the supers are different colors and then mixed up on top of the colonies. The different patterns of colors will help the bees orientate to their own colony.)

Fig. 51. PLATE XVI.

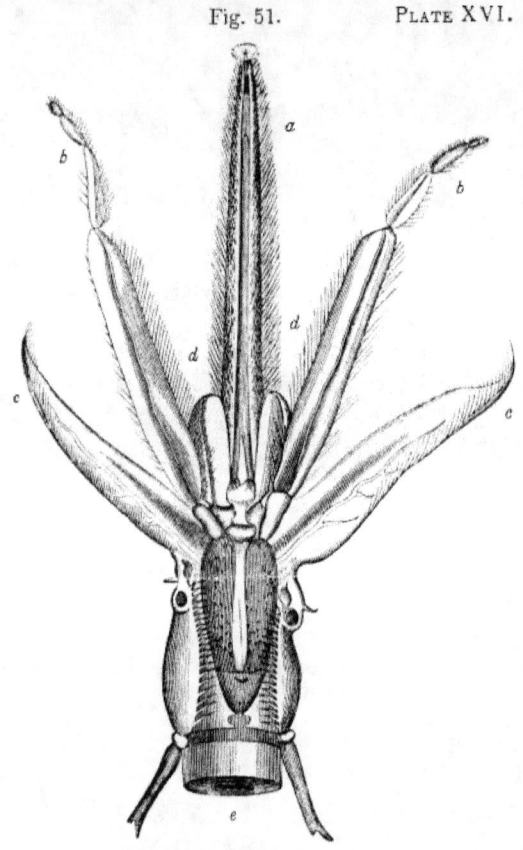

Fig. 52.

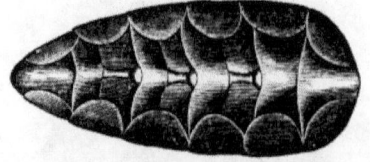

LOSS OF THE QUEEN.

three quarts of bees from the swarm, before they have entered the new hive, which is to be put on the old stand. These bees having the swarming propensity, will supply the place (p. 156) of those which subsequently leave.

If artificial swarming is practiced, the entrances to the hives of the nuclei should be marked with a leafy twig, and; if possible, made to face differently (p. 189) from those of the adjoining stocks. The new colonies should be formed as directed on page 186. If two Apiaries are used, the artificial swarms, may be made in any of the ways previously described, and those colonies which have queens to be impregnated, removed to the second Apiary.[4]

The bees are sometimes so excessively agitated when their queen leaves for impregnation, that they exhibit all the appearance of swarming. They seem to have an instinctive perception of the dangers which await her, and I have known them to gather around her and confine her as though they could not bear to have her leave. If a queen is lost in what the Germans call "her wedding excursion," the bees of an old stock will gradually decline; those of an after-swarm, will either unite with another colony, or speedily dwindle away.

It would be interesting, could we learn how bees become informed of the loss of their queen when she is taken from them, under circumstances that excite the whole colony, we can easily see how they find it out; for as a tender mother, in time of danger, in all anxiety for her helpless children, so bees, when alarmed, always seek first to assure themselves of the safety of their queen. If, however, the queen is very carefully removed, a day, or even more, may elapse, before they realize their loss.[*] How do they first become aware of it? Perhaps some

---

[4] *Langstroth, somewhat, begs the point here. It doesn't matter if you move the colony to another apiary, if the hives are all lined up in neat rows and of the same size and color, the queens may still be lost. Some color near the entrance or other landmark is what is needed. Though, I will have to admit that I have had good success over the years mating queens in divisions above the parent colony with the nuc entrance facing to the rear of the colony. It may be in this case not all colonies had a division on top and thus drifting of the young queens was prevented or reduced.*

[*] "For eighteen hours after the queen was taken away, the usual labors of the hive proceeded as regularly as if she were still present; but no sooner was her loss discovered than all was agitation and tumult—the bees hurried backward and forward over the combs, with a loud noise, rushed in crowds out of the hive, as if going to swarms, and in short, exhibited all the symptoms of bereavement and despair."—BEVAN, p. 24.

dutiful bee, anxious to embrace her mother, makes diligent search for her through the hive. The intelligence that she cannot be found being noised abroad, the whole family is speedily alarmed. At such times, instead of calmly conversing, by touching each other's antennae, they may be seen violently striking them together, and by the most impassioned demonstrations manifesting their agony and despair.

I once removed the queen of a small colony, the bees of which took wing and filled the air, in search of her. Although she was returned in a few minutes, royal cells were found two days later. The queen was unhurt, and the cells untenanted. Was this work begun by some that did not believe the others, when assured that she was safe? or from the apprehension that she might be removed again?

All colonies whose queens are to be impregnated should be watched, that the Apiarian may be seasonably apprised of their loss. Such colonies, if provided with suitable brood-comb, will seldom forsake the hive, if the queen is lost. An old stock which cannot be supplied with a queen or the means of raising one, should be broken up, and the bees added to another colony; a new swarm, unless a queen nearly mature can be given to it (p. 149), should always be broken up. If the new colony is large, it will be better, instead of breaking it up, to give it a queen from some old stock which can easily raise another. If, however, the Apiarian uses movable-comb hives and pursues the nucleus system (p. 188), he will always have queens on hand for all emergencies.

Huber has proved that bees do not ordinarily transport the eggs of the queen from one cell to another. I have, however, in several instances, known them to carry worker-eggs into royal cells. Mr. Wagner put some queenless bees, brought from a distance, into empty

---

(On page 226 Langstroth shows that he knows about the queen scent [pheromone], yet here he becomes fanciful about the loss of a queen—It doesn't normally take very long for a colony to recognize the loss of a queen. Lift off the cover from a hive and listen to the bees, then examine the colony and when you find the queen put her in a queen cage for safekeeping and close up the colony for 30 minutes. Open up the cover again and listen and you will immediately hear the difference. Many bees will be scenting by opening their Nasonov gland (found at the tip end of the abdomen) and these bees will be fanning their wings. You will hear the different noise caused by the fanning of the wings, and with experience you will recognize the smell of the Nasonov pheromone as well.)

LOSS OF THE QUEEN. 219

combs that had lain for two years in his garret. When supplied with brood, they raised their queen in this old comb! Mr. Richard Colvin, of Baltimore, and other Apiarian friends, have communicated to me instances almost as striking.[5]

Having described the precautions necessary to prevent the loss of queens, it remains to show how the bee-keeper can ascertain that a hive is queenless, and how he can remedy such a misfortune. As soon as the bees begin to fly briskly in the Spring, a stock which does not industriously gather pollen,[*] or accept of rye flour, and which refuses clean water, given to it in an empty comb, is almost certain to have no queen, or one that is not fertile—unless it is on the eve of being destroyed by worms, or of perishing from starvation.

A stock is sure to be queenless, if, after taking its first Spring-flight, the bees, by roaming, in an inquiring manner, in and out of the hive (p. 67), show that some great calamity has befallen them. Those that come from the fields, instead of entering the hive with that dispatchful haste so characteristic of a bee returning, well loaded, to a prosperous home, usually linger about the entrance with an idle and dissatisfied appearance, and the colony is restless, late in the day, when other stocks are quiet. Their home, like that of a man who is cursed in his domestic relations, is a melancholy place, and they enter it only with reluctant and slow-moving steps.

And here, if permitted to address a word of friendly advice, I would say to every wife—Do all that you can to make your husband's home a place of attraction. When absent from it, let his heart glow at the thought of returning to its dear enjoyments; as he approaches it, let

---

[5] *The moving of eggs has been reported many times though difficult to prove via experiments. I think that the number of times that I have seen such reports probably makes the case for bees moving eggs. Normally, the workers will eat eggs if given the chance and thus moving them must be triggered by a different behavioral response.*

[*] "Mr. Randolph Peters, of Philadelphia, had a stock which he was satisfied was queenless, as the bees did not carry in pollen for 28 days. I put a queen into the hive, he holding a watch in his hand, and in 3½ minutes from the time she was introduced, a bee was seen to enter with pollen on its legs! We both observed the entrance for some time, and saw many bees carry in pollen."—P. J. MAHAN

(This is too dramatic. A bee can't fly that fast. A foraging trip for pollen would last, on average, about 45 minutes. However, it is known that a queen and brood does stimulate pollen gathering. This is one reason that orchardists have not liked using queenless colonies for pollination even though they were provided at a much cheaper price.)

his countenance involuntarily assume a more cheerful expression, while his joy-quickened steps proclaim that he feels, that there is no place like the cheerful home where his chosen wife and companion presides as its happy and honored Queen.* If your home is not of dear delights, try all the virtue of winning words and smiles, and the cheerful discharge of household duties, and exhaust the utmost possible efficacy of love, and faith, and prayer, before those words of fearful agony,

>"Anywhere, anywhere
>Out of the world!"

are extorted from your despairing lips, as you realize that there is no home for you, until you have passed into that habitation not fashioned by human hands, or inhabited by human hearts.

Although when bees commence their work in the Spring, they usually give reliable evidence either that all is well, or that ruin lurks within, if their first flight is not noticed, it is sometimes difficult, in the common hives, to get at the truth if the bees are driven up among the combs, by smoke, the presence or absence of brood may often be ascertained. If a few imperfect bees are found on the bottom-board, or in front of the entrance, it shows that the hive has a fertile queen.

---

* "The tenth and last species of women were made out of a bee; and happy is the man who gets such a one for his wife. She is full of virtue and prudence, and is the best wife that Jupiter can bestow." —SPECTATOR, No. 209.

I strongly advise giving every movable-comb hive a thorough examination, as soon as the bees begin to work m the Spring.* The combs, with the adhering bees, may be put into a clean hive, and the old one, after being cleansed from everything offensive to the delicate senses of the bees, may be given to another stock.

In making this thorough cleansing of his hives, the Apiarian will learn which require aid, and which can lend a helping hand to others; and any one needing repairs, may be put in order before being needed again. Such hives, if occasionally re-painted, will last for generations, and prove cheaper, in the long run, than any other kind.

If, in the Spring examination, a hive has no queen, it should be supplied, if populous, with one from a weaker stock. If it is small, comb, with hatching bees,† should be given to it from a stronger colony. Or it may change stands with a strong stock, when the bees are actively gathering stores; or bees brought from a distance may be

---

* I would refer those, who think that "*it is too much trouble*" to examine their hives in the Spring, to the practice of the ancient bee-keepers, as set forth by Columella:— "The hives should be opened in the Spring, that all the filth which was gathered in them during the Winter may be removed. Spiders, which spoil their combs, and the worms, from which the moths proceed, must be killed. When the hive has been thus cleaned, the bees will apply themselves to work with the greater diligence and resolution." The sooner those abandon bee-keeping, who consider the proper care of their bees as "too much trouble," the better for themselves and their unfortunate bees.

(Most beekeepers today do not neglect their bees as much as not knowing if they should examine them when the temperature is not perfect on the day that they have available in the Spring. Usually more harm is done by not opening the colony and finding that the colony needs food or more comb space. Certainly, by cleaning off the bottom board the bees will be allowed to expend their efforts at something more productive.)

† That class of bee-keepers who suppose that all such operations are the "new-fangled" inventions of modern times, will be surprised to learn that Columella, 1800 years ago, recommended strengthening feeble stocks, by *cutting* out combs from stronger colonies, containing workers "just gnawing out of their cells."

(But, with these "new-fangled" hives the moving of brood is easy, and a very productive way to equalize colonies. Brood can be moved between hives with absolute impunity. If a colony needs bees very badly then be sure to give them brood that is very close to emerging. Remove a few cappings to determine if the bees are about to emerge. If the colony is weak they will have to keep the brood warm and may not be able to if the amount of brood given is too much, and the new bees do not emerge soon.)

added to it.* If it raises a queen before she can be seasonably impregnated she may be killed, and more brood comb given to them. The smallest stocks may thus be preserved until the drones appear, by which time they may be made as strong as is desired. The stocks deprived of their queens should be managed in the same way. By this device, every queenless stock, however feeble that survives the Winter, may to nursed into profitable strength.[6]

A vigilant eye should be kept upon every colony that has not an impregnated queen; and when its queen is about a week old it should be examined, and if she has become fertile, she will usually be found supplying one of the central combs with eggs. If neither queen nor eggs can be found, and there are no certain indications that she is lost, the hive should be examined a few days later, for some queens are longer in becoming impregnated than others, and it is often difficult to find an unimpregnated one, on account of her adroit way of hiding among the bees.

If the Apiarian relies on artificial swarming, he may deprive his queens of their wings, as soon as they are impregnated.† In a large Apiary, where many swarms might otherwise come off together, this

---

* If a common hive is found, in the Spring, to be very much reduced in numbers, it can be recruited in the last two ways, provided it has a healthy queen. If it has no queen, and is not sufficiently strong to justify giving it one from a weaker stock, the bees should be joined to another colony, and the hive reserved, with its combs, for future swarms. It should, however, be kept out of the reach of the bee-moth, and before it is used again a few of the central combs should be broken out, to see that it is not infested by worms. *(Here, again, we see the wonderful advantages of movable-comb hives. Combs broken out of old box (common) hives were effectively lost.)*

[6] *Modern beekeepers certainly take a different approach to such weak colonies. These colonies just do not deserve such time-consuming care. They should be consolidated with other colonies. Langstroth was just too enamored of every colony, and just could not see it perish. By uniting colonies the joint heat and care of the brood will sometimes allow the colony to expand enough to make a division later.*

† Virgil speaks of clipping the wings of queens, to prevent them from escaping with a swarm. John Mills (1776) quotes the following from an account published of the sheep of Spain:—"The number of bee-hives kept in Spain is incredible. I am almost ashamed to give under my hand, that I knew a parish priest who had five-thousand hives. The bees suck all their honey from the aromatic flowers which enamel and perfume two-thirds of the sheep-walks. This priest cautiously seizes the queens in a small crape fly-catch, and then clips off their wings. He assured me that he never lost a swarm from the day of this discovery to the day he saw me, which was, I think, five years after."—p. 77.

LOSS OF THE QUEEN.

will greatly diminish the labor and perplexity of the bee-keeper. I have devised a way of doing this, so as to designate *the age of the queens*:—With a pair of scissors, let the wings, on one side, of a young queen be carefully cut off: when the hives are examined next year, let one of her two remaining wings be removed, and the last one the third year.[7]

The fertility of queens usually decreases after the second year, and before they die of old age the contents of their spermatheca sometimes become exhausted, and they lay only drone-eggs.[*] Unless, therefore, queens are unusually fertile, it will be safer to remove them after they have entered on their third year.[†]

A young queen, or a sealed royal cell, should be given to a colony, the second day after the old one is removed—for if they raise a queen from the egg, she may find nearly all the cells filled with honey or bee-bread, and the population greatly reduced.

Early in October—when some brood is usually found in every healthy stock, and when all the colonies should be examined, with reference to the coming Winter—if any are found to be queenless, they should he united to other stocks. If, however, the old queens were seasonably removed, and the stocks that raised young ones were properly attended to, few queenless colonies will be found in the Fall. At this season, or as soon as forage fails, such stocks [queenless] may

---

[7] *I use the method of alternating sides in clipping the queen's wings, along with paint. Left wings are cut off about half, or a third, in odd numbered years and right wings in even. The paint color is the 5-year code—Years ending in 0 and 5, Blue; 1 and 6, White; 2 and 7, Yellow; 3 and 8, Red; 4 and 9, Green. I add the clipping because sometimes the paint is lost and then I still know the age of a queen since I do not normally keep queens three years. Some beekeepers feel that clipping causes queens to be superseded; though I do not think there is any evidence for this fear. Usually when they paint or clip a queen the beekeeper then knows exactly when it is superseded. Otherwise, without clipping or marking, they think the queens are living a longer time when they are in fact being superseded without the beekeeper knowing it.*

[*] Posel says, that a queen that has suffered from hunger for 24 hours never recovers her wonted fertility. I shall show, in another place, that after recovering from severe cold, queens cease to lay worker-eggs.

[†] "Queens differ much as to the degree of their fertility. Those are best which deposit their eggs with uniform regularity, leaving no cells unsupplied—as the brood hatches at the same time on the same range of comb, which can be again supplied; the queen thus losing no time in searching for empty cells."—Dzierzon. In bee-life, as well as in human affairs, those who are systematic, ordinarily accomplish the most.

usually be detected by the incessant attempts of other colonies to rob them.

The neglect of a colony to expel its drones, when they are destroyed in other hives, is always a suspicions sign, and generally an indication that it has no queen. *Healthy stocks almost always destroy the drone, as soon as forage becomes scarce.* In the vicinity of Philadelphia, there were only a few days in June, 1858, when it did not rain, and in that month the drones were destroyed in most of the hives. When the weather became more propitious, others were bred to take their place. In seasons when the honey-harvest has been abundant and long protracted, I have known the drones to be retained, in Northern Massachusetts, until the 1st of November. If bees could gather honey and could swarm the whole year, the drones would probably die a natural death.

The importance of preventing the over-production of drones has been corroborated by the discovery of Mr. P. J. Mahan, that those *leaving* the hive have quite a large drop of honey in their stomach—while those *returning* from their pleasure excursions, having digested their dinners, are prepared for a new supply.*

"The drone," says quaint old Butler,[8] "is a gross, stingless bee, that spendeth his time in gluttony and idleness. For however he brave it, with his round velvet cap, his side gown; his full paunch, and his loud voice, yet is he but an idle companion, living by the sweat of others' brows. He worketh not at all, either at home or abroad, and yet

---

* Aristotle (History of Animals, Book IX, Chap. XL), speaks of the *irregular* and *thick* combs built by some stocks, and the superabundance of drones issuing from them. He notices, also, the destruction of the drones when bee-forage fails, and describes their excursions as follows:—"The drones, when they go abroad, rise into the air with a circular flight, as though to take violent exercise, and when they have taken enough, return home, and gorge themselves with honey." Columella says, that the proper time for removing the surplus honey is when the bees expel the drones.

(I have to comment here that Langstroth was certainly very well read. He probably had Latin and Greek training in Theology School, but still he either read the original or some other works where these quotes from the ancient Romans and Greeks were written. I also am continually astonished at how much these ancient beekeepers knew while working with box hives or other non movable-comb hives.)

[8] Langstroth does not properly identify this Butler, though he is probably referring to Charles Butler, who published a book in 1609, "The Feminine Monarchie," Butler was one of the first to recognize that the king bee was actually a queen.

spendeth as much as two laborers: you shall never find his maw without a drop of the purest nectar. In the heat of the day he flieth abroad, aloft and about, and that with no small noise, as though he would do some great act; but it is only for his pleasure, and then returns he presently to his cheer."

It has already been stated (p. 51), that the bee-keepers in Aristotle's time were in the habit of destroying the excess of drones.[9] They excluded them from the hive—when taking their accustomed airing—by contracting the entrance with a kind of basket work. Butler recommends a similar trap, which he calls a *"drone-pot."* The arrangement used in my hives to, prevent swarming will serve also to exclude the drones. Towards dark or early in the morning when clustered, for warmth, in the portico—they may be brushed into a vessel of water, and given to chickens, which will soon learn to devour them. In excluding them from hives having an unimpregnated queen the entrance must be adjusted to let her pass.

It is interesting to notice the actions of the drones when they are excluded from the hive. For a while they eagerly search for a wider entrance, or strive to force their balky bodies through the narrow gateway. Finding this to be in vain, they solicit honey from the workers, and when refreshed, renew their efforts for admission, expressing, all the while, with plaintive notes their deep sense of such a cruel exclusion. The beekeeper, however, is deaf to their entreaties; it is better for him that they should stay without, better for them—if they only knew it—to perish by his hands, than to be starved or butchered by the unfeeling workers. With movable-comb hives, pity and profit may be perfectly reconciled (p.51), by removing all excess of drone-comb from the breeding apartment.*

---

[9] *I think that beekeepers have always been overly concerned with the number of drones. Possibly, in the old common hives where they were not able to control the comb cell size by the foundation used, there might have been excess. In modern hives the numbers are rarely too many. If the beekeeper tries to keep all drones out the colony will just make more drone comb. Thus, it is better to allow some drones.*

* If a number of drones are confined in a small box, they give forth a strong odor: Swammerdam supposed that the queen was impregnated by this scent ('aura seminalis') of the drones.

(Drones most likely do have an odor or pheromone. Though in this case it may well that in a confined box the drones expelled feces and it is the odor of these feces that is

In the Summer of 1853, I discovered that after a queen is taken from a paper cone (p. 159), the bees will run in and out of it for a long time, thus proving that they recognize her peculiar scent. It is this odor which causes them to run inquiringly over our hands, after we have caught a queen, and over any spot where she alighted when her swarm came forth.

This scent of the queen was probably known in Aristotle's time, who says: When the bees swarm, if the king (queen) is lost, we are told that they all search for him, and follow him with their sagacious smell, until they find him." Wildman says: "The scent of her body is so attractive to them, that the slightest touch of her along any place or substance will attract the bees to it, and induce them to pursue any path she takes" [†]

The intelligent bee-keeper will readily perceive not only how the loss of queens may be remedied by the movable-comb hive, but how any operation which in other hives is performed with difficulty, if at all, is in this rendered easy and certain. No hive, however, can make the ignorant or negligent very successful, unless they live in a region where the climate is so propitious, and the honey resources so abundant, that bees will prosper in spite of mismanagement or neglect.

Those who have not the leisure or disposition to manage their own bees, may with my hives, entrust the care of them to competent persons. The business of the gardener seems naturally associated with

---

the "strong odor", since these feces do have a very strong odor to me. When you cage drones for use in instrumental insemination the drones must be "flown" in order to eliminate these feces prior to using the drones to collect semen.)

[†] Before becoming acquainted with these authors, I supposed myself to have made an original discovery. Mr. P. J. Mahan informs me that after handling the queen he has had bees several times alight upon his fingers, when he was a mile or more from his Apiary.

(Again, it would appear that Langstroth was aware of the queen odor or pheromone.)

that of the Apiarian and practical gardeners may find the management of bees, for their employers, quite a lucrative part of their profession. With but little trouble, they can make new colonies, remove the surplus honey, and on the approach of Winter prepare the bees to resist its rigors.

Preface to Chapter XII

If you do not have any idea what is going on in your old box hive then they can become weak through disease, queen failure, the lack of food or any of the problems that can confront a colony. Such was the case with the box hives before the movable-comb hive was invented. When some of these problems happen then the wax worms (wax moths) become much more of a problem. It is with this background that we see Langstroth's fixation about the wax moth. He thought the design of his entrance would eliminate the problem. He was wrong on that point, but was right about the fact that the beekeeper could now recognize the problems with the colony and correct it before the wax worms would consume the combs. They could even physically remove badly infested combs if the wax worms were overpowering the colony. That was not possible before the invention of the movable comb.

You will see in the section on diseases that most of our information on these bacteria was developed in the 20th century and so the diagnostic (bacterial) information is out of date—the management programs to correct or eliminate the problems are not.

CHAPTER XII.

THE BEE-MOTH, AND OTHER ENEMIES OF BEES.—DISEASES OF BEES.

THE Bee-Moth (*Tinea mellonella*) [*Galleria mellonella*][1] is mentioned by Aristotle, Virgil, Columella and other ancient authors, as one of the most formidable enemies of the honey-bee. Modern writers, almost without exception, have regarded it as the plague of their Apiaries; while in this country its ravages have been so fatal, that the majority of cultivators have abandoned bee-keeping in despair. Most of the contrivances devised against it have proved worthless, and not a few have aided its nefarious designs.

Having closely studied its habits, I am able to show how careful bee-keepers may protect their colonies from being ruined by its assaults. The careless will obtain a *"moth-proof"* hive only when the sluggard finds a *"weed-proof"* soil. Before stating how to circumvent the moth, its habits will be briefly described.

Swammerdam speaks of two species of the bee-moth (called in his time the *"bee-wolf"*), one much larger than the other. Linnaeus and Reaumur also describe two kinds—*Tinea cereana* and *Tinea mellonella*. Most writers suppose the former to be the male, and the latter the female of the same species. The following description is abridged from Dr. Harris' Report on the Insects of Massachusetts:

"Very few of the *Tinea* exceed or even equal it in size. In its adult state it is a winged moth, or miller, measuring, from the head to the tip of the closed wings, from five-eighths to three-quarters of an inch in length, and its wings expand from one inch and one-tenth to one inch

---

[1] *Taxonomists have now placed the greater wax moth into its own Genus separated from the "Tinea" meal-moths. The lesser wax moth (Achroia grisella) has also been placed into a different Genus. Most of the related species are stored-products pests that feed on cereals and milled grains. The wax moth evolved from the stored-products pests and instead of feeding on grains it uses pollen and othe debris of the hive to grow and flourish. It can digest wax, but needs pollen to develop.*

and four-tenths. The fore-wings shut together flatly on the top of the back, slope steeply downwards at the sides, and are turned up at the end somewhat like the tail of a fowl. The female is much larger than the male, and much darker colored. There are two broods of these insects in the course of the year. Some winged moths of the first brood begin to appear towards the end of April or early in May—earlier or later, according to climate and season. Those of the second brood are most abundant in August; but some may be found between periods, and even much later."

No writer with whom I am acquainted has given such an exact description of the difference between the sexes, that they can always be readily distinguished. The beautiful wood-cuts of the moths, larva, and cocoons, which I present to my readers, were drawn from nature, by Mr. M. M. Tidd, of Boston, Mass., and engraved by Mr. D. T. Smith, of the same city. A large number of specimens were furnished to Mr. Tidd, and great accuracy has been secured. He seems first to have noticed that the *tongue of the female* projects so as to resemble a beak, while that of the male is very short.·

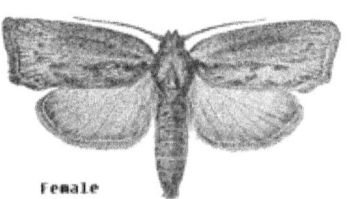
Female

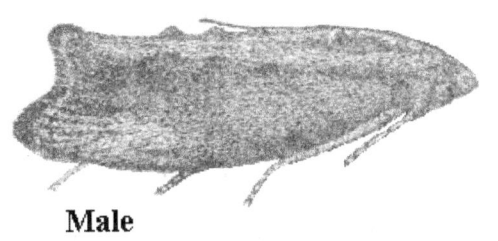
Male

While some males are larger than some females, and some females much lighter colored than the average of males, and occasionally some males as dark as the darkest females, *the peculiarity of the tongue of the female is so marked that she may always be distinguished at a glance.*[2]

· Dr. Harris speaks of the tongue of the moth as "very short, and hardly visible." This is true only of that of the male.

[2] *The rear margin of the wings of the male wax-moth, when at rest, are more curved or scalloped than the female. I think this distinguishing mark is easier to see.*

The tongue of the female is double, and the line of separation is shown in the figure in which she is represented as lying on her back. Both male and female were accurately copied from specimens of the average size and form.

In this sketch, an under-sized male is represented.· His color was so dark that, but for the tongue, he might easily have been mistaken for a female of a different and much smaller species.†

Female

Small Male

These insects are seldom seen on the wing, unless started from their lurking places about the hives, until towards dark. On cloudy days, however, the female may be noticed endeavoring, before sunset, to gain an entrance into the hives. "If disturbed in the daytime," says Dr. Harris, "they open their wings a little, and spring or glide swiftly away, so that it is very difficult to seize or to hold them.‡ In the evening, they take wing, when the bees are at rest, and hover around the hive, till, having found the door, they go in and lay their eggs." "If the approach to the Apiary," says Bevan, "be

---

· The legs are shown in this figure. In the sitting position, they are usually concealed, as in the preceding figures. These drawings appear to better advantage in Plate XIII.

† As all the specimens submitted to Mr. Tidd were taken from two adjoining hives, very late in the Fall, it is possible that the observations at some other season, and in different localities, may confirm the view of those who believe that there are two species.

(The male and female of the wax moth are different, but there are also two species of wax moth—the lesser and greater wax moth. Usually there is only one found in a certain area or locality, but I have seen both species in the same hive.)

Mr. Tidd, while experimenting to ascertain the sexes, found that a female, as soon as she was pinned fast, thrust out her ovipositor, which works with a telescopic motion, and began to feel for some crevice in which to deposit her eggs. Some cracks being made with a small penknife in the wood to which she was fastened, she at once proceeded to fill them with eggs. Her abdomen was then cut off, and the egg-laying process continued as before, while the rest of the body leisurely walked away! The abdomen was now dissected, so as to show the ducts of the ovaries, and, even in this mutilated condition, she thrust out her ovipositor, all the while carefully seeding for appropriate crevices in which to deposit her eggs! I have repeated, with similar results, these experiments, so suggestive of curious speculations as to insect volition.

‡ They are surprisingly agile, both on foot and on the wing, the motions of a bee being very slow, in comparison. "They are," says Reaumur, "the most nimble-footed creatures that I know."

observed of a moonlight evening, the moths will he found flying or running round the hives, waiting an opportunity to enter, whilst the bees that have to guard the entrances against their intrusion, will be seen acting as vigilant sentinels, performing continual rounds near this important post, extending their antennas to the utmost, and moving them to the right and left alternately. Woe to the unfortunate moth that comes in their reach!" "It is curious," says Huber, "to observe how artfully the moth knows how to profit by the disadvantage of the bees, which require much light for seeing objects, and the precautions taken by the latter in reconnoitering and expelling so dangerous an enemy."

"Those that are prevented from getting within the hive, lay their eggs in cracks on the outside; and the little worm-like caterpillars hatched therefrom, easily creep into the hive through the cracks, or gnaw a passage for themselves under the edges of it."·—DR. HARRIS.

"As soon as hatched, the worm encloses itself in a case of white silk, which it spins around its body; at first it is like a mere thread, but gradually increases in size, and, during its growth, feeds upon the cells around it, for which purpose it has only to put forth its head, and find its wants supplied. It devours its food with great avidity, and, consequently, increases so much in bulk, that its gallery soon becomes too short and narrow, and the creature is obliged to thrust itself forward and lengthen the gallery, as well to obtain more room as to procure an additional supply of food. Its augmented size exposing it to attacks flow surrounding foes, the wary insect fortifies its new abode with additional strength and thickness, by blending with the filaments of its silken covering a mixture of wax and its own excrement, for the

---

· If movable-bottom-boards are used, it will be next to impossible to prevent the moth from laying her eggs between them and the edges of the hives. The smallest opening will enable her to thrust in her ovipositor, and place her eggs when her progeny will find an easy admission to the hive.

(There will be cracks in the hive, even if just between supers, and the small first instar larvae can get into the hive. These small larvae can very rapidly crawl into the comb and begin feeding. The larvae are able to exist in the hive until they become large enough for the bees to notice them and then they are removed. If the colony is not very populous sometimes the pressure from too many larvae will overwhelm the bees. This happens more often in the south, or tropics, where more moths survive the winter period, and thus create more pressure on the colonies by laying more eggs.)

external barrier of a new gallery, the *interior* and partitions of which are lined with a smooth surface of white silk, which admits the occasional movements of the insect, without injury to its delicate texture. In performing these

operations, the insect might be expected to meet with opposition from the bees, and to be gradually rendered more assailable as it advanced in age. It never, however, exposes any part but its head and neck, both of which are covered with stout helmets, or scales, impenetrable to the sting of a bee, as is the composition of the galleries that surround it."—BEVAN.

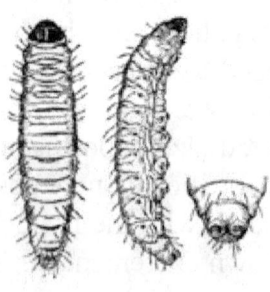

The worm is here given of full size, and with all its peculiarities carefully represented. The scaly head is shown in one of the worms; while the three pairs of claw-like fore legs, and the five pairs of hind ones, which are suckers, are clearly delineated.[3] The tail is also furnished with two of these suckers. The breathing holes are seen on the back.

• This representation of the web, or gallery of the worm, was copied from Swammerdam.
[3] *Like most Lepidoptera larvae (moths and butterflies), these caterpillars have five pairs of pro-legs. These fleshy feet have hooks, known as crochets, but no suckers. The wax worm goes through seven instars, or growth stages, as a larva and then often gouges out a depression in the wood of the frame or hive body where it spins its cocoon to pupate.*

Wax is the chief food of these worms.· When obliged to steal their living among a strong stock of bees, they seldom fare well enough to reach the size which they attain when rioting at pleasure among the full combs of a discouraged population. In about three weeks, the larvae stop eating, and seek a suitable place for encasing themselves in their silky shroud. In hives where they reign unmolested, almost any place will answer their purpose, and they often pile their cocoons one on another, or join them together in long rows. They sometimes occupy the empty combs, so that their cocoons resemble the capping of the honey-cells. In Plate XlX., Fig. 56, Mr. Tidd has given a drawing, accurate in size and form, of a curious instance of this kind. The black spots, resembling grains of gunpowder, are the excrements of the worms. In hives strongly guarded by healthy bees, many a worm, while prying about to find a snug hiding place, is seized by the nape of the neck, and served with an instant writ of ejectment. If a hive is thoroughly made, it runs a dangerous gauntlet, as it passes, in search of some crevice, through the ranks of its enraged foes. Its motions, however, are exceedingly quick, and it is full of cunning devices, being able to crawl backwards, to twist round on itself, to curl up almost into a knot, and to flatten itself out like a pancake. If obliged to leave the hive, it gets under some board or concealed crack, spins its cocoon, and patiently awaits its transformation. In most hives, it readily finds a crack into which it can creep, or a small space between the movable bottom-board and the edges of the hive. It can pass through a *very* small crevice, and as soon as safe from the bees, it will begin to enlarge its cramped tenement, by gnawing into the solid wood. The time required for the larvae to break forth into winged insects, varies with the temperature to which they are exposed, and the season of the

---

· "Larvae fed exclusively on *pure* wax will die, wax being a non-nitrogenous substance, and not furnishing the aliment required for their perfect development."—DÖNHOFF.

This statement agrees with the fact, that the larvae prefer the brood-combs, and that the combs of an old stock are more liable to be devoured than those of a new one.

(The larvae need pollen for development, though Galleria *is one of the few animals that can digest wax. But wax alone does not provide the proteins needed for growth. Sometimes you will see comb honey that has been "ruined" by a small wax worm larvae feeding across the surface, and making a track of open wax. These larvae don't survive but will destroy the salability of the comb. Freezing the comb sections for a few days, which should kill the eggs or small larvae, can control this action.)*

year when they spin their cocoons.· I have known them to spin and hatch in ten or eleven days; and they often spin so late in the Fall, as not to emerge until the ensuing Spring.

The male usually keeps away from the hive, but the female seeks in every way to gain an entrance. If the stock is weak and discouraged, she lays her eggs † among the combs, or inserts them in the cor-

---

· In November (1858), I procured a large number of cocoons for winter observations. From many of them, the moths quickly emerged. In others the larvae slowly changed into pupae or chrysalids; while, in others still, after being exposed for more than two months to a summer temperature, they remained in the worm state. A few were exposed for six weeks to a uniform temperature of over 80°, and only one passed into the winged moth. Some, after being taken out of their cocoons six times, would envelop themselves in a new shroud.

Dr. Dönhoff says, that the larvae become motionless at a temperature of from 38° to 40°, and entirely torpid at a lower temperature. A number which he left all Winter in his summer-house, revived in the Spring, and passed through their natural changes. He appears to have been more successful than myself in inducing them to develop in Winter, by artificial heat; but this may be owing to the fact that he experimented with larvae which greedily ate the food given to them, *and not as I did, with worms which had spun their cocoons.* Further experiments are needed, in order to determine whether dilatory development is peculiar to those reaching maturity late in the Fall, or is caused by the *sudden check* given by cold weather.

"If, when the thermometer stood at 10°, I dissected a chrysalis, it was not frozen, but congealed immediately afterwards. This shows that, at so low a temperature, the vital force is sufficient to resist frost. In the hive, the chrysalides and larvae, in various stages of development, pass the Winter in a state of torpor, in corners and crevices, and among the waste on the bottom-boards. In March or April, they revive, and the bees of strong colonies commence operations for dislodging them."—DÖNOFF.

Some larvae which I exposed to a temperature of 6° below zero, froze solid, and never revived. Others, after remaining for 8 hours in a temperature of about 12° seemed, after reviving, to remain for weeks in a crippled condition.

(Wax worms are not particularly cold hardy, and they do not survive in the duff of the bottom boards during winter, at least in the north. If they are to survive very cold weather, they must be more or less in the cluster area of the hive. The caterpillars are best controlled in stored supers, in the north, by the low over-winter temperatures.)

† "The eggs of the bee-moth (see Plate XIII., Fig. 44) are perfectly round, and very small, being only about one-eighth of a line in diameter. In the ducts of the ovarium, they are ranged together in the form of a rosary. They are not developed consecutively, like those of the queen bee, but are found in the ducts, fully and perfectly formed, a few days after the female moth emerges from the cocoon. She deposits them, usually, in little clusters on the combs. If we wish to witness the discharge of the eggs, it is only necessary to seize a female moth, two or three days old, with finger and thumb, by the head—she will instantly protrude her ovipositor, and the eggs may then be distinctly seen passing along through the semi-transparent duct. (see Plate XII., Fig. 46, C.)

ners or crevices, or among the refuse wax and bee-bread on the bottom board, where her progeny can be concealed and nourished till they am able to reach the combs.

In Plate XX., Fig. 57, Mr. Tidd has faithfully delineated, and Mr. Smith skillfully engraved, the black mass of tangled webs, cocoons, excrements, and perforated combs, which may he found in a hive where the worms have completed their work of destruction.

The entrance of the moth into a hive and the ravages committed by her progeny, forcibly illustrate the havoc which vice often makes when admitted to prey unchecked on the precious treasures of the human heart. Only some tiny eggs are deposited by the insidious moth, which give birth to very innocent-looking worms; but let them once get the control, and the fragrance [*] of the honied dome is soon corrupted, the hum of happy industry stilled, and everything useful and beautiful ruthlessly destroyed.

The honey-bee is not a native of the New World, and, when brought

---

"Last Summer I reared a bee-moth larva in a small box. It spun a cocoon, from which issued a female moth. Holding her by the head, I allowed her to deposit eggs on a piece of honey-comb. Three weeks afterwards, I examined the comb, and found on it some web and two larvae. The eggs were all shriveled and dried up, except a few which were perforated, and from which, I suppose, the larvae emerged. This appears to be a case of true parthenogenesis in the bee moth."—*Translated from* Dr. DÖNOFF *by* S. WAGNER.

As among hundreds of specimens furnished to Mr. Tidd very few males were noticed, I conjectured that the eggs of these females would hatch without impregnation, and took measures to have Dr. Joseph Leidy investigate the subject. It seems, however, that in this matter, our German brethren have the priority.

(I suspect that neither of these researchers realized that it doesn't take many males to produce viable offspring. There is no evidence that wax moths reproduce by parthenogenesis. If they noticed only moths that are trying to enter bee hives these would all be female.)

[*] The odor of the moth and larvae is very offensive.

here, was called by the Indians the white man's fly.* Longfellow, in his "Song of Hiawatha," in describing the advent of the European to the New World, makes his Indian Warrior say of the bee and the white clover:–

> "Wheresoe'er, they move, before them
> Swarms the stinging fly, the Ahmo,
> Swarms the bee, the honey-maker;
> Wheresoe'er they tread, beneath them
> Springs a flower unknown among us,
> Springs the White Man's Foot in blossom."

As the bees flourished for years undisturbed by the moth, it seems probable that it was not brought over in the first hives, but at a much later period. In whatever way it was introduced, it has so multiplied in our propitious climate of hot summers, that few districts are now exempt from its ravages.

Fifty years ago our markets were proportionally better supplied with honey than they now are, and large tubs filled with snow-white combs were a common sight.

Many Apiarians contend that newly-settled countries are most favorable to the bee; and an old German adage runs thus:—

---

* "It is surprising in what countless swarms the bees have overspread the far West, within but a moderate number of years. The Indians consider them the harbingers of the white man, as the buffalo is of the red man, and say that, in proportion as the bee advances, the Indian and the buffalo retire..... They have been the heralds of civilization, steadily proceeding it as it advances from the Atlantic borders; and some of the ancient settlers of the West pretend to give the very year when the honey-bee first crossed the Mississippi. At present it swarms myriads in the noble groves and forests that skirt and intersect the prairies, and extend along the alluvial bottoms of the rivers. It seems to me as if these beautiful regions answer literally to the description of the land of the promise—'a land flowing with milk and honey;' for the rich pasturage of the prairies is calculated to sustain herds of cattle as countless as the sands upon the sea-shore, while the flowers with which they are enameled render them a very paradise for the nectar-seeking bee"—WASHINGTON IRVING, *Tour on the Prairies*, Chap. IX.

> "Bells' ding dong,
> And choral song,
> Deter the bee
> From industry:
> But hoot of owl,
> And 'wolf's long howl,'
> Incite to moil
> And steady toil."

Others affirm that our colonies are too numerous to find sufficient food. That neither of these reasons account for the change, will be subsequently shown. Others lay all the blame on the moth, and others still, on our departure from the old-fashioned mode of keeping bees.

It is undoubtedly true that the moth so super-abounds in many districts, that no profit can be derived from managing bees in the simple way which was once so successful. Often the old bee-keeper, after hiving his swarms, never looked at them again until the Fall, when all the colonies which had too few bees, or were too light to survive the Winter, were condemned to the brimstone pit. Some of the heaviest were also killed for the sake of their honey, and the *very best* were reserved for stock hives.

In a newly-settled country, where weeds are almost unknown, the farmer who plants his corn and "lets it alone," may often harvest a remunerative crop. If, in process of time, as the weeds increase, he continues to plough and plant in the "good old way," he will only be laughed at for complaining that the pestiferous weeds have caused his corn to "run out" And yet, with equal folly, many bee-keepers do not understand why plans which answered when moths were unknown or were very scarce, cannot be made to succeed at the present time.

If the old plans had been rigidly adhered to, the ravages of the moth, destructive as they must have been, would never have been as great as they now are. *The use of patent hives has contributed to fill the land with the devouring pest*. Ever since their introduction, the notion has almost universally prevailed that stocks must not, under any circumstances, be voluntarily destroyed and hence, thousands of colonies, which, under the old system, were mercifully killed, are now left to perish by slow starvation, while thousands more are so feeble in the

Spring that they serve only to breed a host of moths to be the pest of the Apiary.

The truth is, that improved hives, without an improved system of management, have done, on the whole, more harm than good. In no country have they been so extensively used as in our own, and no where has the moth so completely gained the ascendancy. Just so far as they have discourage ordinary bee-keepers from the old plan of "taking up" their weak swarms in the Fall, just so far have they extended "aid and comfort" to the moth. Some of them might, unquestionably, be so managed as, in ordinary cases, to protect the bees against the moth but no hive which does not give the control of the combs, can be relied on for all emergencies. As for many of the complicated contrivances, which have been devised by men ignorant of the first principles of bee-keeping, and the "swindle-traps" of sharpers who, to fill their own pockets would be glad to kill all the bees the world, they not only afford no more security against the moth, than the old box-hive, but are full of fixtures, which serve no end but to annoy the bees and multiply lurking-places for moths and worms. The more they are used, the worse the condition of the bees; just as the more a man uses the nostrums of the lying quack, the farther he gets from health.*

---

* An intelligent man informed me that he paid ten dollars to a "*bee-quack*" professing to have an infallible secret for protecting bees against the moth. After parting with his money, and learning that this secret consisted in "always keeping strong stocks," he felt that he had been as grossly imposed upon, as if, after paying a large sum for an infallible life-preserving secret, he had been turned off with the truism that to live forever, one must keep well!

While freely admitting that the old plan of killing the bees has, in the hands of the ignorant, met with the best success, I am persuaded that a more humane and enlightened system can be made much more profitable. The use of movable frames permitting, as they do, the weakest stocks to be strengthened or united to others, will, I trust, in due time, introduce the happy era when the following epitaph, taken from a German work, might properly be placed over every pit of brimstoned bees :[*]

HERE RESTS
CUT OFF FROM USEFUL LABOR,
A COLONY OF
INDUSTRIOUS BEES,
BASELY MURDERED
BY ITS
UNGRATEFUL AND IGNORANT
OWNER

To the epitaph should be appended Thompson's verses:

"Ah, see, where robbed and murdered in that pit,
Lies the still heaving hive! at evening snatched,
Beneath the cloud of guilt-concealing night,
And fixed o'er sulphur! while, not dreaming ill,
The happy people, in their waxen cells
Sat tending public cares.
Sudden, the dark, oppressive steam ascends,
And, used to milder scents, the tender race,
By thousands, tumble from their honied dome
Into a gulf of blue sulphureous flame!"

The following letter, on the first appearance of the bee-moth in this country, from Dr. Jared P. Kirtland, of Cleveland, Ohio, who is so

---

[*] Killing bees for their honey was, unquestionably, an invention of the dark ages, when the human family had lost—in Apiarian pursuits, as well as in other things—the skill of former ages. In the times of Aristotle, Varro, Columella, and Pliny, such a barbarous practice did not exist. The old cultivators took only what their bees could spare, killing no stocks, except such as were feeble or diseased.

(It is probably difficult to determine if the practice of killing half the hives each year was a dark age invention or when beekeepers had more hives which enabled them to collect more honey by killing the colony. The practice may also have needed a successful poison with which to kill the bees without ruining the honey—sulfur.)

widely known for his interest in Horticultural and Apiarian pursuits, will be read with great interest:

"Cleveland, Feb. 19th, 1859.

"DEAR SIR: —Until 1805, the honey-bee flourished in the United States. At the commencement of the present century, a majority of the farmers and mechanics in the State of Connecticut cultivated the bee. Few, if any, unfavorable contingencies interfered with that pursuit; the simplest form of box-hives was usually employed, though, occasionally a *hollow gum*, and, in a few instances, the conical straw *skep* supplied their place.

"In Autumn, the weak colonies, and such of the old as were depreciating in value, were destroyed by fire and brimstone. The honey thus obtained was sufficiently abundant to satisfy the demand; hence, in those days, caps, drawers and side-boxes, for robbing bees, were not employed.

"During the Spring of the year 1806, I read an article, in the *Boston Patriot*, describing the miller and worm, and their depredations, and representing them as of recent appearance in the vicinity of that city. A few months subsequently, a neighbor informed me that they were depredating extensively on his colonies; and within two years of that time, four-fifths of all the Apiaries in that vicinity were abandoned.[*]

---

[*] Judge Fishback, of Batavia, Ohio, says that the ravages of the moth, in his Apiary, were much more destructive the second season after its appearance, than at any subsequent period. I can only account for this, by supposing that, at first, the bees were ignorant of its nature, and took no special precautions to prevent it for entering their hives. In Europe, where it has been well known for more than two thousand years, its ravages have never been of such a wholesale character. As both worms and moth have a peculiar smell, the bees would soon learn to repel from their hives, a moth smelling so much like the worms that were devouring their combs.

The bees can *learn* to defend themselves against *new* enemies, is proved by the facts related by Huber, of their narrowing their entrances with propolis to keep out the large death-head moth (*Sphinx atropos*), a single one of which can swallow a tablespoonful of honey.

An Aparian, from Ohio, sent me some honey-eating moths, much larger than the bee-moth, which entered weak hives and gorged themselves with honey.

Fig. 53   Plate XVII

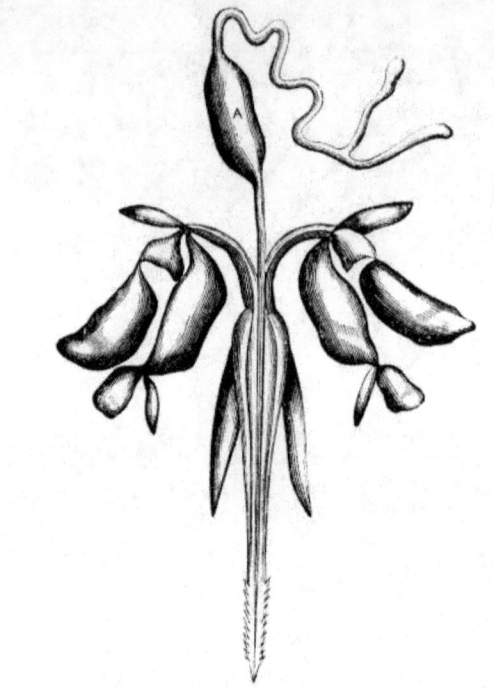

Fig. 54

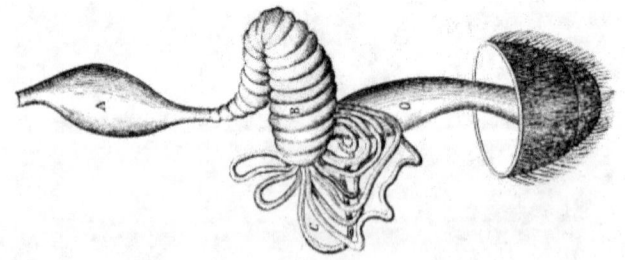

"Since that period, a succession of patent hives, whose originators were ignorant of the habits of the moth, has appeared as its auxiliaries, and the two combined, have nearly exterminated the bee from that section of the country. The efforts of a few individuals, of more than usual perseverance and ingenuity, were occasionally attended with limited success.

"In the Summer of 1810, I resided in the county of Trumbull, Ohio. The moth had not reached this part of the country, and bee-culture was extensively pursued, and with a success I have never witnessed elsewhere. The rich German farmers were on a strife to excel each other in the number of their colonies. Two or three hundred they frequently attained.[4]

"In 1818, I again visited that county, and permanently located there in 1823, and at both periods found that pursuit still prospering. In August, 1828, while visiting a sick family in Mercer Co., Pa., I observed that a large Apiary was suffering severely from the attacks of the worm. The proprietor informed me that it had made its appearance for the first time the present season. Within another year, it spread over all of Northern Ohio, and in the Winter of 1831-2, I learned, from members of the Legislature, that it had reached every part of our State. Similar results followed its progress here, as in the New England States.

"Until the introduction of your system of movable frames, no successful means of counteracting its ravages were devised. I am happy to say that, by the aid of your hives, I have not the least difficulty in meeting it.

"With great respect, yours, &c.,
Jared P. Kirtland."

"Rev. L. L. Langstroth.

---

[4] *Such a large number of colonies in one location give rise to several problems. The first is that the colonies do not make a very good average gain. Secondly, when you have that many colonies in one location the chances of robbing are certainly increased. It may be that not everyone had all their colonies in one place, but the means of transport were not very good and thus you see pictures from that era of apiaries of one to two hundred colonies. The average today is close to 25 colonies per apiary.*

Almost anything hollow will often, for a series of years, be successfully tenanted by bees. To see hives, with large, open cracks, whose owners are ignorant and careless, bidding defiance to the moth, may, at first sight, impair confidence in the value of any precautions. While stocks often flourish in such log-cabin hives, others, in costly "Bee-Palaces," are frequently devoured by the worms—their owner, with all the newest devices in the Apiarian line, being unable to protect them against their enemies, or to explain why some colonies, like the children of the poor, appear almost to thrive upon neglect, while others, like the offspring of the rich, are feeble, apparently in exact proportion to the care lavished on them.[*]

I shall now explain why some stocks flourish in spite of neglect, while others, most cared for, fall a prey to the moth, and shall show how, in suitable hives, and with proper precautions, the moth may be kept from seriously annoying the bees.

A feeble colony being unable to cover its combs, they are often filled with the eggs of the moth, and, frequently, their owner becomes aware of their condition only when their ruin is completed. But how, can the novice know when a stock, in a common hive, is seriously[†] infested with these all-devouring worms? The discouraged aspect of the bees plainly indicates that there is trouble of some kind within, and the bottom-board will be covered with pieces of bee-bread mixed with the *excrement* of the worms, which looks like grains of *gunpowder*.[‡]

Early in the Spring, before the stocks become populous, the bees should be driven up among their combs by smoke, and the bottom-boards cleansed (p. 221). It too frequently happens that, in the com-

---

[*] It is very common to hear bee-keepers speak of having "good luck," or "bad luck," with their bees; and, as bees are managed, success or failure often seems to depend almost entirely upon what is called "luck."

[†] Inexperienced bee-keepers, who imagine that a colony is nearly ruined when they find a few worms, should remember that almost every old stock, however strong or healthy, has some of these enemies lurking about its premises.

[‡] When bees in the Spring prepare their cells for brood, the bottom-board is often covered with small pieces of comb and bee-bread; but if these are not mixed with the black excrement, they are proofs of industry, instead of signs of ruin.

(Since wax moths are related to (or evolved from) the meal moths it is not surprising that the larvae thrive on the debris on the bottom boards. Wax worms will often end up in pollen traps under colonies, as well.)

mon hives, nothing can he effectually done, even when the bee-keeper is aware of the plague within. With movable frames, however, the combs, and all parts of the hives, may be carefully cleansed, and if a stock is weak or queenless, the proper remedies may be easily applied. If a feeble stock cannot be strengthened so as to protect its empty combs, they may be taken away until the bees are numerous enough to need them.

If the bee-moth were so constituted as to require but a small amount of heat for its full development, it would become exceedingly numerous early in the Spring, and might easily enter the hives and deposit its eggs where it pleases: for at this season, not only is there no guard maintained by the bees at night, but large portions of their comb are quite unprotected. How does every fact in the history of the bee, when properly investigated, point with unerring certainty to the wisdom of Him who made it!

Combs having no brood, may be smoked with the fumes of burning sulphur, to kill the eggs or worms of the moth. If kept from the bees, they should be carefully protected, in a dry place, from the moth, and examined: occasionally, to be smoked again if any worms are found.[5]

Directions have been given on page 140 for preventing common hives from swarming so often that they cannot protect their empty combs. If not prevented from over-swarming, in the movable-comb hives, by methods which have been so fully described, some of the combs of the mother-stock may be given to the after-swarms, instead of being left where they may be attacked by the moth.

The most fruitful cause of the ravages of the moth still remains to be described. *If a colony becomes hopelessly queenless, it must, unless otherwise destroyed, inevitably fall a prey to the bee-moth.* By watching, in glass hives, the proceedings of colonies purposely made

---

[5] *The control of the wax moth has become difficult because of restrictions on the use of various chemicals. In the north, where it gets cold, storing empty supers is relatively easy. The only period that needs protection is after the honey is extracted and before it becomes cold in the late Fall. Some beekeepers freeze the supers to kill any larvae and then store the supers in a moth-tight place. Once cold weather sets in there will be little more problem. In the Spring the moths are not very plentiful until the time supers go back onto the colonies.*

In warmer climates storing supers in screen cages exposed to sunlight seems to help control the moths.

queenless, I have ascertained that they make little or no resistance to her entrance and allow her to lay her eggs where she pleases. The worms, after hatching, appear to have their own way, and are even more at home than the dispirited bees.[*]

How worthless, then, to a queenless colony, are all the traps and other devices which, of late years, have been so much relied upon. Any passage which admits a bee is large enough for the moth, and if a single female enters such a hive, she will lay eggs enough to destroy it, however strong. Under a low estimate, she would lay, at least, two hundred eggs in the hive, and the second generation will count by thousands, while those of the third will exceed a million.[†]

---

[*] The fact that queenless stocks do not oppose any effectual resistance to the moths or worms—a fact which I one thought to be a discovery of my own—has for a long time been well known to the Germans. Mr. Wagner informs me "that the best treatises, for many years, speak of this as a settled fact, so that it has become an axiom that, if a colony is overpowered by robber-bees, its owner is not entitled to compensation, *as it was, in all likelihood, queenless, and would certainly have been destroyed by the moth.*

My attention has been recently called to an article in the *Ohio Cultivator* for 1849 page 185, by Micajah T. Johnson, in which, after detailing some experiment, he says:—"One thing is certain—if bees, from any cause, should lose their queen, and not have the means in their power of raising another, the miller and the worms soon take possession. I believe no hive is destroyed by worms while an efficient queen remains in it."

This seems to be the earliest published notice of this important fact by any American observer.

(In the tropics, I have seen small but queen-right colonies overtaken by wax worms. The colonies have to be strong enough to keep constantly removing the larvae. Once the wax worms get numerically ahead they grow very large and overrun the hive.)

[†] This power of rapid increase accounts for Judge Fishbacks's and Dr. Kirtland's facts respecting the rapid dissemination of the moth.

(Such estimates always assume that every larvae and moth lives to reproduce, which just never happens.)

Not only do the bees of a hopelessly queenless hive make no effectual opposition to the bee-moth, but, by their forlorn condition, they positively invite her attacks. She appears to have an instinctive knowledge of their condition, and no art of man can ever keep her out. She will pass by other colonies to get at a queenless one, as if aware that she will find in it the best conditions for the development of her young; and thus the strongest colonies, after losing their queens, are frequently devoured by the worms, while small ones, standing by their side, escape unharmed.[6]

It is certain that a queenless hive seldom maintains a guard at the entrance, and does not fill the air with the pleasant voice of happy industry. Even to our dull ears, the difference between the hum of a prosperous hive and the unhappy note of a despairing one is often sufficiently obvious; may it not be even more so to the acute senses of the provident mother-moth?

Her unerring sagacity resembles the instinct by which birds that prey upon carrion, single out from the herd a diseased animal, hovering over its head with their dismal croakings, or sitting in ill-omened flocks on the surrounding trees, watching it as its life ebbs away, and snapping their blood-thirsty beaks, impatient to tear out its eyes, just glazing in death, and to banquet on its flesh, still warm with the blood of life. Let any fatal accident befall an animal, and how soon will you see them,–

"First a speck and then a Vulture,"

speeding, from all quarters of the heavens, their eager flight to their destined prey, when only a short time before not one could be perceived.

The common hives not only furnish no reliable remedy for the loss of the queen, but, in many cases, their owner cannot be sure that his bees are queenless until their destruction is certain; while not infrequently, after an experience of years, he does not believe that there is such a thing as a queen-bee! Its the Chapter on the Loss of the Queen,

---

[6] *Langstroth gets fanciful here, as he earlier commented that all old stocks have worms in them (Note 2, pg. 242). However, an egg-laying moth will pass empty supers in preference to a living colony, so the female moth can detect different odors.*

full directions have been given for protecting colonies in movable-comb hives, from a calamity which, more than all others the want of food * excepted—exposes them to destruction. When a colony becomes hopelessly queenless, its destruction is certain. Even should the bees retain their wonted zeal in gathering stores and defending themselves against the moth, they must as certainly perish (p. 58) as a carcass must decay, even if it is not assailed by filthy flies and ravenous worms. Occasionally, after the death of the bees, large stores of honey are found its their hives. Such instances, however, though once not uncommon, are now rare; for a motherless hive is almost always assaulted by stronger stocks, which, seeming to have an instinctive knowledge of its orphanage, hasten to take possession of its spoils; or, if it escape the Scylla [7] of these pitiless plunderers, it is dashed upon a more merciless Charybdis, when the miscreant moths find out its destitution. Every year, multitudes of hives are bereft of their queens, most of which are either robbed by other bees or sacked by the moth, or both robbed and sacked, while their owner imputes all the mischief to something else than the real cause.

To one acquainted with the habits of the moth, the bee-keeper who is constantly lamenting its ravages, seems almost as much deluded as a farmer would be who after diligently searching for his missing cow, and finding her nearly devoured by carrion worms, should denounce these worthy scavengers as the primary cause of her untimely end.

The bee-moth is the only insect known to feed on wax. It has, for thousands of years, supported itself on the labors of the bee, and there is no reason to suppose that it will ever become exterminated. In a state of nature, a queenless hive or one whose inmates have died, being of no further account, the mission of the moth is to gather up its fragments that nothing may be lost.*

---

* Colonies which are almost starved become almost as indifferent to the attacks of the moth as those which have no queen.

[7] *Scylla, a dangerous rock on the Italian side of the Strait of Messina, facing Charybdis, a whirlpool on the Sicillian side, both personified in classical mythology as female monsters; between Scylla and Charybdis, between two evils or dangers, either one of which can be safely avoided only by risking the other. In modern times we would say, "Between a rock and a hard place," maybe this is the origin of that saying.*

* In the times of Aristotle and Columella, the ravages of the moth were kept under by a judicious system of management. It may be seriously questioned whether its *extermina-*

From these remarks, the bee-keeper will see the means on which he must rely, to protect his colonies from the moth. Knowing that strong stocks which have a fertile queen, can take care of themselves in almost any kind of hive, he should do all that he can to keep them in this condition. They will thus do more to defend themselves than if he devoted the whole of his time to fighting the moth.

It is hardly necessary, after the preceding remarks, to say much upon the various contrivances to which so many resort, as a safeguard against the bee-moth. The idea that gauze-wire doors, to be shut at dusk and opened again at morning, can exclude the moth, will not weigh much with those who have seen them on the wing, in dull weather, long before the bees have ceased their work. Even if they could be excluded by such a contrivance, it would require, on the part of those using it, a regularity almost akin to that of the heavenly bodies.

An ingenious device has been employed for dispensing with such close supervision, by governing the entrances of all the hives by a long lever-like hen-roost, so that they may be regularly closed by the crowing and cackling tribe when they go to bed at night, and opened again when they fly from their perch to greet the merry morn. Alas! that so much skill should be all in vain! Some chickens are sleepy, and wish to retire before the bees have completed their work, while others, from ill-health or laziness, have no taste for early rising, and sit moping on their roost, long after the cheerful sun has purpled the glowing east. Even if this device could entirely exclude the moth, it could not save a colony which has lost its queen. The truth is, that most of the contrivances on which we are instructed to rely, are equivalent to the lock put upon the stable door after the horse has

---

tion in any Apiary would be desirable, unless it could be destroyed *everywhere else*. The bees would soon forget all about it, and if again exposed to its attacks, similar results might follow to those described on p. 240; for unless the bees know how to protect themselves, no art of man can save them, as is clearly seen in queenless hives, where they will not attend to their combs. Aristotle says, "that good bees expel the moths and worms, but others, from slothfulness, *neglect their combs*, which then perish." His *bad* bees were doubtless those which had no fertile queen.

(Langstroth does not understand the difference between learning and instinct. The bees will remove any intruder from their hive, and they are not likely to forget that behavior pattern unless pests and intruders are removed from ever entering all colonies.)

been stolen; or, to attempts to banish the chill of death by warm covering, or artificial heat.[8]

Let me not be understood as asserting that there are no means of protecting the common hives from the ravages of the bee-moth. *If bee-keepers will be careful to place their hives where the young queens are not in danger of being lost (p. 214), they will lose comparatively few of their colonies.* The knowledge of this fact will enable the Apiarian to contend successfully against the moth, let him use what hive he will. He will, undoubtedly, lose many colonies which have become queenless, from other causes than the close proximity of their hives, and which might be easily saved in movable-comb hives; but his losses will not be of such a wholesale character as utterly to dishearten him in his attempts to keep bees.[9]

The prudent bee-keeper, remembering that "prevention is better than cure," will take unwearied pains to destroy, as *early* in the season as he can, the larvae of the moth. The destruction of a single female worm

---

[8] *Langstroth sometimes sends a mixed message on wax moth control. In one case he talks about closing up the entrance so that only a single bee can exit to prevent the moths from entering, and then he talks about cracks in the hive where the moth can lay eggs. I feel that very few moths actually enter via the entrance of a hive, but rather lay their eggs in the cracks between the supers. The wax moth has been selected via evolution to not try and face the guard bees—they lay their eggs in the edges and cracks of the hive. The very small 1st instar larvae crawl into the hive and escape detection. They enter the comb where they grow until they reach a size that attracts the bees.*

[9] *Langstroth is really using a "soft sell" approach here. Not insisting that his hive would be the only one to succeed against the wax moth, or loss of a queen. But if the reader was not already sold on the merits of his hive before this point, he quietly adds that in the movable-comb hive you would **know** when it became queenless, whereas in a common hive you might not. It is this knowledge of the condition within the hive that makes the movable comb hive the powerful tool that it is.*

may thus be more effectual that the slaughter of hundreds at a later period.* If the common hives are used, the worms will usually be found where the hive rests upon the bottom-board. Such hives should be propped up on both ends with strips of wood, about three-eighths of an inch thick, and a piece of woolen-rag put between the bottom-board and the back of the hive. The fall-grown worm retreating to this warm hiding-place to spin its cocoon, may be easily caught, and effectually dealt with. Only provide some hollow, easily accessible to the worms when they wish to spin, and to yourself when you want them, and as bees in good condition will not permit them to spin among the combs, you can easily entrap them. If the hive has lost its queen, and the worms have gained possession of it, break it up, instead of reserving it as a moth-breeder, to infest your Apiary.

In the movable-comb hive, blocks of a peculiar construction (Plates III., VI., Figs. 11, 17) are used, both to entrap the worms and exclude the moth. The only place where she can get into these hives, is at the bee-entrance, and as abundant ventilation can be given, independent of this, it may be contracted to suit all possible emergencies, and thus force her to pass ever a space which, by continually narrowing, is more and more easily defended by the bees. These traps are slightly elevated, so that the heat and smell of the hive pass under them through small openings, into which the moth can enter, but which do not admit her to the hive. These openings, which resemble the crevic-

---

* Few, who have not seen their ravages by lifting out a comb, are aware how many young bees fall a prey to the worm as it burrows in the comb.

Mr. M. Quinby, of St. Johnsville, New York, whose common-sense treatise on "*The Mysteries of Bee-Keeping*" will richly repay perusal, is of opinion that the larger number of imperfect bees carried out of the hive in the Spring, have been destroyed by the worms. He thinks that enough are often thus lost from a single hive to make a moderate swarm of bees.

This estimate will not seem extravagant, if we take into account the number of breeding-cells which are destroyed, and the large vacancies which are often made by the bees in cutting out the webs and cocoons of the moth.

Dr. Kirtland, in an article in the *Ohio Farmer*, Dec. 1857, alluding to the times before the advent of the bee-moth, says: "In those halcyon days of bee-raising, swarms often came out earlier, and in larger numbers, than in recent times. It was no unusual occurrence for a Spring swarm to fill the hive with stores and young brood so rapidly, as to allow it, also, to throw off a swarm sufficiently early for the latter to lay up stores for Winter."

es between the common hives and their bottom-boards, she will enter, rather than attempt to force her way through the guards; and, finding here the nibblings of comb and bee-bread, its which her young can flourish, she deposits her eggs where they may be reached and destroyed. All this is on the supposition that the hive has a healthy queen, and that the bees have no more comb than they can warn and defend; for it there is no guard, or only a feeble resistance, she will penetrate to the heart of the citadel to deposit her seeds of mischief.

These blocks have also grooves which communicate with the *interior* of the hives, and which appear to the prowling worm, in search of a comfortable nest, the very place—so warm and secure—in which to spin its web, and "bide its time." When the hand of the bee-master lights upon it, it finds that it has been caught in its own craftiness.

All such contrivances, instead of helping the *careless* bee-keeper, will but give him greater facilities for injuring his bees. Worms will spin undisturbed under the blocks, and moths lay their eggs: his traps only affording them more effectual aid. If such incorrigibly careless persons will persist in the folly of keeping bees, they should use only smooth blocks, which, by regulating the entrance to the hives, will assist the bees in defending themselves against all enemies which seek admission to their castle.·

· In Plate V., Fig. 16, a small entrance is shown in front of the hives *above* the frames. If the lower one is closed, and the bees of a feeble colony are allowed to use this, it will be kept *warm* by the heat rising to the top of the hive, and will be guarded even in cool nights. Such an entrance may, in many cases, be found a great protection against the moth.

(Since we now know that the small larvae enter the hive, trying to keep the moth out of the hive is not essential or even very practical. It is true that a smaller entrance will allow the bees to more easily defend the colony against all intruders, but since we no longer have special ventilation ports built into our hives the bees need a larger entrance in the summer to move air in and out and also to allow bees to forage more readily.)

If the worms, by any means, get the ascendancy in movable-comb hives, the frames should be removed, (p. 243), and the worms destroyed. If proper care has been exercised, such and operation will be seldom needed.· Shallow vessels of sweetened water, placed on the hives after sunset, will often entrap many of the moths. they are so fond of sweets, that I have caught them sticking fast to pieces of moist sugar candy. Whey and sour milk are said to destroy them.[†]

I shall close what I have to say upon the bee-moth, with an extract from that accomplished scholar, and well-known enthusiast in bee-culture, Henry K. Oliver, of Massachusetts:

"The ravages of all the other enemies [‡] of the bee are but a baby bite to the destruction caused by the bee-moth. They are a paltry-looking, insignificant little gray-haired pestilent race of wax-and-honey-eating and bee-destroying rascals, that have baffled all contrivances that ingenuity has devised to conquer or destroy them.

"Your committee would be very glad to be able to suggest any effectual means by which to assist the honey-bee and its friends against the inroads of this foe, whose desolating ravages are more despondingly referred to than those of any other enemy.

"He who shall be successful in devising the means of ridding the bee-world of this destructive and merciless pest, will richly deserve to be crowned 'King Bee', in perpetuity; to be entitled to a never-fading wreath of budding honey flowers, from sweetly breathing fields, all murmuring with bees; to be privileged to use, during his natural life, 'night tapers from their waxen thighs,' (best wax candles, two to the

---

· Old combs are much the most liable to suffer from the moth. In movable-comb hives, no combs need remain so long in the hive as to have their value seriously impaired.

(Again we see the value of the movable-comb hive as a single comb can be removed. Many beekeepers do not have a regular pattern of replacing combs regardless if they have wax worms or not. They should be replaced when they become so darkened that you cannot see light through them—probably in about 5 years or so. Wax worms thrive better on dark combs simply because they have more pollen or other nutrients, and the darker combs tend to have these. These older combs also tend to harbor more diseases and pesticide residues.)

[†] Devices for *burning* the moth date back to the times of Columella, who recommends placing near the hives, at night, a brazen vessel, with a light burning in it. to destroy the moths resorting to it.

[‡] Report on Bees, to the Essex County Agricultural Society, 1851.

pound!); to have an annual offering, from every bee-master, of ten pounds each of very best virgin honey; and to a body guard, for protection against all foes, of thrice ten thousand workers, all armed and equipped as Nature's law directs. Who shall have these high honors?"

It seems almost incredible that such puny animals as mice should venture to invade a hive of bees; and yet they often slip in when cold compels the bees to retreat from the entrance. Having once gained admission, they build a warm nest in their confortable abode, eat up the honey and such bees as are too much chilled to offer resistance, and fill the premises with such a stench, that the bees, on the arrival of warm weather, often abandon their polluted home. On the approach of cold weather, the entrances of the hives should be so contracted that a mouse cannot get in.*

That various kinds of birds are fond of bees, every Apiarian knows to his cost. The King-bird (*Tyrannus musicapa*), which devours them by scores, is said—when he can have a choice—to eat only the drones; but as he catches bees on the blossoms—which are never frequented by these fat and lazy gentlemen—the industrious workers must often fall a prey to his fatal snap. There is good reason to suspect that this gourmand can distinguish between an empty bee in search of food, and one which, returning laden to its fragrant home, is in excellent condition to glide—already sweetened—down his voracious maw.[10]

If—as in the olden time of fables—birds could be moved by human language, it would be worth whale to post up, in the vicinity of our Apiaries, the old Greek poet's address to the swallow:

> "Attic maiden, honey fed,

---

* If, as the weather grows cold, the bees are allowed to use only the upper entrance (p. 250), it will be almost impossible for the mice to effect a lodgment.

(There are various mouse guards that have been designed to work with the lower entrances of modern hives. Most work fairly well if put in place at the right time. The cost of the labor to install them is the greatest deterrent. I have found that the antirobbing screens, described in Bee Culture, vol. 110, pg. 92, [1982], also keep mice out fairly well, but not completely.)

[10] *Any of the insectivorous birds can eat bees, it is rare that they cause much damage. A critical problem could arise in a queen-mating yard.*

On the other hand, I have seen the bee-eaters of Eastern Europe, (Merops sp.) sometimes become a serious problem.

>   Chirping warbler, bears't away
> Thou the busy buzzing bee,
>   To thy callow brood a prey?
> Warbler, thou a warbler seize?
>   Winged, one with lovely wings?
> Guest thyself, by Summer brought,
>   Yellow guests whom Summer brings?
> Wilt not quickly let it drop?
>   'Tis not fair; indeed, 'tis wrong,
> That the ceaseless warbler should
>   Die by mouth of ceaseless song."

No Apiarian ought ever to encourage the destruction of birds, because of their fondness for his bees. Unless we can check the custom of destroying, on any pretense, our insectivorous birds, we shall soon, not only be deprived of their aerial melody among the leafy branches, but shall lament, more and more, the increase of insects, from whose ravages nothing but these birds can protect us. Let those who can enjoy no music made by these winged choristers of the skies, except that of their agonizing screams as they fall before their well-aimed weapons, and flutter out their innocent lives before their heartless gaze, drive away, as far as they please from their cruel premises, all the little birds that they cannot destroy, and they will, eventually, reap the fruits of their folly, when the caterpillars weave their destroying webs over their leafless trees, and insects of all kind riot in glee on their blasted harvests.[*]

The toad is a well-known devourer of bees. Sitting, towards evening, under a hive, he will sweep into his mouth, with his swiftly-darting tongue, many a late returning bee, as it falls, heavily laden, to the ground; but as he is also a diligent consumer of various injurious insects, he can plead equal immunity with the insectivorous birds.

---

[*] "The farmers of Europe having learned, by repeated observation, that, without the aid of mischievous birds, their work would be sacrificed to the more destructive insect race, forgive the trespasses of such birds, as we forgive those of cats and dogs. The respect shown to birds by any people, seems to bear a certain ratio to the antiquity of the nation. Hence, the sacredness with which they are regarded in Japan, where the population is so dense that the inhabitants would feel that they could ill afford to divide the produce of their fields with the birds, unless they were convinced of their usefulness."— *Atlantic Monthly for* 1859, p.325.

Chirping warbler, bears't away
Thou the busy buzzing bee,
　To thy callow brood a prey?
Warbler, thou a warbler seize?
　Winged, one with lovely wings?
Guest thyself, by Summer brought,
　Yellow guests whom Summer brings?
Wilt not quickly let it drop?
　'Tis not fair; indeed, 'tis wrong,
That the ceaseless warbler should
　Die by mouth of ceaseless song."

No Apiarian ought ever to encourage the destruction of birds, because of their fondness for his bees. Unless we can check the custom of destroying, on any pretense, our insectivorous birds, we shall soon, not only be deprived of their aerial melody among the leafy branches, but shall lament, more and more, the increase of insects, from whose ravages nothing but these birds can protect us. Let those who can enjoy no music made by these winged choristers of the skies, except that of their agonizing screams as they fall before their well-aimed weapons, and flutter out their innocent lives before their heartless gaze, drive away, as far as they please from their cruel premises, all the little birds that they cannot destroy, and they will, eventually, reap the fruits of their folly, when the caterpillars weave their destroying webs over their leafless trees, and insects of all kind riot in glee on their blasted harvests.[*]

The toad is a well-known devourer of bees. Sitting, towards evening, under a hive, he will sweep into his mouth, with his swiftly-darting tongue, many a late returning bee, as it falls, heavily laden, to the ground; but as he is also a diligent consumer of various injurious insects, he can plead equal immunity with the insectivorous birds.

---

[*] "The farmers of Europe having learned, by repeated observation, that, without the aid of mischievous birds, their work would be sacrificed to the more destructive insect race, forgive the trespasses of such birds, as we forgive those of cats and dogs. The respect shown to birds by any people, seems to bear a certain ratio to the antiquity of the nation. Hence, the sacredness with which they are regarded in Japan, where the population is so dense that the inhabitants would feel that they could ill afford to divide the produce of their fields with the birds, unless they were convinced of their usefulness."—
Atlantic Monthly for 1859, p.325.

out of the mire; and then, being let go, away he trots, more afeard than hurt, leaving the smeared swain in a joyful fear."

Ants, in some places, are so destructive, that it becomes necessary to put the hives on stands, whose legs are set in water.· My limits forbid me to speak of wasps, hornets, millipedes (or wood-lice), spiders, and other enemies of bees. If the Apiarian keeps his stocks strong, they will usually be their own best protectors, and, unless they are guarded by thousands ready to die in their defense, they are ever liable to fall a prey to some of their many enemies, who are all agreed on this one point, at least—that stolen honey is much sweeter than the slow accumulations of patient industry.

DISEASES OF BEES.

Bees are subject to but few diseases which deserve special notice. The fatal effects of dysentery have already been alluded to (p 90). "The presence of this disorder," says Bevan, "is indicated by the appearance of the excrement which, instead of a reddish yellow, exhibits a muddy black color, and has an intolerably offensive smell. Also, by its being voided upon the floors and at the entrance of the hives, which bees, in a healthy state, are particularly careful to keep clean."·

Various opinions have prevailed as to the causes of this disease. All Apiarians are agreed that dampness in the hives, especially if the bees are long confined, is sure to produce it. Feeding bees late in the Fall on liquid honey—which they have not time to seal over, and which sours by attracting moisture—should be avoided; also, all unnecessary disturbance of colonies in the Winter, which, by exciting

---

· Small ants often make their nests about hives, to have the benefit of their warmth, and neither molest the bees nor are molested by them.

(On a worldwide basis, ants are one the major enemies of honey bees. In the tropics they are especially a problem and bee hives are often hung from trees or, as Langstroth indicates, put on stands with the feet of the stands in water or oil.)

· I have discovered a kind of dysentery which confines its ravages to a few bees in a colony. Those attacked are at first excessively irritable, and sting without any provocation. In the latter stages of this complaint, they may often bee seen on the ground, stupid and unable to fly, their abdomens unnaturally distended with an offensive yellow matter. I can assign neither cause nor cure for this disease.

(The disease he describes fits the parasite Nosema apis *quite well.)*

them, causes an excessive consumption of food. Populous stocks, well stored with honey, in hives so ventilated as to keep the combs dry, will seldom suffer severely from this disease.

The disease called by the Germans "*foul brood*," is of all others the most fatal (p. 19) to bees. The sealed brood die in the cells, and the stench from their decaying bodies seems to paralyze the bees.†

There are two species of foul-brood, one of which the Germans call the dry, and the other, the moist or fetid. The dry appears to be only partial in its effects, and not contagious, the brood simply dying and drying up in certain parts of the combs. In the moist, the brood, instead of drying up, decays, and produces a noisome stench, which may be perceived at some distance from the hive.*

If the Spring or Summer, when the weather is fine and pasturage abounds, the following cure for foul-brood is recommended by a German Apiarian:—"Drive out the bees into any clean hive, and shut them up in a dark place without food for twenty-four hours; prepare for them a clean hive, properly fitted up with comb from healthy col-

---

† Dzierzon thinks that this disease was produced in his Apiary by feeding bees on "American honey" (honey from the West India Islands). As this honey does not ordinarily produce it, he probably used some taken from colonies having the disease. Such honey is always infectious.

*(The Caribbean Islands have been quite free of foulbrood diseases, so Langstroth may have been right. The spores of American foulbrood are almost always transmitted through honey. After the colony dies it gets robbed out and thus the disease transmitted to another colony. Very often this robbing happens early in the spring when the strong colonies go in search of food and there is no nectar yet, they find the dead colony and return with the honey **and** disease.)*

Mr. Quinby informs me that he has lost as many as 100 colonies in a year from this pestilence. It has never made its appearance in my Apiaries, and I should regard its general dissemination through our country as the greatest possible calamity to beekeeping.

* As Aristotle (*History of Animals, Book LX., Chap 40*) speaks of a disease which is accompanied by a disgusting smell of the hive, there is reason to believe that foul-brood was common more than two thousand years ago.

(Langstroth has quite simply described the differences between European (EFB) and American foulbrood (AFB) diseases. While EFB does smell a little it is nothing like AFB. The dead and dying larvae of EFB look almost dried up, while the ropy, decaying larvae of AFB are certainly "moist." The larvae or pupae that are killed by AFB die after the cell is sealed, whereas the larvae killed by EFB usually die when they are still young and they have not yet spun a cocoon or the cell capped.)

onies, transfer the bees into it, and confine them two days longer, feeding them with pure honey."

My readers are indebted to Mr. Samuel Wagner for a translation of Dzierzon's mode of treating foul-brood:

"I admit that I can furnish no prescriptions by which a diseased colony may be forthwith cured. Nay, I consider it highly improbable that a colony, in which the disease has made marked progress, can be cured by any medicaments. The removal of the putrid and infectious matter, already so abundant in the cells, must at least be simultaneously effected—this seems to be altogether impracticable. Nevertheless, there would be much gained if we could neutralize or destroy the virus in the bees themselves, and also render the infected honey harmless. A beekeeping friend recently informed me that, if such honey be somewhat diluted with water, and then well boiled and skimmed, it may he safely used in feeding bees. Suspected honey should invariably be boiled and skimmed before it is fed to bees.[12] For the hive itself, chloride of lime might prove an efficient disinfectant. I simply let the hives, which contained diseased colonies, stand exposed to sun and air for two seasons, and stock them thereafter without experiencing a return of the malady.[13]

"On the whole, the disease has now lost its terrors for me. Though my bees may reintroduce it from neighboring Apiaries or other foreign sources, I no longer apprehend that it will suddenly break out in a number of my colonies, or spread rapidly in any of my Apiaries, because I shall hereafter avoid feeding foreign or imported honey, even if in an unfavorable year, it should become necessary to reduce the number of my stocks to one-half or one-fourth of the usual complement.

"But when the malady makes its appearance in only two or three of the colonies, and is discovered early (which may readily be done in hives having movable combs), it can be arrested and cared without damage or diminution of profit. *To prevent the disease from spread-*

---

[12] *Most beekeepers just use sugar syrup to avoid the difficulty of boiling honey and water to kill any AFB spores.*

[13] *The fact that AFB spores will persist for years in the comb would lead us to believe that Dzierzon at this point was talking about European foulbrood.*

ing in a colony, there is no more reliable and efficient process THAN TO STOP THE PRODUCTION OF BROOD, *for where no brood exists, none can perish and putrefy*. The disease is thus deprived both of its ailment and its subjects. The healthy brood will mature and emerge in due time, and the putrid matter REMAINING in a few cells will dry up and be removed by the workers. All this will certainly result from a *well-timed removal of the queen* from such colonies. If such removal becomes necessary in the Spring or early part of Summer, a supernumerary queen is thereby obtained, by means of which an artificial colony may be started, which will certainly he healthy if the bees and brood used, be taken from healthy colonies. Should the removal be made in the latter part of Summer, the useless production of brood will at once be stopped, and an unnecessary consumption of honey prevented. Thus, in either case, we are gainers by the operation. If we have a larger number of colonies than it is desired to winter, it is judicious to take the honey from the colonies deprived of their queens, immediately after all the brood has emerged, as they usually contain the greatest quantity of stores at that time. If the disease be not malignant foul-brood, the colony may be allowed to remain undisturbed after it has bred a new queen, and, in most instances, such colonies will subsequently be found free from disease. I have, indeed, ascertained the singular fact that, if both bees and combs be removed from an infected hive, and healthy bees and pure comb be placed therein, these will speedily be infected with foul-brood; whereas, when *the queen* of an incipiently infected colony is removed, or simply confined in a cage, and the workers are still sufficiently numerous to remove all impurities, the colony will speedily be restored to a healthy condition. It thus seems as though the bees can become accustomed to the *virus* which usually adheres so perniciously to the hive.[14]

"*Foul-brood*, indeed, is a disease exclusively of the larvae, and not of the emerged bees, or of brood sufficiently advanced to be nearly

---

[14] *Shaking bees onto new combs was one of the only ways to save the bees prior to the use of the antibiotics sulfathiazole, terramycin and now tylosin. Because Dzierzon keeps talking about caging the queen to cure the disease it would appear to me that he is often mixing the two foulbrood diseases. Since they were not well separated until early in the 20th century that confusion may be understandable. Caging would work for EFB but not for AFB. Since the European foulbrood disease bacteria does not produce long-lived spore, the disease often is abated by a disruption in the brood cycle.*

ready to emerge. Hence, the cause of the disease may exist already in the food provided for the larvae, and have its seat in the chyle-stomach[15] of the nursing bees, though these latter may not themselves be injuriously affected thereby.

"Though the colonies treated in this manner generally appear to be free from infection during the ensuing season, and the brood proceeding from the eggs of a queen subsequently given to them, or from those of one reared by themselves, is healthy, maturing and emerging in due time still, the disease, in most instances, reappears in the following Summer. It is, indeed, possible that the bees may have re-introduced it from foreign sources, but it is not unlikely, also, that the infectious matter really remained latent in the hive. The bees do not usually remove all the putrid matter from the cells, but let some portions remain in the corners after it has become dry, merely covering it with a film of wax or propolis, through which, subsequently, when circumstances favor its action, the *virus* may exert a malignant influence and cause a revival of the disease. Hence, when I do not break up such colonies altogether in Autumn, and transfer the bees to new hives or other colonies with pure combs, I invariably regard them with suspicion, as unreliable, and keep them under strict surveillance at least a year longer.

"I also use these suspected colonies, by preference, for the production of queens with which to supply queenless colonies or start artificial swarms—successively removing from them the young queens as soon us they prove to be fertile or I have occasion to use them. In this way, I make such a colony furnish three or four—nay, sometimes, by inserting sealed royal cells, even five or six young queens. But, in such operations, I invariably take the bees and brood for the artificial swarms, from colonies which are unquestionably free from the disease. For this purpose, I select strong colonies having young and vigorous queens, and which are consequently able to furnish the required supplies without any serious diminution of population, when the sea-

---

[15] *Chyle comes from the Latin Chymos or juice. It may be that the idea was that there was a stomach that produced the brood food rather than the brood glands in the head and thorax.*

The spores of American foulbrood bacteria are transmitted via the nurse bees that have fed on diseased honey, or have cleaned a diseased cell.

son is at all favorable to the multiplication of stocks. In such seasons, strong colonies, in good condition, with a vigorous queen in the prime of life, can easily supply brood and bees sufficient for four swarms."—*Bienenzeitung*, 1857, *No.* 4.

CHAPTER XIII.

ROBBING, AND HOW PREVENTED.

BEES are so prone to rob each other that, unless great precautions are used, the Apiarian will often lose some of his most promising stocks. Idleness is with them, as with men, a fruitful mother of mischief. They are, however, far more excusable than the lazy rogues of the human family; for they seldom attempt to live on stolen sweets when they can procure, a sufficiency by honest industry.

As soon as they can leave their hives in the Spring, if urged by the dread of famine, they begin to assail the weaker stocks. In this matter, however, the morals of our little friends seem to be sadly at fault; for, often those stocks which have the largest surplus—like some rich oppressors—the most anxious to prey upon the meagre possessions of others.

If the marauders, who are ever prowling about in search of plunder, attack a strong and healthy colony, they are usually glad to escape with their lives from its resolute defenders. The beekeeper, therefore, who neglects to feed his needy colonies, and to assist such as are weak or queenless (p. 221), must count upon suffering heavy losses from robber-bees.[1]

It is sometimes difficult for the novice to discriminate between the honest inhabitants of a hive, and the robbers which often mingle with them. There is, however, an air of roguery about a thieving bee which, to the expert, is as characteristic as are the motions of a pickpocket to a skillful policeman. Its sneaking look, and nervous, guilty agitation, once seen, can never be mistaken. It does not, like the laborer carrying

---

[1] *The act of feeding bees can stimulate robbing. Either the feeders leak a little or syrup is spilled. In any event, the beekeeper should be careful when putting any type of feeder into, or on, a colony. In preventing robbing from starting, it is the ounce of prevention that is important. Some actions taken by beekeepers can be OK at certain times of the year and not at others. For example, leaving honey supers uncovered will not be important during the honey flow but can become a disaster when there is no nectar coming in. An entrance feeder can induce robbing in a colony that already needs help, so be extra careful if you use an entrance feeder.*

home the fruits of honest toil, alight boldly upon the entrance-board, or face the guards, knowing well that, if caught by these trusty guardians, its life would hardly be worth insuring. If it can glide by without touching any of the sentinels, those within—taking for granted that all is right—usually permit it to help itself.

Bees which lose their way, and alight upon a strange hive, can be readily distinguished from these thieving scamps.[2] The rogue, when caught, strives to pull away from his executioners, while the bewildered unfortunate shrinks into the smallest compass, submitting to any fate his captors may award.

These dishonest bees are the "*Jerry Sneaks*" of their profession, and, after following it for a time, lose all taste for honest pursuits. Constantly creeping through small holes, and daubing themselves with honey, their plumes assume a smooth and almost black [*] appearance, just as the hat and garments of a thievish loafer acquire a "seedy" aspect. "Honesty is as good policy" among bees as among men, and, if the pilfering bee only knew its true interests, it would be safely laboring amid the smiling fields, instead of risking its life for a taste of forbidden sweets.

It is said, that bees occasionally act the part of highway robbers, by waylaying a humble-bee as it returns to its nest with a well-stored sac. Seizing the honest fellow, they give him to understand that they want his honey. If they killed him, they would never be able to extract his spoils from their deep recesses; they, therefore, bite and tease him, after their most approved fashion, all the time singing in his ears, "Your honey or your life," until he empties his capacious receptacle, when they release him and lick up his sweets.

---

[2] *It is the darting and uncertain flight that distinguishes the robbing bee. Often working the edges of the entrance looking for a place where there are no guard bees. The orientation flights in early afternoon sometimes look like robbing bees except that these circular flight patterns are usually some feet from the entrance, whereas the robbing bee is just a few inches away.*

[*] Dzierzon thinks that these black bees, which Huber has describes as so bitterly persecuted by the rest, are nothing more than thieves. Aristotle speaks of "a *black* bee which is called a *thief.*"

(*These robber bees lose some of their hair in the fighting that often takes place during these robbing encounters. This gives them a darker or black appearance.*)

Bees sometimes carry on their depredations upon a more imposing scale. Having ascertained the weakness of some neighboring colony, they sally out by thousands, eager to engage in a pitched battle. A furious onset is made, and the ground in front of the assaulted hive is soon covered with the bodies of innumerable victims.[3] Sometimes the baffled invaders are compelled to sound a retreat; too often, however, as in human contests—right proving but a feeble barrier against superior might–the citadel is stormed, and the work of rapine forthwith begins. And yet, after all, matters are not so bad as at first they seemed to be, for often the conquered bees, giving up the unequal struggle, assist the victors in plundering their own hive, and, are rewarded by being incorporated into the triumphant nation. The poor mother however, remains in her pillaged hive, some few of her children–faithful to the last–staying with her to perish by her side amid the ruins of their once happy home.[*]

If the bee-keeper would not have his bees so demoralized that their value will be seriously diminished, he will be *exceedingly careful* (p. 199) to prevent them from robbing each other. If the bees of a strong stock once get a taste of forbidden sweets, they will seldom stop until they have tested the strength of every hive. Even if all the colonies are able to defend themselves, many bees will be lost in these encounters, and much time wasted; for bees, whether engaged in robbing, or battling against the robbery of others, lose both the disposition and the ability to engage in useful labors.·

---

[3] *Langstroth does not indicate how many bees can be killed, or how long that this warfare can go on. I have seen whole apiaries essentially ruined by the bees robbing and killing each other. Once robbing gets started in an apiary it is hard to stop, and the beekeeper ends up getting a lot of stings as well.*

[*] "Bees, like men, have their different dispositions, so that even their loyalty will sometimes fail them. An instance not long ago came to our knowledge, which probably few bee-keepers will credit. It is that of a hive which, having early exhausted its store, was found, on being examined one morning, to be utterly deserted. The comb was empty, and the only symptom of life was the poor queen herself, 'unfriended, melancholy, slow,' crawling over the honeyless cells, a sad spectacle of the fall of bee-greatness. Marius among the ruins of Carthage—Napoleon at Fontainebleau—was nothing to this."—*London Quarterly Review*.

· If the Apiarian would guard his bees against dishonest courses, he must be exceedingly careful, in his various operations, not to leave any combs where strange bees can find them (see note, p. 172); for, after once getting a taste of stolen honey, they will hover

By keeping the movable entrance-blocks of my hives very close together when a colony is feeble, if thieves try to slip in, they are almost sure to be overhauled and put to death; and if robbers are bold enough to attempt to force an entrance, as the bottom-board slants forward, it gives the occupants of the hive a decided advantage. If any succeed in entering, they find hundreds standing in battle-array, and fare as badly as a forlorn hope that has stormed the walls of a beleaguered fortress, only to perish among thousands of enraged enemies.

By putting these blocks before the entrance of a hive which has ceased to offer any effectual resistance, the dispirited colony will often recover heart, and drive off their assailants.[4]

When bees are actively engaged in robbing, they sally out with the first peep of light, and often continue their depredations until it is so late that they cannot find the entrance to their hive. When robbing has become a habit, they are sometimes so infatuated with it as to neglect their own brood![5]

The cloud of robbers arriving and departing need never be mistaken for honest laborers carrying, with unwieldy flight, their heavy burdens to the hive. These bold plunderers, as they *enter* a hive, are almost as

---

round him as soon as they see him operating on a hive, all ready to pounce upon it and snatch what they can of its exposed treasures.

Some bee-keepers question whether a bee that once learns to steal ever returns to honest courses. I have known the value of an Apiary to be so seriously impaired by the bees beginning early in the season to rob each other, that the owner was often tempted to wish that he had never seen a bee.

[4] *The use of the anti-robbing screen on the entrance of the hive works with great success in preventing robbing. The greatest benefit to the beekeeper is that the bees do not get so defensive that the sting anything or anybody. See* [Gleanings in] *Bee Culture Vol. 110, pg. 92 (1982), on how to make and use the anti-robbing screens.*
[5] *Robbing can occur at any time, but happens mainly in the Fall when brood rearing is also on the decline.*

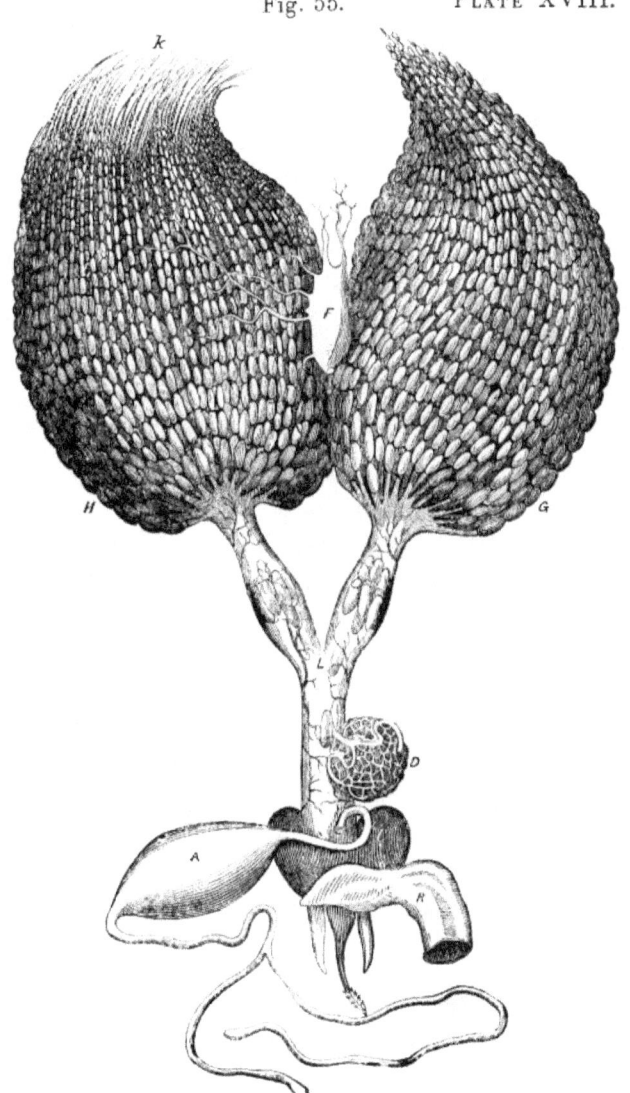

Fig. 55.　　　Plate XVIII.

hungry looking as Pharaoh's lean kine,[6] while, on *coming out*, they show by their burly looks that, like aldermen who have dined at the expense of the city, they are stuffed to their utmost capacity.

When robbing-bees have fairly overcome a colony, the attempt to stop them—by shutting up the hive or by moving it to a new stand—if improperly conducted, is often far more disastrous than to allow them to finish their work. The air will be quickly filled with greedy bees, who, unable to bear their disappointment, will assail, with almost frantic desperation, some of the adjoining stocks. In this way, the strongest colonies are sometimes over-powered, or thousands of bees slain in the desperate contest.

When an Apiarian perceives that a colony is being robbed, he should contract the entrance, and, if the assailants persist in forcing their way in, he must close it entirely. In a few minutes the hive will be black with, the greedy cormorants, who will not abandon it till they have attempted to squeeze themselves through the smallest openings. Before they assail a neighboring colony, they should be thoroughly sprinkled with cold water, which will make them glad to return to their homes.

Unless the bees that were shut up can have an abundance of air, they should be carried to a cool and dark place. Early the next morning they may be examined,[†] and, if necessary, united to another stock.

There is a kind of pillage which is carried on so secretly as often to escape all notice. The bees engaged in it do not enter in large num-

---

[6] *Kine is an old word used for cows or cattle.*

· "In Germany, when colonies in common hives are being robbed, they are often removed to a distant location, or put in a dark cellar. A hive, similar in appearance, is placed on their stand, and leaves of wormwood and the expressed juice of the plant are put on the bottom-board. Bees have such an antipathy to the odor of this plant, that the robbers speedily forsake the place, and the assailed colony may then be brought back.

"The Rev. Mr. Kleine says, that robbers may be repelled by imparting to the hive some intensely powerful and unaccustomed odor. He effects this the most readily by placing in it, in the evening, a small portion of *musk*, and on the following morning the bees, it they have a healthy queen, will boldly meet their assailants. These are nonplussed by the unwonted odor, and, if any of them enter the hive and carry off some of the coveted booty, on their return home, having a strange smell, they will be killed by their own household. The robbing is thus soon brought to a close."—S. WAGNER

[†] It will usually be found that a stock which is overpowered by robbers has no queen, or one that is diseased (p. 244, note).

bers, no fighting is visible, and the labors of the hive appear to be progressing with their usual quietness. All the while, however strange bees are carrying off the honey as fast as it is gathered. After watching such a colony for some days, it occurred to me, one evening, as it had an un-hatched queen, to give it a fertile one. On the next morning, rising before the rogues were up, I had the pleasure of seeing them met with such a warm reception, that they were glad to make a speedy retreat.

May not the fertile mother give to each hive (p. 203) its distinguishing scent? And may not a hive without such a queen be so pleased (p. 226) with the odor of other bees, as to let them do what they will with its stores? As bees are seldom engaged in raising young queens, except in the swarming season, when honey is so plenty that they are not inclined to rob, this may, if my conjectures are correct, account for the scarcity of this kind of pillage.

CHAPTER XIV.

DIRECTIONS FOR FEEDING BEES.

Few things in practical bee-keeping are more important than the feeding of bees; yet none have been more grossly mismanaged or neglected. Since the sulphur-pit has been discarded, thousands of feeble colonies starve in the Winter or early Spring; while often, when an unfavorable summer is followed by a severe Winter and late Spring, many persons lose most of their stocks, and abandon bee-keeping in disgust.

In the Spring, the prudent beekeeper will no more neglect to feed his destitute colonies, than to provide for his own table. At this season, being stimulated by the returning warmth, and being largely engaged in breeding, bees require a liberal supply of food, and many populous stocks perish, which might have been saved with but trifling trouble or expense.·

> "If e'er dark Autumn, with untimely storm,
> The honey'd harvest of the year deform;
> Or the chill blast from Eurus' mildew wing,
> Flight the fair promise of returning Spring;
> Full many a hive, but late alert and gay,
> Droops in the lap of all-inspiring May."–EVANS

When bees first begin to fly in the Spring, it is well to feed them a

---

· "If the Spring is not favorable to bees, they should be fed, because that is the season of their greatest expense in honey, for feeding their young. Having plenty at that time, enables them to yield early and strong swarms."—EVANS. *(Again, this is a good reason to have 80-90 kbs. of honey on the colony over winter. Beside the buffering effect of the extra honey it provides the food in the spring for rearing the young brood.)*

A bee-keeper, whose stocks are allowed to perish after the Spring has opened, is on a level with a farmer whose cattle are allowed to starve in their stalls; while those who withhold from them the needed aid, in seasons when they cannot gather a supply, resemble the merchant who burns up his ships, if they have made an unfavorable voyage.

Columella gives minute instructions for feeding needy stocks, and quotes approvingly the directions of Hyginus—whose writings are no longer extant– that his matter should be most carefully (*"diligentissime"*) attended to.

little, even when they have abundant stores, as a small addition to their hoards encourages the production of brood. Great caution, however, should be used to prevent robbing, and as soon as forage abounds, the feeding should be discontinued. If a colony is *overfed*, the bees will fill their brood-combs, so as to interfere with the production of young, and thus the honey given to them is worse than thrown away.

The overfeeding of bees resembles, in its results, the noxious influences under which too many children of the rich are reared. Pampered and fed to the full, how often does their wealth prove only a legacy of withering curses, as, bankrupt in purse and character, they prematurely sink to dishonored graves.

The prudent Apiarian will regard the feeding of bees—the little given by way of encouragement excepted—as an evil to be submitted to only when it cannot be avoided, and will much prefer that they should obtain their supplies in the manner so beautifully described by him whose inimitable writings furnish us, on almost every subject, with the happiest illustrations:

> "So work, the honey bees,
> Creatures that, by a rule in Nature, teach
> The art of order to a peopled kingdom.
> They have a king and officers of sorts,
> Where some, like magistrates, correct at home,
> Others, like merchants, venture trade abroad;
> Others, like soldiers, armed in their stings
> Make boot upon the Summer's velvet buds;
> Which pillage they, with merry march, bring home
> To the tent royal of their emperor,
> Who, busied in his majesty, surveys
> The singing masons building roofs of gold;
> The civil citizens kneading up the honey;
> The poor mechanic porters crowding in
> The heavy burdens at his narrow gate;
> The sad-eyed justice, with his surly hum,
> Delivering o'er, to executors pale,
> The lazy, yawning drone." —SHAKESPEARE'S *Henry V., Act I., Scene 2.*

FEEDING.

Impoverished stocks, if in common hives, may be fed by inverting the hives and pouring a teacup full of honey among the combs in which the bees are clustered.[1] A bee deluged by sweets, when away from home, is a sorry spectacle; but what is thus given them does no harm, and they will lick each other clean, with as much satisfaction as a little child sucks its fingers while feasting on sugar candy. When the bees have taken up what has been poured upon them, the hive may be replaced, and the operation repeated, at intervals, as often as is needed. If the stock is in a movable-comb hive, the food may be put into an empty comb, and placed where it can be easily reached by the bees.

If a colony has too few bees, its population must be replenished (p. 221) before it is fed. If it has but a small quantity of brood-combs, unless fed very moderately, it will fill the cells with honey instead of brood. If the Apiarian wishes the bees to build new comb, the food must be given so regularly as to resemble natural supplies, or they will store it in the cells already built.

To build up small colonies by *feeding*, requires more care and judgment than any other process in bee-culture, and will rarely be required by those who have movable-comb hives. It can only succeed when everything is made subservient to the most rapid production of *brood.* [2]

By the time the honey-harvest closes, all the colonies ought to be strong in numbers; and, in favorable seasons, their aggregate resources

---

[1] *Here we see how far beekeeping has come with the movable-comb hives. It would be impossible to feed very many hives if you had to turn them upside down and pour honey or syrup onto the bees and combs.*

[2] *If possible, the easiest way to make colonies stronger is to add frames of brood, taken from strong colonies, to the weak colonies. Bees will accept any brood regardless of its source. By equalizing colonies many problems of swarming at a later time are averted as well. The strong colonies can make up the loss of a frame or two of brood very quickly. It is, however, important to give the weak colonies brood that is just about to emerge. That way the newly emerged bees can give instant support to the colony.*

should be such that, when an equal division is made, there will be enough food for all. If some have more and others less than they need, an equitable division may usually be effected in movable-comb hives. Such an agrarian procedure would soon overthrow human society; but bees thus helped, will not spend the next season in idleness; nor will those which were deprived of their surplus, limit their gatherings to a bare competency.

Early in October—in northern latitudes, by the middle of September—if forage is over, all feeding required for wintering bees should be carefully attended to. If delayed, to a later period, the bees may not have sufficient time to seal over their honey, which, by attracting moisture and souring, may expose them (p. 256) to dysentery.[3] Such colonies as have too few bees to winter well, should be added to other stocks.

West India honey is, ordinarily, the cheapest liquid bee-food. To remove its impurities, and prevent it from souring or candying in the cells, it should have a little water added to it, and, after boiling a few minutes, should be set to cool; the scum on the top should then be removed. A mixture of three lbs. of honey, two of brown sugar, and one of water, prepared as above, has been used by me (p. 257) for many years, without injury to my bees.

It is desirable to get through with feeding as rapidly as possible*,· as the bees are so excited by it, that they consume more food than they otherwise would. In my hives, the feeder may be put over one of the holes of the honey-board, into which the heat ascends. The bees can then get their food without being chilled in cold weather, and its smell is not so likely to attract robber bees. To make a cheap and convenient feeder (see Plate XI, Fig. 26), take any wooden box holding at

---

[3] *The wax capping does not prevent moisture from entering honey within the cells. It slows up the process a little but does not prevent the exchange of water. Combs of honey removed from a colony and stored above 50% relative humidity will eventually ferment.*

The lack of time to ripen the honey in the Fall can be a problem in wintering bees, whether from feeding sugar syrup, or from a late honey flow, if bad weather does not allow the bees to ripen the honey. When the un-ripened honey is eaten it can cause the bees to have dysentery and the colony may die. For this reason, if I plan on feeding bees in the Fall, I try to get the syrup on the colonies by the middle of September. This is usually about a month before the weather gets very cold here in Michigan.

*Feeding stocks, driven late in the Fall into empty hives, unless combs (p. 71) can be given to them, will seldom pay expenses.

FEEDING. 271

least two quarts; about two inches from one end put a thin partition, coming within half an inch of the top; cut a hole in the bottom of the small apartment, so that when the feeder is put over any hole, the bees can pass into it and get access to the division holding the food. The joints of the feeding apartment should be made *honey-tight*, by running into the corners a mixture (p. 78) of wax and rosin; and if the sides are washed with the same hot mixture, the wood, absorbing no honey, will keep sweet. The lid should have a piece of glass, to show when the feeder needs replenishing, and a hole, for pouring in the food, made and closed like those admitting the bees to the spare honey receptacles. Some clean straw, cut short enough to sink readily, as the bees consume the honey, will prevent them from being drowned.·

Water is indispensable to bees when building comb or raising brood. They take advantage of any warm Winter day (see Chapter on Wintering Bees) to bring it to their hives; and, in early Spring, may be seen busily drinking around pumps, drains, and other moist places. Later in the season, they sip the dew from the grass and leaves.

Every careful bee-keeper will see that his bees are well supplied with water.[†] If he has not some sunny spot where they can safely obtain it, he will furnish them with shallow wooden troughs, or vessels filled with floats or straw, from which—sheltered from cold wind and warmed by the genial rays of sun–they can drink without risk of drowning.

Bees seem to be so fond of salt, that they will alight upon our hands to lick up the saline perspiration. "During the early part of the

---

· If such a box is covered thickly with cotton or wool, so as to retain the ascending heat, it may be used all Winter as a honey or water-feeder.

(Subsequent adaptations of this "hive-top feeder" by C. C. Miller allow the bees to come up through an opening, from the hive below. The feeder has wire screens that go into the syrup for the bees to hang onto. These feeders can hold 1-2 gallons of syrup.)

Columella recommends wool, soaked in honey, for feeding bees. When the weather is not too cold, a saucer, bowl, or vessel of any kind, filled with straw, will make a convenient feeder. *(Most modern feeders use the fact that heat rises to the top of the hive, which allows for feeding by the bees in cool or cold weather. An inverted pail (with a small-screened hole) placed inside an empty super on the topbars of the colony makes an excellent feeder.)*

[†] An old Grecian bee-keeper says, "that if the weather is such that the bees are prevented from flying for only a few days, the brood will perish from want of water."

(The dryness of the climate would have a bearing on how soon, and how much, water they need. Brood rearing does require water.)

breeding season," says Dr. Bevan, "till the beginning of May, I keep a constant supply of salt and water near my Apiary, and find it thronged with bees from early morn till late in the evening. About this period, the quantity they consume is considerable, but afterwards they seem indifferent to it.[4] The eagerness they evince for it at one period of the season, and their indifference at another, may account for the opposite opinions entertained respecting it."

The Rev. Mr. Weigel, of Silesia, recommends plain sugar-candy as a substitute for liquid honey. If bees can get access to it, without being chilled, they will cluster on it, and, when supplied with water, will gradually eat it up. Four pounds of candy[*] will, it is said, sustain a colony having scarcely any winter stores. It is cheaper than liquid food, and less liable, to sour in the cells.

If the common hives are inverted, and sticks of candy placed gently between the combs where the bees are clustered, they may be easily fed in the coldest weather. In my hives, if the spare honey-board, or cover, is elevated on strips of wood, about an inch and a half above the frames, and the candy laid on them just above the clustered bees, it will be accessible to them in the coldest weather. It may also be gently put between the combs, in an upright position among the bees.·

Mr. Wagner has furnished me with the following interesting facts, translated by him from the *Bienenzeitung*:

---

[4] *During a nectar flow there is usually an abundance of water for the brood since the bees have to remove the water in order to make the nectar into honey. Water still may be needed if the area where the bees are has very low relative humidity, or the nectar flow is light.*

[*] To make candy for bee-feed: add water to the sugar, and clarify the syrup with eggs; put about a teaspoonful of cream of tartar to about 20 lbs. of sugar, and boil until the water is evaporated. To know when it is done, dip you finger first into cold water and then into the syrup. If what adheres is brittle when chewed, it is boiled enough. Pour it into shallow pans, slightly greased, and, when cold, break it into pieces of suitable size. After boiling, balm, or any other flavor agreeable to bees, may be put into the syrup.

(Candy boards are still commonly used. The sugar needs to be heated to a temperature of 240° F. to make a soft-ball candy. Beekeepers use these as insurance against the bees running out of food during the winter when it is almost impossible to feed the bees any other way. The soft-ball candy can also be poured into wax-paper lined bread tins to make cakes that are put onto the top bars of the colony. This way you do not have to have special boards.)

· By sliding a few sticks of candy under their frames, a small colony may be fed in warm weather, without tempting robbers by the smell of liquid honey. If a small quantity of liquid food is needed in Summer, loaf sugar dissolved in water, having little smell, is best.

FEEDING. 273

"'The use of sugar-candy for feeding bees,' says the Rev. Mr. Kleine, 'gives to bee-keeping a security which it did not possess before. Still, we must not base over-sanguine calculations on it, or attempt to winter very weak stocks, which a prudent Apiarian would at once unite with a stronger colony. I have used sugar-candy for feeding, for the last five years, and made many experiments with it, which satisfy me that it cannot be too strongly recommended, especially after unfavorable Summers. Colonies well furnished with comb, and having plenty of pollen, though deficient in honey, may be very profitably fed with candy, and will richly repay the service thus rendered them.

"'Sugar-candy, dissolved is a small quantity of water, may he safely fed to bees late in the Fall, and even in Winter, if absolutely necessary. It is prepared by dissolving two pounds of candy in a quart of water, and evaporating, by boiling, about two gills [5] of the solution; then skimming and straining through a hair sieve. Three quarts of this solution, fed in Autumn, will carry a colony safely through the Winter, in an ordinary location and season. The bees will carry it up into the cells of such combs as they prefer, where it speedily thickens and becomes covered with a thin film, which keeps it from souring.[6]

"'*Grape-sugar*,[7] for correcting sour wines, is, now extensively made from potato-starch, in various, places on the Rhine, and has been highly recommended, for bee-feed. It can be obtained at a much lower price than cane-sugar, and is better adapted to the constitution of the bee, as it constitutes the saccharine matter of honey, and hence, is frequently termed honey-sugar. "'It may be fed either diluted with boiling-water, or in its raw state, moist, as it comes from the factory. In the latter condition, bees consume it slowly, and, as there is not the waste that occurs when candy is fed, I think it is better winter-food.' [8]

"'The Rev. Mr. Sholz, of Silesia, recommends the following as a substitute for sugar-candy in feeding bees:

---

[5] *A gill was an old U. S. Customary measure equal to ¼ pint or 4 ounces.*
[6] *I am confident that the bees made this into honey before winter.*
[7] *Grape-sugar = dextrose = glucose. Glucose is one of the two simple sugars found in honey; fructose (levulose) is the other.*
[8] *In modern times many beekeepers have switched to high-fructose corn syrup as a winter-feed for bees. It may be fed in almost the same way. It is used because it is cheaper than other sugars, as well.*

"Take one pint of honey and four pounds of pounded lump sugar; heat the honey, without adding water, and mix it with the sugar, working it together to a stiff doughy mass.[9] When, thus thoroughly incorporated, cut it into slices, or form it into cakes or lumps, and wrap them in a piece of coarse linen and place them in the frames. Thin slices, enclosed in linen, may be pushed down between the combs. The plasticity of the mass enables the Apiarian to apply the food in any manner he may desire. The bees have less difficulty in appropriating this kind of food than where candy is used, and there is no waste.

Mr. Kleine grates [*] candy, for a winter bee-food into cells previously dampened with sweetened water."

It is, impossible to say how *much* honey will be needed to carry a colony safely through the Winter. Much will depend (see Chapter on Wintering Bees) on the way in which they are wintered, whether in the open air or in special depositories, where they are protected against the undue excitement caused by sudden and severe atmospheric changes; much, also, on the length of the Winters, which vary so mach in different latitudes, and the forwardness of the ensuing Spring. In some of our Northern States, bees will often gather nothing for more than six months, while, in the extreme South, they are seldom deprived of all natural supplies for as many weeks. In all our Northern and Middle States, if the stocks are to be wintered out of doors, they should have at least twenty-five pounds [*] of honey.

---

[9] *We would now call this "queen-cage candy" and is made in almost the same way except that we use powdered sugar instead of plain sugar. You should not make this queen-candy with honey, since honey may contain spores of American foulbrood disease.*

[*] *Granulated* loaf-sugar would probably make a good bee-feed, and, by wetting the combs after it has been sifted into them, it might easily be made to stay into the cells. Neither sugar nor candy can be used by bees unless they have water to dissolve them.

I have seen bees flock by thousands around the mills where the Chinese sugar-cane (*Sorghum*) was being ground. The value, as a bee-food, of the raw juice and the syrup should be carefully tested.

[*] In movable-comb hives, the amount of stores may be easily ascertained by actual inspection. The weight of hives is not always a safe criterion, as old combs are heavier than new ones, besides being often over-stored (p. 82) with bee bread.

(A colony of bees uses from one to two pounds of honey per week in the early part of winter, and toward Spring more brood rearing is being carried on and the use increases. I still like to have 80-90 lbs. on a colony here in the North, as the extra amount helps to buffer the colony against sudden temperature shifts. Any extra honey is used for spring brood rearing.

FEEDING. 275

All attempts to derive profit from selling cheap honey fed to bees, have invariably proved unsuccessful. The notion that they can change *all sweets*, however poor their quality, into *good honey* [†] on the same principle that cows secrete milk from any acceptable food, is a complete delusion.

It is true that they can make white comb from almost every liquid sweet, because wax being a natural secretion of the bee, can be made from all saccharine substances, as fat can be put upon the ribs of an ox by any kind of nourishing food. But the quality of the comb has nothing to do with its contents; and the attempt to sell, as a prime article, inferior honey, stored in beautiful comb, is as truly a fraud as to offer for good money, coins which, although pure on the outside, contain a baser metal within. The quality of honey depends very little, if any, upon the secretions of the bees; and hence, apple-blossom, white clover, buckwheat, and most other varieties of honey, have each its peculiar flavor.[‡]

The evaporation[*] of its watery particles is the only well-marked

---

[†] When the bees are rapidly storing their combs, they disgorge the contents of their honey-sacs as soon as they return from the fields. That the honey undergoes *no* change during the short time it remains in their sacs cannot positively be affirmed, but that it can undergo only a *very slight* change is evident from the fact that the different kinds of honey or sugar-syrup fed to the bees can be almost as readily distinguished, after they have sealed them up, as before.

The Golden Age of bee-keeping, in which bees are to transmute inferior sweets into such balmy spoils as were gathered on Hybla or Hymettus, is as far from prosaic reality as the visions of the poet, who saw–

"*A golden hive, on a golden bank,*
Where golden bees, by alchemical prank,
Gather gold instead of honey."

[‡] "That bees *gather* honey, but do not *secrete it*, is argued from the fact that bee-keepers find cells filled with honey (in new swarms) on the first or second day."— *Aristotle.*

(Such statements by Aristotle that bees gather honey probably confused Linneaus in the naming of the honey bee, as Apis mellifera, *which means honey gatherer. Linneaus latter changed the name to* Apis mellifica, *or honey maker. This later name, while more accurate, has been discarded by taxonomists because of the priority rules in zoological nomenclature.)*

[*] If a strong colony is put on a platform scale, it will be found, during the height of the honey harvest, to gain a number of pounds on a pleasant day. Much of this weight, however, will be lost in the night from the evaporation of the newly-gathered honey, the water from which often runs in a stream from the bottom-board. The Rev. Levi Wheaton, of North Falmouth, Mass., is of the opinion that ventilation will greatly aid the bees in evaporating the

change that honey appears to undergo,[10] from its natural state in the nectaries of the blossoms, and bees are very unwilling to seal it over until it has been brought to such a consistency that it is in no danger of becoming acid in the cells.‡

Even if cheap honey could be "made over" by the bees so as to be of the best quality, it would cost the producer, taking into account the amount consumed (p 71) in elaborating wax, almost, if not quite, as much as the market price of white clover honey; and, if he feeds his bees after the natural supplies are over, they will suffer from filling up their brood cells.†

The experienced Apiarian will fully appreciate the necessity of preventing his bees getting a taste of forbidden sweets, and the inexperienced, if incautious, will soon learn a salutary lesson. Bees were intended to gather their supplies from the nectaries of flowers, and, while following their natural instincts, have little disposition to meddle with property that does not belong to them; but, if their incautious

---

water from their unsealed honey. The thorough upward ventilation which I now give to my hives may, therefore, contribute to increase the yield of honey.

[10] *About half of the nectars are pure sucrose and thus are changed into the two simple sugars glucose and fructose in the process of making honey. Other nectars are mixtures of fructose and glucose, or some are mixtures of all three.*

‡ Aristotle notices this fact, which I once thought a discovery of my own. The remarks of this wonderful genius on the generation of bees show that he appreciated the difficulties which, until of late, have so much perplexed modern Apiarians. After discussing this topic, he says; "All pertaining to this subject has not yet been sufficiently ascertained; but, if it ever should be, then we must place more confidence in our observations than in our reasonings. Theory, however, as far as it conforms to facts observed, is worthy of credit." Have we not here the inductive system of well guarded and as well expressed as ever it was by Bacon?

† The following is my recipe for a beautiful *liquid* honey, which the best judges have pronounced one of the most luscious articles they ever tasted. Put two pounds of the purest white sugar in as much hot water as will dissolve it; take one pound of strained white clover honey—any honey of good flavor will answer—and add it warm to the syrup, thoroughly stirring them together. As refined loaf sugar is a pure and inodorous sweet, one pound of honey will give its flavor to two pounds of sugar, and the compound will be free from that smarting taste which pure honey often has, and will usually agree with those who cannot eat the latter with impunity. Any desired flavor may be added to it.

Although no profit can be realized from inducing bees to store this mixture in boxes or glasses, the amateur may choose, in bad seasons, or in districts where the honey is poor, to secure in this way choice specimens for his table.

(*Under FDA pure food laws this product may not be sold, as this is adulteration of pure honey.*)

FEEDING. 277

owner tempts them with liquid food, especially at times when they can obtain nothing from the blossoms, they become so infatuated with such easy gatherings as to lose all discretion, and will perish by thousands if the vessels which contain the food are not furnished with floats, on which they can safely stand to help themselves.

As the fly was not intended to banquet on blossoms, but on substances in which it might easily be drowned, it cautiously alights on the edge of any vessel containing liquid food, and warily helps itself; while the poor bee, plunging headlong, speedily perishes. The sad fate of their unfortunate companions does not in the least deter others who approach the tempting lure, from madly alighting on the bodies of the dying and the dead, to share the same miserable end! No one can understand the extent of their infatuation, until he has seen a confectioners shop assailed by myriads of hungry bees. I have seen thousands strained out from the syrups in which they had perished; thousands more alighting even upon the boiling sweets; the floors covered and windows darkened with bees, some crawling, others flying, and others still, so completely besmeared as to be able neither to crawl nor fly— not one in ten able to carry home its ill-gotten spoils, and yet the air filled with new hosts of thoughtless comers.

I once furnished a candy-shop in the vicinity of my Apiary, with gauze-wire windows and doors, after the bees had commenced their depredations. On finding themselves excluded, they alighted on the wire by thousands, fairly squealing with vexation as they vainly tried to force a passage through the meshes. Baffled in every effort, they attempted to descend the chimney, reeking with sweet odors, even although most who entered it fell with scorched wings into the fire, and it became necessary to put wire-gauze over the top of the chimney also.*

As I have seen thousands of bees destroyed in such places, thousands more hopelessly struggling in the deluding sweets, and yet increasing thousands, all unmindful of their danger, blindly hovering over and alighting on them, how often have they reminded me of the

---

* Manufacturers of candies and syrups will find it to their interest to fit such guards to their premises; for, if only one bee in a hundred escapes with its load, considerable loss will be incurred in the course of the season. *(This is also true of honey extracting houses. The extracting often occurs after the honey flow and the bees will try to rob the honey.)*

infatuation of those who abandon themselves to the intoxicating cup. Even although such persons see the miserable victims of this degrading vice falling all around them into premature graves, they still press madly on, trampling, as it were, over their dead bodies, that they too may sink into the same abyss, and their sun also go down in hopeless gloom.

The avaricious bee that, despising the slow process of extracting nectar from "every opening flower," plunges recklessly into the tempting sweets, has ample time to bewail its folly. Even if it does not forfeit its life, it returns home with a woe-begone look, and sorrowful note, in marked contrast with the bright hues and merry sounds with which its industrious fellows come back from their happy rovings amid "budding honey-flowers and sweetly-breathing fields."

CHAPTER XV.

THE APIARY— PROCURING BEES TO STOCK IT— TRANSFERRING BEES FROM COMMON TO MOVABLE-COMB HIVES.

An intimate acquaintance with the honey resources of the country is highly important to those desirous of engaging largely in bee-culture. While, in some localities, bees will accumulate large stores, in others, only a mile or two distant, they may yield but a small profit.[*]

Wherever the Apiary is established, great pains should be taken to protect the bees against high winds.[†] Their hives should be placed where they will not be annoyed by foot passengers or cattle, and should never be very near places where sweaty horses must stand or pass. If managed on the swarming plan, it is very desirable that they should be in full sight of the rooms most occupied, or at least where the sound of their swarming will be easily heard.

In the Northern and Middle States, the hives should have a south-eastern exposure, to give the bees the benefit of the sun when it will be most conducive to their welfare. By using my movable stands (Plate V., Fig. 16), the hives may be made to face in any desired direction. The plot occupied by the Apiary should be in grass, mowed frequently, and kept free from weeds. Hives are too often placed where many bees perish by falling into the dirt, or among the tall weeds and grass, where spiders and toads find their choicest lurking-places.

---

[*] "While Huber resided at Cour, and afterwards at Vivai, his bees suffered so much from scanty pasturage, that he could only preserve them by feeding, although stocks that were but two miles from him were, in each case, storing their hives abundantly."— *BEVAN.*

[†] By tacking a piece of muslin to the alighting-board and the projecting parts of the stand (Plate V., Fig. 16), the bees, as they slack up, will alight on the cloth—to escape being bruised or blown away—and thus easily gain their hives. In windy situations, thousands of bees (p. 186) may be thus saved.

(I feel that wind protection is far more important in the winter, though bees will be benefited with a good wind-break at any time of the year. Cold, winter winds pull off any residual heat from the outside of hives.)

Covered Apiaries, unless built at great expense, afford little or no protection against extreme heat or cold, and much increase the risk losing the queens.[1]

In the Summer, no place is so congenial to bees as the shade of trees, if it is not too dense, or their branches so low as to interfere with their flight. As the weather becomes cool, they can easily be moved to any more desirable Winter location. If colonies are moved in the line of their flight, and *a short distance at a time*, no loss of bees will be incurred; but, if moved only a few yards, *all at once*, many will often be lost. By a *gradual process*, the hives in an Apiary may, in the Fall, be brought into a narrow compass, so that they can be easily sheltered from the bleak Winter winds. In the Spring, they may be gradually returned to their old positions.[*]

PROCURING BEES TO STOCK AN APIARY.

The beginner will ordinarily find it best to stock his Apiary with swarms of the current year, thus avoiding, until he can prepare himself to meet them, the perplexities which often accompany either natural or artificial swarming. If new swarms are purchased, unless they are large and early, they may only prove a bill of expense. If old stocks are purchased, such only should be selected as are healthy and populous. If removed after the working season has begun, they should be brought from a distance of at least two miles (p. 156).

If the bees are not all at home when the hive is to be removed, blow a little smoke into its entrance, to cause those within to fill themselves with honey, and to prevent them from leaving for the fields. Repeat this process from time to time, and in about half an hour all will have re-

---

[1] *In Europe where bee-houses are more common they usually use a different colored front for each hive, thus allowing the hives to be placed close together. It doesn't mean that they don't have some drifting it is just that the colors reduce the problem. These houses also provide some extra insulation and combined heat during the winter.*

[*] By removing the strongest stocks in an Apiary the first day, and others not so strong the next, and continuing the process until all were removed, I have safely changed the location of my Apiary, when compelled to move my bees in the working season. On the removal of the last hive, but few bees returned to the old spot. The change, as thus conducted, strengthened the weaker stocks.

turned.² If any are clustered on the outside, they may be driven within by smoke.

The common hives may be prepared for removal by inverting them and tacking a coarse towel over them, or strips of lath may be laid over wire-cloth, and brads driven through them into the edges of the hive.

Confine the hive, so that it cannot be jolted, to a bed of straw in a wagon with springs, and be sure, before starting, that it is *impossible* for a bee to get out. The inverted position of the hive will give the bees what air they need, and guard their combs from being loosened It will be next to impossible, in warm weather, to move a hive which contains much new comb.

New swarms may be brought home in any old box which has ample ventilation. A tea-chest, with wire-cloth on the top, sides, and bottom-board, will be found very convenient. The bees may be shut up in this box as soon as they are hived. *New swarms require even move air than old stocks*, being full of honey, and closely clustered together. They should be set in a cool place, and if the weather is very sultry, should not be removed until night. Many swarms are suffocated by the neglect of these precautions. The bees may be easily shaken out from this temporary hive (p. 139).

When movable-comb hives are sent away to receive swarm, two strips of wood, with small pieces nailed to them to go between the frames and keep them apart, should be laid over the frames.³ The cover, or honey-board, should then be *screwed* fast, and, if, the strips are of proper thickness, one-eighth of an inch air-space will be left all around the hive, which, with the other ventilators, will give air enough.⁴ If an old stock, in hot weather, is to be moved any distance in such a hive, it will be advisable to fasten wire-cloth in front of the portico, so that the

---

[2] *This is a useful technique to prevent bees from getting poisoned by pesticides if you find that the chemicals are being applied when the bees are flying to the sprayed field. This method works best in late afternoon when you may not have to keep the bees home for a long period. By using a sprinkling watering can you can "syrup" quite a few colonies in a short time.*

[3] *With modern Hoffman self-spacing frames this procedure is not needed.*

[4] *Since most modern Langstroth hives do not have special ventilation, beekeepers that migrate, or move bees for pollination, in warm weather, will often put screens on the top of the hives while they are being moved. Moving at night also helps keep the bees cooler.*

bees can leave their combs (p. 91) and cluster there.[5] Hives with movable frames should be arranged in such a position that the frames run from *front to rear*, and not from *side to side*, in the carriage. My glass hives ought never to be sent off for swarms.

Inexperienced persons will seldom find it profitable to begin bee-keeping on a large scale. By using movable-comb hives, they can rapidly increase their stocks after they have acquired skill, and have ascertained, not simply that money can be made by keeping bees, *but that they can make it*. While large profits can be realized by careful and experienced bee-keepers, those who are otherwise will be almost sure to find their outlay result only in vexatious losses. An Apiary neglected or mismanaged is worse than a farm overgrown with weeds or exhausted by ignorant tillage; for the land, by prudent management, may again be made fertile, but the bees, when once destroyed, are a total loss.

TRANSFERRING BEES FROM COMMON TO MOVABLE-COMB HIVES

This process may be easily effected whenever the weather is warm enough for bees to fly.[*] It is conducted as follows: Drive the bees into a forcing-box (p. 154), which put on their old stand, and carry the parent hive to some place where you cannot be annoyed by other bees. Have on hand tools for prying off a side of the hive; a large knife for cutting out the combs; vessels for the honey; a table or board, on which to lay the brood combs; cotton-twine or tape, for fastening them into the frames; and water for washing off, from time to time, the honey which will stick to your hands. Having selected the working combs, carefully cut them rather large, so that they will just *crowd* into the frames, and retain their places in their natural position until the bees have time to fasten them. It will be well to wind some twine or tape, which should be subsequently removed, around the upper and lower slats of the

---

[5] *We use the anti-robbing screen, described earlier (p. 264, note 4), with the top exit covered with a thin strip of wood which allows the bees extra ventilation during moving.*

[*] It has frequently been done in Winter, for purposes of experiment, by removing the bees into a warm room.

frames, as an additional security. Small pieces of empty comb may be fastened with melted wax and resin (p. 72).*

When the hive is thus prepared, the bees may be put into it and confined, water being given to them, until they have time to make all secure against robbers.

When the weather is cool, the transfer should be made in a warm room, to prevent the brood from being fatally chilled. An expert Apiarian can easily complete the whole operation—from the driving of the bees to the returning of them to their new hive– in about half an hour, and with the loss of very few bees, old or young. The best time for transferring bees is about ten days after a swarm has issued or been forced from the old hive. The brood will then be sealed over, and able to bear considerable exposure.[6]

Until the feasibility of transferring bees by movable frames had been thoroughly tested, I felt irreconcilably opposed to any attempt to dislodge them from their previous habitations. The process, as it has been ordinarily conducted, has resulted in the wanton sacrifice of thousands of stocks.

Dr. Kirtland thus speaks of the results of transferring some of his colonies to the movable-comb hives: "I had three stocks transferred to an equal number of Mr. Langstroth's hives. The first had not swarmed in two years, and had long ceased to manifest any industry; the others had never swarmed. All the hives were filled with black and filthy

---

* The Rev Levi Wheaton prefers to use combs for guides, and confines them by a thin strip of wood sprung between the uprights of the frames, so as to press against the lower edges of the combs.

Mr. Wm. W. Cary, in transferring, uses strips three-eighths of an inch wide and one-eighth thick, cut from any springy wood, and half an inch longer than the depth of the frames. He fastens them together in pairs, with strings which keep them just far enough apart to pass over the tops and bottoms of the frames. Two pairs will be needed for each frame, and they must be removed after the combs are firmly secured by the bees, which will be done in two or three days.

(I generally use large rubber bands, or string, to hold the combs into the frames. The bees will remove either in just a few days.)

[6] *I have found that eggs and larvae can stand much more cold than pupae. However, the eggs and larvae need to be kept from drying out if the humidity is lower than about 60 percent. Often when grafting larvae for queen rearing, beekeepers keep their grafting rooms very warm. It is not necessary, but keeping up high humidity is important.*

comb, candied honey, concrete bee-bread[7], and an accumulation of the cocoons and larvae of the moth. Within twenty-four hours, each colony became reconciled to its new tenement, and began to labor with far greater activity than any of my old stocks... I have now no stronger colonies than these, which I considered of little value till my acquaintance with this new hive."— *Ohio Farmer*, Dec. 12, 1857.

---

[7] *Old dried pollen, or bee-bread, can be removed from the cells of the comb. The question may be if it is worth the effort. First, get the comb, and pollen, wet and let it stand for a while. Then dry the comb and pollen and most of the pollen can be shaken out of the comb. I would only do this on fairly new comb that had not been used in a couple of years, and thus the pollen had dried. Otherwise it would be best to recycle the comb.*

CHAPTER XVI.

HONEY.

THAT honey is a vegetable product, was known to the ancient Jews, one of whose Rabbins asks: "Since we may not eat bees, which are *unclean*, why are we allowed to eat honey?" and replies: "Because bees do not *make* honey, but only *gather* it from plants and flowers." [1]

Bees often obtain a saccharine substance from the honey-dews, which are found on the foliage of many trees, and are sometimes merely an exudation from their leaves, though oftener a discharge from the bodies of small aphides or "plant-lice."[*]

Messrs. Kirby and Spence, in their interesting work on Entomology, have given a description of the honey-dew furnished by the aphides:

"The loves of the ants and the aphides have long been celebrated; you will always find the former very busy on those trees and plants on which the latter abound; and, if you examine somewhat more closely, you will discover that the object of the ants, in thus attending upon the aphides, is to obtain the saccharine fluid secreted by them, which may well be denominated their milk. This fluid, which is scarcely inferior to honey in sweetness, issues in limpid drops from the abdomen of these insects, not only by the ordinary passage, but also by two setiform tubes, placed one on each side just above it. Their sucker being inserted in the tender bark, is, without intermission, employed in absorbing the sap, which, after it has passed through these organs, they

---

[1] *This was a common error early in history, and probably why Linneaus first named the honey bee* Apis mellifera *(honey gatherer) and then later changed it to* Apis mellifica *(honey maker). The current rules of zoological nomenclature prevent us from using the correct Latin name because of priority.*

[*] The Abbé *Boissier de Sauvages*, in "1672 described very fully and accurately these two species of honey-dew. The first kind, he says, has the same origin with the manna of the ash and maple trees of Calabria and Briancon, where it flows plentifully from their leaves and trunks, and thickens in the form in which it is usually seen.' I have received specimens of a honey-dew from California, which is said to fall from the oak trees in stalactites of considerable size.

keep continually discharging.[2] When no ants attend them, by a certain jerk of the body, which takes place at regular intervals, they ejaculate it to a distance."

"Mr. Knight once observed a shower of honey-dew descending in innumerable small globules, near one of his oak trees. He cut off one of the branches, took it into the house, and, holding it in a stream of light admitted through a small opening, distinctly saw the aphides ejecting the fluid from their bodies with considerable force, and this accounts for its being frequently found in situations where it could not have arrived by the mere influence of gravitation. The drops that are thus spurted out, unless interrupted, by the surrounding foliage, or some other interposing body, fall upon the ground; and the spots may often be observed, for some time, beneath and around the trees, affected with honey-dew, till washed away by the rain. The power which these insects possess of ejecting the fluid from their bodies, seems to have been wisely instituted to preserve cleanliness in each individual fly, and, indeed, for the preservation of the whole family; for, pressing as they do upon one another, they would otherwise soon be glued together, and rendered incapable of stirring. On looking steadfastly at a group of these insects (*Aphides salicis*) while feeding on the bark of the willow, their superior size enabled us to perceive some, of them elevating their bodies and emitting a transparent substance in the form of a small shower:

> "'Nor scorn ye now, fond elves, the foliage sear,
> When the light aphids, arm'd with puny spear,
> Probe each emulgent vein, till bright below,
> Like falling stars, clear drops of nectar glow.'—EVANS.

"Honey-dew usually appears upon the leaves as a viscid transparent substance, as sweet as honey itself, sometimes in the form of globules, at others resembling a syrup. It is generally most abundant from the middle of June to the middle of July—sometimes as late as September.

---

[2] Many of the Homoptera tap into a plant's phloem, which has a transpiration pressure that forces the sap through the gut so fast that it does not allow time for the insect to digest any of the nutrients. Thus, these bugs have developed a by-pass system that allows for digestion for some of the sap, while most of the sap goes straight out as honeydew.

"It is found chiefly upon the *oak*, the *elm*, the *maple*, the *plane*, the *sycamore*, the *lime*, the *hazel* and the *blackberry*; occasionally also on the *cherry*, *currant*, and other fruit trees. Sometimes only one species of trees is affected at a time. The oak generally affords the largest quantity. At the season of its greatest abundance, the happy, humming noise of the bees may be heard at a considerable distance, sometimes nearly equaling in loudness the united hunt of swarming."—BEVAN.

In some seasons, bees gather large supplies from these honey-dews, but it is usually abundant only once in three or four years. The honey obtained from it, though seldom light-colored, is generally of a good quality.[3]

The quality of honey varies very much: some kinds are bitter, and others very unwholesome, being gathered from poisonous flowers. A Mandingo African informed a lady of my acquaintance that his countrymen eat none that is *unsealed* until it has been boiled. In some of our Southern States, all that is unsealed is rejected. The noxious properties of honey gathered from poisonous flowers would seem to be mostly evaporated (p. 276) before it is sealed over by the bees. The boiling, however expels them still more effectually, for some persons cannot eat even the best, when raw, with impunity. When honey is taken from the bees, it should be put where it will be safe from all intruders, and not exposed to so low a temperature as to candy in the cells. The little red and the large black ants are extravagantly fond of it, and will carry off large quantities if within their reach. Old honey is more wholesome than that freshly gathered by the bees.[*]

---

[3] *Some of the largest honey yields are from areas that have regular honey-dew flows. The Black Forest area of Europe is a good example of such an area.*

[*] The following extract from the work of Sir J. More, London, 1707, will show the extravagant estimate which the old writers set upon bee-products.

"Natural wax is altered by distillation into an oyl of marvellous vertue: it is rather a Divine medicine than humane, in wounds or inward diseases, it worketh miracles. The bee helpeth to cure all your diseases, and is the best little friend a man has in the world..... Honey is of subtil parts, and therefore doth pierce as oyl, and easily passeth the parts of the body; it openeth obstructions, and cleareth the heart and lights of those humors which fall from the head; it purgeoth the foulness of the body cureth phlegmatick matter, and sharpeneth the stomach; it purgeth those things which hurt the clearness of the eyes, breedeth good blood, stirreth up natural heat, and prolongeth life; it keepeth all things uncorrupt which are put into it, and is a sovereign medicament, both for outward and inward maladies; it helpeth the greif of the jaws, the kernels growing within the mouth, and the squinancy; it is drank against the biting of a serpent or a mad dog; it is good for such as have eaten mush-

To drain honey from virgin combs, bring it to the boiling point in any clean vessel, and, when cool, the wax will float on the top, and the honey may be strained and poured into bottles or jars, which should be tightly covered, to exclude the air.[4] Should it candy, these may be put into cold water and brought to the boiling-point, when the honey will be as nice as ever. Combs which contain bee-bread should be kept separate from the others, as the honey from them is of an inferior quality.·

Empty comb which cannot be used in the hive or spare honey-boxes (p. 71), may be put into water and boiled, when the pure wax will float upon the top, and harden if poured into cold water. If melted again, and run into vessels slightly greased, the impurities will settle at the bottom. Combs which have been so long used by bees for breeding that they will not readily part with their wax, may be put into a coarse woolen bag, with a flat-iron on the top to make it sink, and boiled until the wax has risen to the top of the kettle. *Very old brood-combs* are seldom worth rendering into wax.

New swarms, unless very large, ought not to be admitted to the surplus honey receptacles until they have been hived three or four days. Old stocks should have access to them quite early in the season. If the hives stand in the sun, and the weather is warm, ample ventilation[†] should be given, while bees are storing honey.

The surplus honey may be taken from my hives in a great variety of ways:

(1st.) The hive may be made so long that it can be taken from the ends on frames; and if these ends be separated from the main body of the hive by movable or permanent partitions, the purest honey will be

---

rooms, for the falling sickness, and against the surfeit. Being boiled, it is lighter of digestion, and more nourishing."

[4] *Today we would not think of heating our honey to the point of boiling. Heating darkens the honey by converting some of the sugars into furfurals. Many of the volatile components of honey are also lost because of heating. These are the aroma and flavorings of the honey. Some counties will reject honey that is heated above 40° C (105° F.).*

· In Russia and Germany, very little honey is sold in the comb. Purchasers in this country should beware of the inferior *West India* honey, which is often sold in cans as a superior article, for two or three times its cost.

[†] My hives admit of such complete ventilation, that they may be safely put anywhere except where there is a *pent* heat.

Fig. 56. PLATE XIX.

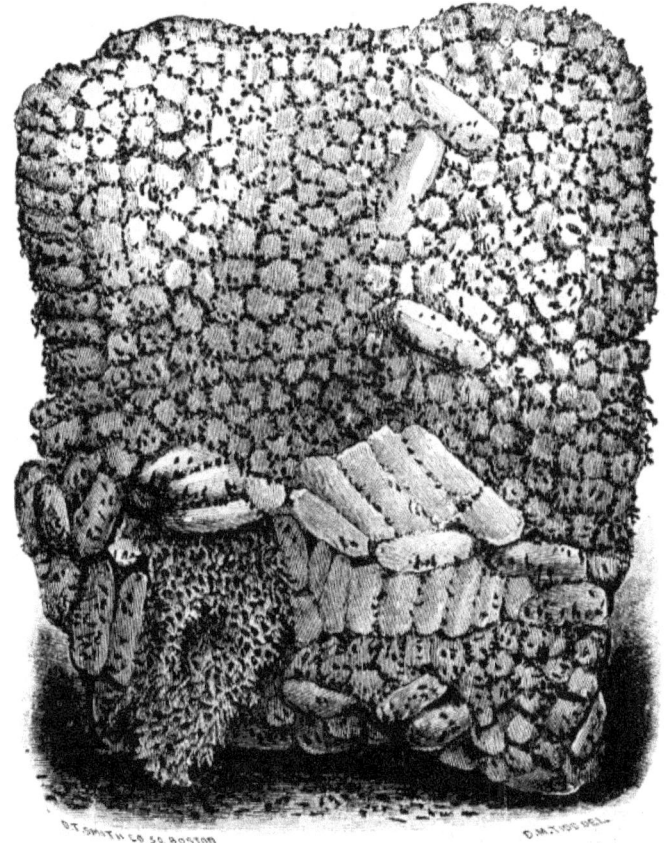

deposited in them. The partitions should be kept about a quarter of an inch from the top and bottom, to allow the bees to pass freely into the ends.·

(2d.) The surplus honey may be stored in large or small frames, put in an upper box or hive (see Plates III., V., and VII, Figs. 9, 16, and 20). Such a box,† when full, may, by a little smoke, be easily removed, and the bees driven from it. Its contents may be sold in gross, or by the single frame.

In all my hives, any additional storage-room may be given, which the season or locality can ever require. The experienced beekeeper well knows that bees will make much more honey in a large box, than in several small ones whose united capacity is the same. In small boxes, they cannot so well maintain their animal heat, and their effective force is thus often wasted at the height of the honey-harvest, when time is, to the last degree, precious.‡

No metallic slides are needed for removing surplus honey-boxes. By blowing smoke into them, before they are taken off, most of the bees will retreat to the main hive, and, if removed early in the morning, or late in the afternoon, and planed on a sheet fastened to the hive, the bees, attracted by the hum of their companions, will speedily leave them, but not until they have swallowed all that they can hold.

· Such a hive, holding a dozen frames in the central apartment, and six in each of the end ones may be cheaply made. The side apartments may be rabbeted so as to receive short frames running from the ends to the partitions, or long ones from front to rear.

† In a favorable season, I have taken two such boxes, each holding over fifty pounds, from a non-swarming hive, and, in good locations, still larger returns may often be realized. The boxes may be set over the main hive, and, as the bees can pass into them without being obliged to travel over the combs, the unusual height will not annoy them.

‡ I am not aware that the attention of Apiarians has ever been called to the loss incurred by compelling bees to store their surplus honey in small receptacles. The bee-keeper cannot afford to sell honey stored in small receptacles, except at a considerable advance over its value in large boxes. By movable frames, the usual objections to large boxes are removed, as honey may be conveniently taken from them for sale or use.

(The newer plastic comb honey boxes [Half-comb© cassettes] seem to allow the bees to enter more easily than some of the earlier designs. Regardless, honey produced in section boxes will be less than the amount produced via the extracted honey method. Some beekeepers put thin-wax foundation into regular frames and then cut the honey sections from them. Such "cut-comb" honey allows the bees to work in the supers the same as any other. This method of producing comb honey also will generally produce less honey since the bees must produce the wax that is needed to store the honey.)

When gorged, they are very reluctant to fly, and this is the reason they are so long in leaving when boxes are carried from the hive. The sooner the bees are driven from them the better, and care must be taken to protect them from robbers, who would soon carry their contents to their own hives. If any of the frames contain brood, they may be returned to the bees. Should the queen be in the box, many bees will refuse to leave it until she is returned to the hive.

(3rd.) Glass vessels, of almost any size or form, make beautiful receptacles for the spare honey; they should have a piece of comb fastened in them, and should be covered with something warm if the weather is cool.*

(4th) If small boxes are used for surplus honey, the one shown in Plate X., Fig. 24, the dimensions of which are given in the Explanation of Hives, will probably be found the simplest, cheapest, and best.†

To remove surplus honey stored in small receptacles, slowly pass a thin knife or *spatula* under the box, to loosen its attachments to the hive; then, before raising it enough to allow any bees to escape, blow smoke under it, and, when they have gorged themselves, it may be safely removed, the hole from the hive being closed or covered with another box. The few bees remaining is the receptacle that is taken off, will quickly fly to their hive. Those who are very timid, may use a slide to prevent any bees from escaping from the hole. Smoke, however, is altogether preferable.

While the most timid may, with proper instructions, safely remove

---

* Honey, stored in tumblers just large enough to receive one comb, may be placed in an elegant form upon the table. While all small receptacles waste the time of the bees, the shallow cells, so many of which must be made in any *cylindrical* vessel, require as large a consumption of time and materials for their covers and bottoms as those which hold more than twice as much honey. *(The closest we come to these glass tumblers are the plastic Halfcomb® cassettes that allow you to see the comb and honey very clearly.)*

† Such a box, which should be furnished either with guides or pieces of comb, will hold three store-combs, weighing together over four pounds, and, by removing a glass, one may be cut out without disturbing the others.

If all the joints of a box are made air-tight by a melted mixture of wax and resin, the bees will be saved much labor in stopping them with propolis; and, when the entrance is closed and covered with the same mixture, the honey may be transported without leakage, even if the combs are broken. Boxes containing honey should be very carefully packed, and lifted without the slightest jarring.

honey, even from the main hive (p. 169), a child ten years old may learn to take off small boxes or glasses.

CHAPTER XVII.

BEE-PASTURAGE — OVER-STOCKING.

EVERY bee-keeper should carefully acquaint himself with the honey-resources of his own neighborhood. My limits will allow me to mention only some of the most important plants from which bees draw their supplies. Since Dzierzon's discovery of the use which may be made of rye flour, early blossoms, producing pollen only, are not so important.[1]

All the varieties of willow abound in both bee-bread and honey, and their early blossoming gives them a special value:

> "First the gray willow's glossy pearls they steal,
> Or rob the hazel of its golden meal,
> While the gay crocus and the violet blue,
> Yield to their fexile trunks ambrosial dew."—EVANS.

The sugar-maple (Acer saccharinus) yields a large supply of delicious honey, and its blossoms, hanging in graceful fringes, will be alive with bees.

Of the fruit trees, the apricot, peach, plum, cherry, and pear, are great favorites but none furnishes so much honey as the apple.[2]

The dandelion, whose blossoms furnish pollen and honey, when the yield from the fruit trees is nearly over, is worthy of a high rank among honey-producing plants.

The tulip tree (Liriodendron), often called "poplar" and "white wood," is one of the greatest honey-producing trees in the world. As

---

[1] *Honey bees will collect almost any dry powdered substance in the spring. That does not mean that they receive any nutrition from these gatherings.*

[2] *Some modern nectar analyses would probably disagree with Langstroth as pear has a very low percentage of sugar, and bees will often leave it for other crops.*

In my experience, the tart cherry is probably as good a nectar producer as apple, but the weather is often not very favorable for gathering nectar when cherries are in bloom.

its blossoms expand in succession, new swarms will sometimes fill their hives from this source alone. The honey, though dark,* is of a good flavor. This tree often attains a height of over one hundred feet, and its rich foliage, with its large blossoms of mingled green and yellow, make it a most beautiful sight.

The linden, or basswood (*Tilia americana*) yields an abundance of white honey of a delicious flavor, and, as it blossoms when both the swarms and parent-colonies are usually populous, the weather settled, and other bee forage scarce, its value to the bee-keeper is very great.†

> "Here their delicious task, the fervent bees
> In Swarming millions tend: around, athwart,
> Through the soft air the busy nations fly,
> Cling to the bud, and with inserted tube,
> Suck its pure essence, its ethereal soul."—THOMSON.

This majestic tree, adorned, so late in the season, with beautiful clusters of fragrant blossoms,, is well worth attention as an ornamental shade-tree. By adorning our villages and country residences with a fair allowance of tulip, linden, and such other trees as are not only beautiful to the eye, but attractive to bees, the honey-resources of the country might, in process of time, be greatly increased.

The common locust is a very desirable tree for the vicinity of an Apiary, yielding much honey when it is peculiarly needed by the bees. In many districts, locust and basswood plantations would be valuable for their timber alone.³

Hives in the vicinity of extensive beds of seed-onions will speedily become very heavy; the offensive odor of the freshly-gathered honey

---

* The honey of Hymettus, which has been so celebrated from the most ancient times, is of a fair golden color. The lightest-colored honey is by no means always the best.
† Judge Fishback says that nearly all his surplus honey is gathered from the linden. A correspondent of the *Bienenzeitung,* in Wisconsin, states that, in 1853, several of his hives increased in weight one hundred pounds each, while this tree was in blossom.
³ *I think it is interesting that beekeepers made their section-comb boxes out of basswood for so many years. Maybe it was just a small percent of the trees that were cut, but still it was an important nectar-bearing tree. With the advent of plastic boxes, for comb honey, this trend is no longer true.*

disappears before it is sealed over by the bees.

Of all the sources from which bees derive their supplies, white clover is usually the most important. It yields large quantities of very pure white honey, and wherever it abounds, the bee will find a rich harvest. In most parts of this country, it seems to be the chief reliance of the Apiary.[4] Blossoming at a season of the year when the weather is usually both dry and hot, and the bees gathering its honey after the sun has dried off the dew, it is ready to be sealed over almost at once. This clover ought to be much more extensively cultivated than it now is. The Hon. Frederick Holbrook, of Brattleboro', Vermont, one of New England's ablest practical farmers and writers on agricultural subjects, thus speaks of its value:

"Red-top, red clover, and white clover seeds, sown together, produce a quality of hay universally relished by stock. My practice is, to seed all dry,* sandy, and gravely lands with this mixture. The red and white clover pretty much make the crop the first year; the second year, the red clover begins to disappear, and the red-top to takes its place; and after that, the red-top and white clover have full possession, and make the very best hay for horses or oxen, milch cows or young stock that I have been able to produce. The crop per acre, as compared with herds-grass (timothy) is not so bulky; but, tested by weight and by spending quality in the Winter, it is much the more valuable."

For years I sought in vain to procure a cross between the red and white clover, having the honey and hay-producing properties of the red, with a short blossom, into which the domestic bee-might insert its proboscis. Such a variety, originating in Sweden, has been imported by Mr. B. C. Rogers, of Philadelphia. It grows as tall as the red clover,

---

[4] *The whole upper Mid-West of the U. S. once was called the White Clover Honey Region by the Bureau of Entomology, Apiculture Section. The predominance of white clover* (Trifolium repens) *as a major source of honey probably persisted until the 1920's when alfalfa replaced most of the clovers as the preferred hay crop. While alfalfa may be an excellent honey crop the farmers rarely allow it to bloom. If alfalfa is allowed to bloom it will produce a yield of about a 100 pounds per colony.*

* Mr. Wagner says: "The yield of honey from various plants and trees depends not only on the character of the season, but on the kind of soil, in which they grow. Marshy meadows are inferior to those of a drier soil for bee-pasturage. White clover growing in the latter will be visited by bees, when that growing in the former is neglected by them."

bears many blossoms on a stalk, in size resembling the white, and, while it answers admirably for bees, is said to be preferred by cattle to almost any other kind of grass. It is known by the name of Alsike, or Swedish white clover.[5]

Mr. Wagner thus speaks of it:

"The views of the value of Swedish white clover, presented by reports from twelve different agricultural societies in the district of Dresden, are the result of careful experiments, made in localities differing greatly in soil and exposure. We recapitulate the chief points:

"1. That Swedish white clover is not so liable as red clover to suffer from cold and wet weather. 2. That on dry and sandy soils it is not so certain or valuable a crop as common white clover, but succeeds admirably on more loamy soils, and, on such, surpasses either of the other kinds. 3. That, in any rotation, it may safely follow the common red clover. 4. That the yield per acre of the first mowing is not inferior to that of the red clover, but that, ordinarily, the aftermath, or rowen,[6] is not so abundant. 5. That, for soiling purposes, it should not be mown till it is in full blossom. 6. That, when cured, it is, as hay, a highly nutritious fodder, and is preferred, by cattle, and milch cows, to that made from red clover. 7. That the aftermath is followed by a dense and excellent growth, furnishing most valuable pasturage till late in the season. 8. That it yields an abundance of seed, easily threshed out by flail or machine, three or four days after mowing. 9. That Swedish white clover is fed to most advantage after it has fully matured its blossoms; whilst red clover, if allowed to stand to this stage, will have already lost a considerable portion of its nutritive properties.

"E. Fürst, the editor of the *Frauendorfer Blatter*, says that this clover is pre-eminent, both in quality and quantity of product, and is especially valuable for the continued succulency of the stalk, even when the plant is in full bloom. It requires a less fertile soil than the red clover, and is less liable to be thrown out by frost in Winter. It also yields a heavier second crop than the common white clover."

The blossoms of buckwheat often furnish, late in the season, a very valuable bee-food.[*]

---

[5] *Alsike clover* (Trifolium hybridum) *will indeed produce a fine crop of honey. My first big crop of surplus clover honey was from Alsike, produced on my great-grandfather's farm, but the crop was sown by my uncle.*

[6] *The second crop of hay; today we usually refer to this as the 2nd cutting, or 3rd, etc.*

[*] This honey is usually gathered when the atmosphere is moist, and in wet seasons, is somewhat liable to sour in the cells. Honey gathered when the atmosphere is dry is usually of the thickest consistency.

Buckwheat is uncertain [†] in its honey-yielding qualities, and, in some seasons, hardly a bee will be seen upon large fields of it.[7] Our best agriculturists are agreed that, on many soils, it is a very profitable crop, and every Apiary ought to have some in its vicinity.[‡]

The Canada thistle yields copious supplies of very pure honey, after the white clover has begun to fail. If farmers will tolerate its growth, it is interesting to know that it can be turned to so good an account.

The raspberry furnishes a most delicious honey. In flavor it is superior to that from the white clover, while its delicate comb almost melts in the mouth. The sides of the roads, the borders of the fields, and the pastures of much of the "hill-country" of New England, abound with the wild red raspberry, and, in such favored locations, numerous colonies of bees may be kept. When it is in blossom, bees hold even the white clover in light esteem.[8] Its drooping blossoms protect the honey from moisture, and they can work upon it when the weather is so wet that they can obtain nothing from the upright blossoms of the clover. As it furnishes a succession of flowers for some weeks, it yields a sup-

---

[†] The secretion of honey in plants, like the flow of the sap from the sugar-maple, depends on a variety; of causes, many of which elude our closest scrutiny. In some seasons the saccharine juices abound, while in others they are so deficient that bees can obtain scarcely any food from fields all white with clover. A change in the secretion of honey will often take place so suddenly, that the bees will, in a few hours, pass from idleness to great activity.

(Langstroth again talks about honey from plants, rather than nectar, which the bees make into honey.

We still do not know very much about nectar secretion. It may be the last frontier as far as beekeeping knowledge.)

[7] *Buckwheat went from a major crop to almost non-existent in the U.S. It has rebounded to some extent, though it will probably never return to its peak. The varieties today are still variable in their nectar secretion.*

[‡] Dzierzon says: "In the stubble of Winter grain, buckwheat might be sown, whereby ample forage would be secured to the bees, late in the season, and a remunerating crop of grain garnered besides. This plant, growing to rapidly and maturing so soon, so productive in favorable seasons, and so well adapted to cleanse the land, certainly deserves more attention from farmers than it receives; and its more frequent and general culture would greatly enhance the profits of bee-keeping. Its long-continued and frequently-renewed blossoms yield honey so abundantly, that a populous colony may easily collect fifty pounds in two weeks, if the weather is favorable."

(I think it is interesting that these people were so observant about the various crops. We know today that buckwheat is allellopathic, that is, it prevents other plants (weeds) from growing, "...so well adapted to cleanse the land." Our experience with buckwheat is that, in mid-summer, it will be blooming in less than three weeks after sowing.)

[8] *Raspberry nectar has one of the highest percentages of sugar of all the flowering plants. In addition, the flower keeps replenishing the nectar throughout the day, so when a bee removes the nectar another bee can get more after a little time passes.*

ply almost as lasting as the white clover. The precipitous and rocky lands, where it most abounds, might be made almost as valuable as some of the vine-clad terraces of the mountain districts of Europe.[9]

"Dr. Bevan suggests the use of lemon-thyme as an edging for garden walks and flower beds. No material good, however, can be done to a large colony by the few plants that can be sown around a bee-house. The bee is too much of a roamer to take pleasure in trim gardens.[*] It is the wild tracts of heath and furze,[10] the broad acres of bean-fields and buckwheat, the lime [11] avenues, the hedge-row flowers, and the clover meadows, that furnish her haunts and fill her cells. To those who wish to watch their habits, a plot of bee-flowers is important, and we knew not the bee that could refuse the following beautiful invitation of Professor Smythe:

"Thou cheeful Bee! come, freely come,
And travel round my woodbine bower;
Delight me with thy wandering hum,
    And rouse me from my musing hour:
Oh! try no more those tedious fields;
Come, taste the sweets my garden yields:
The treasures of each blooming mine,
The buds, the blossoms—all are thine!
And, careless of this noontide heat,
    I'll follow as thy ramble guides,
To watch thee pause and chafe thy feet,
    And sweep them o'er thy downy sides;
Then in a flower bell nestling lie,
And all thy envied ardor ply!
Then o'er the stem, though fair it grow,
With touch rejecting, glance and go.
O Nature kind! O laborer wise!
    That roam'st along the Summer's ray,
Glean'st every bliss thy life supplies,
    And meet'st prepared thy wintry day!

---

[9] *In the cut-over forest hills of New England in the middle of the 19th Century there must have been a lot of raspberry. Today these same hills have mature trees and the raspberry will have been greatly diminished, or is gone.*

[*] I should almost as soon expect, from a small grass-plot, to furnish food for a herd of cattle, as to provision bees from garden plants.

[10] *Any plant of the leguminous genus* Ulex, *especially U. europaeus, a low, much-branched, spiny shrub with yellow flowers, common on waste lands in Europe.*

[11] *Lime is the same as linden or basswood* (Tilea sp.). *The names lime and linden are more often used in Europe and basswood in the U.S. There are several species of* Tilea.

Go, envied, go—with crowded gates
The hive thy rich return awaits;
Bear home thy store in triumph gay,
And shame each idler of the day!"
London Quarterly Review.

If there is any plant which would justify cultivation exclusively for bees, it is the borage (*Borago officinalis*).[12] It blossoms continually from June until severe frost, and, like the raspberry, is frequented by bees even in moist weather. The honey from it is of a superior quality, and an acre would support a large number of stocks.

The golden-rod (*Solidago*) affords a late and very valuable pasturage for bees, yielding, in some regions and seasons, an important part of their Winter stores.[13] Some of the earlier-flowering varieties are of no value to bees; but those which blossom in September abound in honey of a superior quality.

The numerous species of asters, lining, in many districts, the road-sides and the borders of fields, are almost as valuable to the bees as the golden-rod. Where these two plants abound, bees should not be fed until they have passed out of bloom, as light but populous stocks will often obtain from them all the Winter stores they need.

The following catalogue of bee-plants, which might easily be enlarged, is taken from Nutt, an English Apiarian:

"Alder, almond, althea frutex, alyssum, amaranthus, apple, apricot, arbutus, ash, asparagus, aspen, aster, balm, bean, beach, betony, blackberry, borage, box, bramble, broom, bugloss (viper's), buckwheat, burnet, cabbage, cauliflower, celery, cherry, chestnut, chickweed, clover, cole or coleseed, coltsfoot, coriander, crocus, crowfoot, crown imperial, cucumber, currants, cyprus, daffodil, dandelion, dogberry, elder, elm, endive, fennel, furze, golden-rod, gooseberry, gourd, hawthorn, hazel, heath, holly, hollyhock (*trumpet*), honeysuckle, honeywort (*cerinthe*), hyacinth, hyssop, ivy, jonquil, kidney bean, laurel, laurustinus, lavender, leek, lemon, lily (*water*), lily (*white*), lime, linden (bass-wood), liquidamber, liriodendron, locust, lucerne, mallow (marsh), marigold (*French*) marigold (*single*), maple, marjoram (*sweet*), mellilot, melons, mezereon, mignionette, mustard, nasturtium, nectarine, nettle (white), oak, onion, orange,

---

[12] *Borage is now becoming a specialty oil crop, and large acreage has been planted in Canada for the special oil it contains.*
[13] *Goldenrods and asters furnish some of the largest amounts of pollen collected by a colony during the year. Part of the reason for this is that the colony is relatively populous in the Fall, as opposed to the period when there are Spring flowers blooming.*

ozier, parsnip, pea, peach, pear, peppermint, plane, plum, poplar, poppy, primrose, privet, radish, ragweed, raspberry, rosemary (*wild*), roses (*single*), rudbeckiae, saffron, sage, saintfoin, St. John's wort, savory (*winter*), snowdrop, snowberry, stock (*single*), strawberry, sunflower, sycamore, squash, tansy (*wild*), tare, teasel, thistles, thyme (*lemon*), thyme (*wild*), trefoil, turnip, vetch, violet (*single*), wallflower (*single*), woad, willow-herb, willow tree, yellow weasel-snout."

OUR COUNTRY NOT IN DANGER OF BEING OVERSTOCKED WITH BEES

If the opinions commonly entertained on the danger of overstocking are correct, bee-keeping must, in this country, be always an insignificant pursuit.

It is difficult to repress a smile when the owner of a few hives, in a district where as many hundreds might be made to prosper, gravely imputes his ill-success to the fact, that too many bees are kept in his vicinity. If, in the Spring, a colony of bees is prosperous and healthy, it will gather abundant stores, in a favorable season, even if hundreds equally strong are in its immediate vicinity; while, if it is feeble, it will be of little or no value, even if it is in "a land flowing with milk and honey," and there is not another stock within a dozen miles of it.

As the great Napoleon gained many of his victories by having an overwhelming force at the right place, in the right time, so the bee-keeper must have strong colonies, when numbers can be turned to the best account. If his stocks become strong only when they can do nothing but consume what little honey has been previously gathered, he is like a farmer who suffers his crops to rot on the ground, and then hires a set of idlers to eat him out of house and home. There is probably *not* a *square mile* in this whole country which is overstocked with bees, unless it is so unsuitable for bee-keeping as to make it unprofitable to keep them at all. Such an assertion may seem unguarded, but I am happy to be able to confirm it by the following letter from Mr. Wagner, showing the experience of the largest cultivators in Europe:

"Dear Sir:—In reply to your inquiry respecting the over-stocking of a district, I would say, that the present opinion of the correspondents of the Bienenzeitung, appears to be that it cannot readily be done. Dzierzon says, in practice at least, 'it never is done,' and Dr, Radlkofer, of Munich, the President of the second Apiarian Convention, declares that his apprehensions on that score were dissipated by observations which he had opportunity and occasion to make when on his way home from the Convention. I have numerous accounts of Apiaries in pretty close proximity, containing from 200 to 300 colonies each.[14] Ehrenfels had a thousand hives, at three separate establishments, indeed, but so close to each other that he could visit them all in half an hour's ride; and he says that in 1801, the average net yield of his Apiaries was two dollars per hive. In Russia and Hungary, Apiaries numbering from 2,000 to 6,000 colonies, are said not to be infrequent; and we know that as many as 4,000 hives are oftentimes congregated, in Autumn, at, one point on the heaths of Germany. Hence, I think we need not fear that any district of this country, so distinguished for abundant natural vegetation and diversified culture, will very speedily be overstocked. Particularly, after the importance of having stocks populous early in the Spring comes to be appreciated. A week or ten days of favorable weather at that season, when pasturage abound, will enable a strong colony to lay up an ample supply for the year, if its labor be properly directed. "Mr. Kaden, one of the oldest contributors to the Bienenzeitung, in the number for December, 1852, noticing the communication from Dr. Rallkofer, says: 'I also concur in the opinion that a district of country cannot be overstocked with bees, and that, however numerous the colonies, all can procure sufficient sustenance, if the surrounding country contain honey-yielding plants and vegetables, in the usual degree. Where utter barrenness prevails, the case is different, of course, as well as rare.'

"The Fifteenth Annual Meeting of German Agriculturists was held in the city of Hanover, on the 10th of September, 1852, and in compliance with the suggestions of the Apiarian Convention, a distinct section devoted to bee-culture was instituted. The programme propounded sixteen questions for discussion, the fourth of which was as follows:

"'Can a district of country embracing meadows, arable land, orchards, and forests; be so overstocked with bees, that these may no longer find adequate sustenance, and yield a remunerating surplus of their products?'

---

[14] *There are numerous photographs of apiaries with 100-300 colonies in the U.S. in the late 1800's and early 1900's. This was a time when transportation of bees and honey was not easily accomplished. Currently in this country the average number of colonies is about 25/apiary. The number of colonies a beekeeper places in an apiary is a trade off between convenience of having bees in one location as opposed to the average trip of a bee being of greater distance to find a flower that has not had its nectar already removed by bees from another colony from the same apiary..*

"This question was debated with considerable animation. The Rev. Mr. Kleine—nine-tenths of the correspondents of the Bee-Journal are Clergymen—president of the section, gave it as his opinion that 'it was hardly conceivable that such a country could be overstocked with bees.' Counsellor Herwig, and the Rev. Mr. Wilkens, on the contrary, maintained that 'it might be overstocked.' In reply, Assessor Heyne remarked that, whatever might be supposed possible, as an extreme case, it was certain that, as regards the kingdom of Hanover, it could not be even remotely apprehended that too many Apiaries would ever be established; and that, consequently, the greatest possible multiplication of colonies might safely be aimed at and encouraged. At the same time, he advised a proper distribution of Apiaries.'

"I might easily furnish you with more matter of this sort, and designate a considerable number of Apiaries in various parts of Germany, containing from twenty-five to five hundred colonies. But the question would still recur, do not these Apiaries occupy comparatively isolated positions? and, at this distance from the scene it would obviously be impossible to give a perfectly satisfactory answer.

"According to the statistical tables of the kingdom of Hanover, the annual production of bees-wax in the province of Lunenberg is 300,000 lbs., about one-half of which is exported; and, assuming one pound of wax as the yield of each hive,[15] we must suppose that 300,000 hives are annually '*brimstoned*' in the province; and assuming further, in view of casualties, local influences, unfavorable seasons, &c., that only one-half of the whole number of colonies maintained, produce a swarm each every year, it would require a total of at least 600,000 colonies (141 to each square mile) to secure the result given in the tables. The number of square miles stocked, even to this extent, in this country, are, I suspect, few and far between.' It is very evident that this country is far from being overstocked; nor is it likely that it ever will be.

"A German writer alleges that the bees of Lunenberg pay all the taxes assessed on their proprietors, and leave a surplus besides.' The importance attached to bee-culture accounts, in part, for the remarkable fact that the people of a district so barren, that it has been called the 'Arabia of Germany,' are, almost

---

[15] *In a recent analysis that I made on commercial beekeepers in the U.S. (Amer. Bee J. v. 130:405-407), the amount of beeswax produced by the "average" beekeeper on a surplus of 90 lbs./colony was 1.7 lbs of wax. However, a comparison cannot be made to the amount of honey produced in Germany at that time since the method of rendering the honey was very much different before extractors and movable-combs, as "brimstoned" means the colonies were killed with sulfur fumes, and all the combs and honey melted. Later you will see beeswax and honey figures from other areas that would indicate between 10 and 15 pounds of honey per colony were secured from these colonies.*

without exception, in easy and comfortable circumstances. Could not still more favorable results be obtained in this country, under a rational system of management, availing itself of the aid of science art, and skill?

"But I am digressing. My design was, to furnish you with an account of bee-culture as it exists in an entire district of country, in the hands of the common peasantry. This, I thought, would be more satisfactory, and convey a better idea of what may be done on a large scale, than any number of instances which might he selected of splendid success in isolated cases.—Very truly yours,
SAMUEL WAGNER"
REV. L. L. LANGSTROTH.

I am persuaded that, even in the poorest parts of New England, there are but few districts which could not be made to yield as large returns as the province of Lunenberg, even if the old-fashioned plan of management was adhered to. The following interesting statements have been furnished to me by Mr. Wagner:

"'When a large flock of sheep,' says Oettl, 'is grazing on a limited area, there may soon be a deficiency of pasturage. But this cannot be asserted of bees, as a good honey-district cannot readily be overstocked with them. To-day, when the air is moist and warm, the plants may yield a superabundance of nectar; while to-morrow, being cold and wet, there may be a total want of it. When there is sufficient heat and moisture, the saccharine juices of plants will readily fill the nectaries, and will be quickly replenished when carried off by the bees. Every cold night checks the flow of honey, and every clear, warm day re-opens the fountain. *The flowers expanded to-day must be visited while open; for if left to wither, their stores are lost*. The same remarks will apply substantially in the case of honey-dews. Hence, bees cannot, as many suppose, collect to-morrow what is left ungathered to-day, as sheep may graze hereafter on the pasturage they do not need now. Strong colonies and large Apiaries are in a position to collect ample stores when forage suddenly abounds, while, by patient, persevering industry, they may still gather a sufficiency, and even a surplus when the supply is small, but more regular and protracted.'

"The same able Apiarian, whose golden rule in bee-keeping is, to keep none but strong colonies, says that, in the lapse of twenty years since he established his Apiary, there has not occurred a season in which the bees did not procure adequate supplies for themselves, and a surplus besides. Sometimes, indeed, he came near despairing, when April, May, and June were continually cold, wet, and unproductive; but in July, his strong colonies speedily filled their garners, and stored up some treasures for him; while, in such seasons, small colonies could not even gather enough to keep them from starvation.

"Mr. A. Braun states, in the *Bienenzeitung*, September, 1854, that he has a mammoth hive furnished with combs containing at least 184,230 cells,[*] and placed on a platform scale, that its weight may readily he ascertained at stated periods. On the 18th of May, it gained eighteen pounds and a half. On the eighteenth of June, a swarm weighing seven pounds issued from it, and the following day it gained over six pounds in weight. Ten days of abundant pasturage would enable such a colony to gather a large surplus, while five times the number of equally favorable opportunities would be of small avail to a feeble stock.

"The island of Corsica paid to Rome an annual tribute of 200,000 lbs. of wax, which presupposes the production of from two to three million pounds of honey yearly. The island contains 3790 square miles.

"According to Oettl (p.389), Bohemia contained 160,000 colonies in 1853, from a careful estimate, and he thinks the country could readily support four times that number. The kingdom contains 20,200 square miles.

"In the province of Attica, in Greece, containing forty-five square miles, and 20,000 inhabitants, 20,000 hives are kept, each yielding, on an average, thirty pounds of honey and two pounds of wax.

"East Friesland, a province of Holland, containing 1,200 square miles, maintains an average of 2,000 colonies per square mile.(HEUBEL, *Bienenzeitung*, 1854, p. 11.)

"According to an official report, there were in Denmark, in 1838, eighty-six, thousand and thirty-six colonies of bees. The annual prod-

---

[*] Such a hive would hold about three bushels. Wildman says that "a clergyman set a well-stocked hive of bees on a tub turned bottom up, after having made a hole through the bottom, and took from the tub four hundred and twenty pounds of honey."

uct of honey appears to be about 1,841,800 lbs. In 1855, the export, of wax from that country was 118,379 lbs.

"In 1856, according to official returns, there were 58,964 colonies of bees, in the kingdom of Wurtemberg.

"In 1857, the yield of honey and wax in the empire of Austria was estimated to be worth over seven millions of dollars."

Doubtless, in these districts, where honey is so largely produced, great attention is paid to the cultivation of crops which, while in themselves profitable, afford abundant pasturage for bees.

Although bees will fly, in search of food, over three miles,[*] still, if it is not within a circle of about two miles in every direction from the Apiary, they will be able to store but little surplus honey.[†] If pasturage abounds within a quarter of a mile from their hives, so much the better; there is no great advantage, however, in having it close to them, unless there is a great supply, as bees, when they leave the hive, seldom alight upon the neighboring flowers. The instinct to fly some distance seems to have been given them to prevent them from wasting their time in prying into flowers already despoiled of their sweets by

---

[*] "Mr. Kaden, of Mayence, thinks that the range of the bee's flight does not usually extend more than three miles in all directions. Several years ago, a vessel, laden with sugar, anchored, off Mayence, and was soon visited by the bees of the neighborhood, which continued to pass to and from the vessel from down to dark. One morning, when the bees were in full flight, the vessel sailed up the river. For a short time, the bees continued to fly as numerously as before; but gradually the number diminished, and, in the course of half an hour, all had ceased to follow the vessel, which had, meanwhile, sailed more than four miles."—*Bienenzeitung,* 1854, p. 83.

(More recent studies would agree with most of these statements. Bees have been recorded to fly up to 6 miles under special circumstances. Most bees fly under two miles, and only rarely will a colony's bees forage at three miles.)

[†] "Judging from the sweep that bees take from the side of a railroad train in motion, we should estimate their pace at about thirty miles an hour. This would give them four minutes to reach the extremity of their common range.

(Honey bees have a top speed of about 25 mph.)

"Mr. Cotton saw a man in Germany who kept all his numerous stocks rich by changing their places as soon as the honey-season varied. 'Sometime he sends them to the moors, sometimes to the meadows, sometimes to the forest, and sometimes to the hills. In France—and the same practice has existed in Egypt from the most ancient times—they often put hundreds of hives in a boat, which floats down the stream by night and stops by day.'"—*London Quarterly Review.*

previous gatherers.

In all my arrangements, I have aimed to save every step for the bees that I possibly can. With the alighting-board properly arranged, and covered, in windy situations, with cotton cloth (p. 279), bees will be able to store more honey, even if they have to go a considerable distance for it, than they otherwise could from pasturage nearer at hand. Many beekeepers utterly neglect all suitable precautions to facilitate the labors of their bees, as though they imagined them to be miniature locomotives, always fired up, and capable of an indefinite amount of exertion. A bee cannot put forth more than a certain amount of physical effort, and a large portion of this ought not to be spent in contending against difficulties from which it might easily be guarded. They may often be seen panting after their return from labor, and so exhausted as to need rest before they enter the hive.

Dzierzon's [*] experience as to the *profits* of bee-keeping has already been given (p. 21). With proper management, five dollars worth of honey may, on an average of years, be obtained for each stock that is wintered in good condition. The worth of the new colonies I set off against the labor of superintendence, cost of hives, and interest on the capital invested.

A careful man, who, with my hives, will begin bee-keeping on a prudent [*] scale, enlarging his operations as his skill and experience increases, will find, in any region where honey commands a good price,

---

[*] "It is by no means easy to devise a rule for estimating the profits of bee-culture, whether we regard the number of colonies or the number of square miles. He is not the best Apiarian who obtains the largest yield from a single hive, but keeps only one or two. By very judicious and careful management, a hundred colonies might yield a large profit, yet fall far short of what three hundred would have yielded in the same location and same season, with much less supervision and attention. He is not the most successful farmer who produces the most extraordinary yield from a single rod of ground, but he who secures the amplest crops from an extensive area, well cultivated. The swarming system may be very advantageous in certain localities, in spite of its manifest wastefulness; though, in other localities, it would, because of that unavoidable wastefulness, render bee-keeping a decidedly losing business, since the system involves a vast expenditure of honey for the production and maintenance of brood, which scarcely matures before it is doomed to the brimstone-pit, leaving to its owner often a smaller quantity of honey than the swarm would have produced if taken up three weeks after it was hived.

"Confine the queen of an artificial swarm, so as to prevent her from depositing eggs in the combs, and the colony will, in a short time in the gathering season, accumulate much larger stores of honey than one whose queen is left at liberty, though equal in age and population. Thus, also, a colony having a very prolific queen, will, even in favorable seasons, lay up much less honey, unless ample storeroom is given them, than one whose queen lays fewer eggs. From these and similar facts, which might be enumerated, it is evident that a very large number of particulars must be taken into consideration when endeavoring to form some general rule for estimating the profits of bee-culture."—DZIERZON.

The old-fashioned bee-keeper should know well the honey-recourses of his district, in order to decide upon the best time for "taking up" his bees. If bees are smothered, it will be found decidedly advantageous to remove and destroy their queens, at least three weeks before taking their honey. In this way, the production of brood and consumption of honey will be checked, and the combs will be in a much better condition for melting.

[*] Bee-keepers cannot be too cautious in entering largely upon new systems of management, until they have ascertained, not only that they are good, but that *they* can make a good use of them. There is, however, a golden mean between the stupid conservatism that tries nothing new, and the rash experimentation, on an extravagant scale, which is so characteristic of the American people.

that the preceding estimate is a moderate one. In favorable localities, a much larger profit may be realized.

CHAPTER XVIII.

THE ANGER OF BEES — REMEDIES FOR THEIR STINGS.

The gentleness of bees, when properly managed, makes them wonderfully subject to human control. When gorged with honey, they may be taken up by handfuls, and suffered to run over the face, and may even have their glossy backs gently smoothed as they rest on our persons; and all the feats of the celebrated Wildman may be safely imitated by experts, who, by securing the queen, can make the bees hang in large festoons from their chin, without incurring any risk of being taken by the beard.

> "Such as the spell, which round a Wildman's arm,
> Twin'd in dark wreaths the fascinated swarm;
> Bright o'er his breast the glittering legions led,
> Or with a living garland bound his head.
> His dextrous hand, with firm yet hurtless hold,
> Could seize the chief; know by her scales of gold,
> Prune'mid the wondering train her filmy wing,
> Or o'er her fold, the silken fetter fling."

M. Lombard, a skillful French Apiarian, narrates the following interesting occurrence, to show how peaceable bees are in swarming time, and how easily managed by those who have both skill and confidence:

"A young girl of my acquaintance, who much afraid of bees, was completely cured of her fear by the following incident: A swarm having come off; I observed the queen alight by herself at a little distance from the Apiary. I immediately called my little friend, that I might show her the queen, she wished to see her more nearly; so, after having caused her to put on her gloves, I gave the queen into her hand. We were in an instant surrounded by the whole bees of the swarm. In this emergency, I encouraged the girl to be steady, bidding her be silent and fear nothing, and remaining myself close by her. I then made her stretch out her right hand, which held the queen, and covered her head and shoulders with a very thin handkerchief. The swarm soon fixed on her hand, and hung from it, as from the branch of a tree. The little girl was delighted above measure at the novel sight and so entirely free from all fear, that she bade me uncover her face. The spectators were charmed with the interesting spectacle. At length I brought a hive, and, shaking the swarm from her hand, it was lodged in safety, and without inflicting a single wound."

A practical acquaintance with the principles set forth in this Treatise, will render it unnecessary, *under any circumstances, to provoke to fury a colony of bees*. When thoroughly aroused, by the overturning, or violent jarring of their hive, or by the presence of a sweaty horse, or any offensive animal, they are terribly vindictive and severe, and even

An Unfortunate Bee-ing

dangerous consequences may ensue. As our domestic animals may, by ill-treatment, be roused to such fury as to endanger our lives, so the most peaceful family of bees may be quickly taught to attack any living thing that approaches their domicile.

When a colony of bees is unskillfully dealt with, they will "compass about" their assailant with savage ferocity; and wo be to him, if they can creep up his clothes, or find a single unprotected spot on his person. He will fare as badly as the "UNFORTUNATE *Bee-ing*," so ludicrously depicted in "Hood's Comic Sketches."

Those who have much to do with bees, should wear a *bee-hat*, unless they are proof against the venom of their stings; for, while tens of thousands will continue their pursuits without annoying those who do not molest them, a few *dyspeptic* bees (p 256), will come buzzing around their ears, determined to sting, without the slightest provocation. Even these, however, retain some touch of grace, amidst all their desperation. Like the scold, whose elevated voice gives timely warning to escape the sound of her tongue, so a bee bent on mischief, by raising its note far above the peaceable pitch, gives timely warning that danger is impending. Even then, if it has not been provoked to madness, it will seldom sting, unless it can plant its weapon on the face of its victim, and, if possible, near the eye; for, like all the stinging tribe, it has an intuitive perception that this in the most vulnerable spot.[1] If the head is quietly lowered, and the face covered with the hands, they will follow a person, often for rods, all the time sounding their war-note in his ears, and daring the sneaking fellow to allow them to catch but a glimpse of his coward face.

Cotton, quoting from Butler, who, in these remarks, follows mainly Columella, says:

"Listen to the words of an old writer:— 'If thou wilt have the favour of thy bees, that they sting thee not, thou must avoid such things as offend them: thou must not be unchaste or uncleanly; for impurity and sluttiness (themselves being most chaste and neat) they utterly abhor; thou must not

---

[1] Bees, more than likely, recognize the changes in pattern from light to dark. This is the reason that eyes are often attacked. This is a major reason that a bee veil is worn—to protect the eyes. In addition the contrast of a dark veil along with a light (white) bee suit focus the attention on the veil where the bees normally can't reach the skin. Bees will also attack dark hair or fur, as over evolutionary time the bee's major pests have been dark furry animals, such as bears.

come among them smelling of sweat, or having a stinking breath, caused either through eating of leeks, onions, garlick, and the like, or by any other means, the noisomeness whereof is corrected by a cup of beer, thou must not be given to surfeiting or drunkenness; thou must not come puffing or blowing unto them, neither hastily stir among them, nor resolutely defend thyself when they seem to threaten thee; but softly moving thy hand before thy face gently put them by; and lastly, thou must be no stranger unto them. In a word thou must be chaste, cleanly, sweet, sober, quiet, and familiar; so will they love thee, and know thee from all others. When nothing hath angered them, one may safely walk along by them; but if he stand still before them in the heat of the day, it is a marvel but one or other spying him, will have a cast at him.'·

"Above all, never blow [†] on them; they will try to sting directly, if you do.

"If you want to catch any of the bees, make a bold sweep at them with your hand; and if you catch them without pressing them, they will not sting. I have so caught three or four at a time. If you want to do anything to a single bee, catch him 'as if you loved him,' between your finger and thumb, where the tail joins on to the body and he cannot hurt you."

If a person is attacked by angry bees, not the slightest attempt should be made to act on the offensive; for, if a single one is struck at, others will avenge the insult; and if resistance is continued, hundreds, and at last, thousands, will join them.[2] The assailed party should

---

· Many persons image themselves to be quite safe, if they stand at a considerable distance from the hives; whereas, cross bees delight to attack those whose more distant position makes them a surer mark to their long-sighted vision, than persons who are close to their hives.

(Bee's vision is no better, or worse, at 6 feet than it is at 100. Persons at a distance, by chance, may have different colored clothing or are attacked for other reasons.)

[†] While bees resent the *warm* breath exhaled *slowly* from the lungs, I have ascertained, that they will run from a blast of cold air blown upon them by the mouth of the operator, almost as quickly as from smoke. Before employing smoke, I often used a pair of bellows.

[2] *We now know that bees leave an alarm pheromone behind when they sting. This attracts more bees to the same spot as the first sting. The chemicals in the pheromone are somewhat volatile and dissipate in time, but usually not soon enough that the person won't get stung again. The chemicals that make up the alarm pheromone are known, e.g., one of them is isoamyl acetate, the same chemical you smell in a banana.*

quickly retreat to the protection of a building, or if none is near, should hide in a clump of bushes, and lie perfectly still, with his head covered, until the bees leave him. When no bushes are at hand, they will generally give over the attack, if he lies still on the grass, with his face to the ground.

Those who are alarmed if a bee enters the house, or approaches them in the garden or fields, are ignorant of the important fact, that *a bee at a distance from its hive, never volunteers an attack.* Even if assaulted, they seek only to escape, and never sting, unless they are hurt.

If they were as easily provoked away from home, as when called to defend those sacred precincts, a tithe of the merry gambols in which our domestic animals indulge, would speedily bring about them a swarm of infuriated enemies; we should no longer be safe in our quiet rambles among the green fields; and no jocund mower could whet or swing his peaceful scythe, unless clad in a dress impervious to their stings. The bee, instead of being the friend of man, would, like savage wild beasts, provoke his utmost efforts for its extermination.

Let none, however, take encouragement from the contract between the contrast of bees at home and abroad, to reserve all their pleasant ways for other places than the domestic roof; for, towards the members of its own family the bee is all kindness and devotion; and while, among human beings, a *mother* is often treated by her own children with disrespect or neglect, among bees she is always waited upon with reverence and affection.

It is true, that if any members of a colony become unable to perform their share of labor, they are dragged from the hive by their pitiless companions. It is, however, a necessary law of their economy, that those who *cannot* work, shall not eat; nor is there anything in the nature of a bee, that can be benefited by nursing the sick, while the

---

Africanized bees react to the alarm pheromone much more actively than do the European bees. This is one reason that more of these bees sting when their nest is disturbed.

Fig. 57. PLATE XX.

noblest traits of humanity are often developed by the incessant care bestowed upon the weak and helpless.

Huber has demonstrated, that bees have an exceedingly acute sense of smell, and that unpleasant odors quickly excite their anger.* Long before his time, Butler said, "Their smelling is excellent, whereby, when they fly aloft into the air, they will quickly perceive anything under them that they like, even though it be covered." They have, therefore, a special dislike to those whose habits are not neat,† and who bear about them a perfume not in the least resembling

> "Sabean odors
> From the spicy shores of Araby the blest"

A sweaty horse is detested by bees, and, when assailed by them, is often killed; as, instead of running away, like most other animals, it will plunge and kick until it falls overpowered. The Apiary should be fenced in, to prevent, horses and cattle from molesting the hives.

The sting of a bee, upon some persons, produces very painful and even dangerous effects.[3] I have often noticed that, while those whose systems are not sensitive to the venom, are rarely molested by bees, they seem to take a malicious pleasure in stinging those upon whom their poison produces the most virulent effect. Something in the secretions of such persons may both provoke the attack and render its consequences more severe.

---

* Strong perfumes, however pleasant to us, are disagreeable to bees; and Aristotle observes, that they will sting those scented with them. I have known persons ignorant of this fact to be severely treated by bees.

† Some persons, however cleanly, are assaulted by bees as soon as they approach their hives. It is related of a distinguished Apiarian that, after a severe attack of fever, he was never able to be on good terms with his bees. That they can readily perceive the slightest differences in smell, is apparent from the fact that any number of colonies, fed from a common vessel, will be gentle towards each other, while they will assail the first strange bee that alights on the feeder.

(Bees will often attack members of another colony at a feeder. It appears to be the same response as guarding against robbers.)

[3] *About 0.5% of the people are hypersensitive to stings. This is because they have been stung at least once before and the immunoglobin E (IgE) is stimulated and over produced, and at the next sting the person may go into anaphylactic shock and could die. Hypersensitivity can occur with any foreign substance, e.g., antibiotics.*

The smell of their own poison produces a very irritating effect upon bees. A small portion of it offered to them on a stick, will excite their anger.* "If you are stung," says old Butler, "or anyone in the company—yea, though a bee has stricken but your clothes, especially in hot weather—you were best be packing as fast as you can, for the other bees, smelling the rank flavor of the poison, will come about you as thick as hail."

REMEDIES FOR THE STING OF A BEE.

If only a few of the host of cures, so zealously advocated, could be made effectual, there would be little reason to dread being stung.

The first thing to be done after being stung, is to pull the sting out of the wound as *quickly as possible*. When torn from the bee, the poison-bag, and all the muscles which control the sting, accompany it; and it penetrates deeper and deeper into the flesh, injecting continually more and more poison into the wound. If extracted at once, it will very rarely produce any serious consequences. After the sting is removed, the utmost care should be taken not to irritate the wound by the *slightest rubbing*. However intense the smarting, and the disposition to apply friction to the wound, *it should never be done*, for the moment that the blood is put into violent circulation, the poison is quickly diffused over a large part of the system, and severe pain and swelling may ensue. On the same principle, by severe friction, the bite of a mosquito, even after the lapse of several days, may be made to swell again. As most of the popular remedies are *rubbed in*, they are worse than nothing.

If the mouth is applied to the wound, unpleasant consequences may follow; for, while the poison of snakes, affecting only the circulating system, may be *swallowed* with impunity, the poison of the bee acts with great power on the organs of digestion.[4] Distressing headaches

---

* When bees thrust out their stings in a threatening manner, a minute drop of poison can be seen on their points, some of which is occasionally flirted into the eye of the Apiarian, and causes severe irritation.

[4] *Langstroth is wrong here. Bee's venom causes mast cell degranulation and the release of histamines that cause the pain at the site of the sting. There also may be immunoglobin antibodies brought into play that are systemic, but not in the digestive system.*

are often produced by it, as any one who has been stung or has tasted the poison, very well knows.*

Mr. Wagner says: "The juice of the ripe berry of the common coral honeysuckle (*Lonicera caprifolium*) is the best remedy I have ever used for the sting of bees, wasps, hornets, &c. The berries or the expressed juice may be preserved in a bottle well closed, and will keep their efficacy more than a year."

The milky juice of the white poppy is highly recommended. An old German writer states that it will instantaneously allay the pain and prevent swelling.

Others recommend the juice of tobacco as a sovereign panacea. Relief has unquestionably been found, by different persons, from each of these remedies, and there is as little reason to expect that one remedy will answer for all as that the same disease can always be cured by the same medicines.

In my own case, I have found cold water to be the best remedy for a bee-sting. The poison being very volatile, is quickly dissolved in it; and the coldness of the water has also a powerful tendency to check inflammation.[5]

The leaves of the plantain, crushed and applied to the wound, are a very good substitute when water cannot at once be procured. Bevan recommends the use of spirits of hartshorn, and says that, in cases of severe stinging, its internal use is also beneficial.†

Some of the symptoms of anaphylactic shock can include nausea, and this may be the reason Langstroth felt the digestive system was involved.

* An old writer says; "If bees, when dead, are dried to powder, and given to either man or beast, this medicine will often give immediate ease in the most excruciating pain, and remove a stoppage in the body when all other means have failed." A tea made by pouring boiling water upon bees has recently been prescribed, by high medical authority, for violent strangary; while the poison of the bee, under the name of *apis*, is a great homeopathic remedy.

[5] *Modern allergists would concur. An ice-cube or cold is about the best treatment for a sting. Aspirin taken about 30 minutes prior to a visit to the apiary will also help reduce the localized swelling.*

† It may be some comfort to novices to know that the poison will produce less and less effect upon their system. Old bee-keepers, like Mithridates, appear almost to thrive upon poison itself. When I first became interested in bees, a sting was quite a formidable thing, the pain being often very intense, and the wound swelling so as sometimes to

Timid Apiarians, and all who suffer severely from the sting of a bee, should by all means protect themselves with a bee-dress. The great objection to such a dress, as usually made, is, that it obstructs clear vision, so highly important in all operations, besides producing such excessive heat and perspiration, as to make one using it peculiarly offensive to the bees. I prefer what I call a bee-hat (Plate XI., Fig. 25), of entirely novel construction. It is made of wire-cloth, the meshes of which are too fine to admit a bee, but coarse enough to allow a free circulation of air, and to permit distinct sight. The wire-cloth should be first sewed together like a hat, and made large enough to go very easily over the head; its top may be of cotton cloth, and the same material should be fastened around its lower edge. If the top is made of sole leather, it would serve a better purpose. A piece of wire-cloth one foot wide, by two and a half feet long, will make a good fit for most persons. With such a hat, there is no danger from waspish bees, and its cape may be tucked under the coat, or so securely fastened, as to defy all assailants.

The hands may be protected by India-rubber gloves, such as are

---

obstruct my sight. At present, the pain is usually slight, and if the sting is quickly extracted, no unpleasant consequences ensue, even if no remedies are used. Huish speaks of seeing the bald head of Bonner, a celebrated practical Apiarian, covered with stings, which seemed to produce upon him no unpleasant effects. The Rev. Mr. Kleine advises beginners to suffer themselves to be stung frequently, assuring them that, in two seasons, their system will become accustomed to the poison!

An Old English Apiarian advises a person who has been stung, to catch as speedily as possible another bee, and make it sting on the same spot. Even an enthusiastic disciple of Huber might hesitate to venture on such a singular homeopathic remedy; but as this old writer had stated, what I had verified in my own experience, that the oftener a person was stung the less he suffered from the venom, I determined to make trial of his prescription. Allowing a sting to remain until it had discharged all its poison, I compelled another bee to insert its sting, as nearly as possible, in the same spot. I used no remedies of any kind, and had the satisfaction, in my zeal for new discoveries, of suffering more from the pain and swelling than for years before.

(Modern science knows about an increase in the fraction of the immunoglobins (IgG) from repeated stinging. This fraction (IgG) blocks the venom and thus keeps the venom from stimulating the IgE fraction. It is the IgE fraction that reacts with the mast cells causing de-granulation and localized swelling. In the case cited by Langstroth, the first sting probably used up most of the readily available IgG fraction, thus allowing the IgE to produce its effect.

Most beginners do not get stung enough, and as a result the IgE fraction is increased without the help of the IgG fraction, which appears to be slower to develop.)

now in common use. These gloves, while impenetrable to the sting of a bee, do not materially interfere with the operations of the Apiarian. As soon, however, as the bee-keeper acquires confidence and skill, he will much prefer to use nothing but the bee-hat, even at the expense of an occasional sting on his hands. If the hands are wet with honey, they will seldom be stung.

Woolen gloves are objectionable, as everything rough or hairy has an extremely initiating influence upon bees. This is probably owing to the fact that, in a state of nature, bears, foxes, and other hairy animals, are their principal enemies. No sooner do they feel the touch of anything rough or hairy, than they dart out their stings.

Butler says: "They use their stings against such things as have outwardly some offensive excrement, such as hair or feathers, the touch where of provoketh them to sting. If they alight upon the hair of the head or beard, they will sting if they can reach the skin. When they are angry, their aim is most commonly at the face, but the bare hand, that is not hairy, they will seldom sting, unless they be much offended."

Preface to Chapter XIX.

This chapter is interesting in its historical context. Langstroth devotes this entire chapter to a bee that he had never seen, as it had not yet been established in this country. However, he was convinced by his reading of all of the accounts of the Italian bee that this was the honey bee that he wanted in his apiary. I am sure he was influenced by Samuel Wagner in this interest, as he wrote the letter that is incorporated, along with footnotes by him as well. As you will read in the final footnote, they were importing Italian bees as the book went to press. Langstroth (and Wagner) were certainly correct in their assessment, as the Italian bee has become the dominant race within North America.

CHAPTER XIX.

THE ITALIAN HONEYBEE.

ARISTOTLE speaks of three different species[1] of the honey-bee, as well known in his time. The *best variety* he describes as "μιχρα, ξτρογγυλη χαι ποιχιλη"—that is, small and round in size and shape, and variegated in color.

Virgil (*Georgicon*, lib. IV., 98) speaks of two kinds as flourishing in his time; the better of the two, he thus describes:

"Elucent aliae, et fulgore coruscant,
Ardentes auro, et paribus lita corpora guttis.
Haec pottor soboles; hinc coeli tempore certo
Dulcia melia premes."

The better variety, it will be seen, he characterizes as spotted or variegated, and of a beautiful golden color.

The attention of bee-keepers has recently been called to this variety of the honey-bee, which, after the lapse of more than two thousand years, still exists distinct and pure from the common kind. The following letter from Mr. Wagner will show the importance attached to this species, by some of the most skillful and successful Apiarians in Europe:

"YORK, Pa., August 5, 1856.

"MY DEAR SIR:—The first account we have of the Italian bees, as a distinct race or variety, is that given by Capt. Baldenstein, in the *Bienenzeitung*, 1848, p. 26.* Being stationed in Italy, during part of the Napole-

---

[1] *Today we would say race of bees or sub-species—all of the same species, Apis mellifera.*

* The Rev. E. W. Gilman, of Bangor Maine, has recently directed by attention to Spinola's "*Insectorum Liguriae species novae aut rariores*," from which it appears, that Spinola accurately described all the peculiarities of this bee, which he found in Piedmont, in 1805. He fully identified it with the bee described by Aristotle, and calls it the *Ligurian Bee*, a name now very generally adopted in Europe. (*This Italian or Ligurian bee is now given the scientific name of* Apis mellifera ligustica, *as a separate sub-species*

onic wars, he noticed that the bees in the Lombardo-Venetian district of Valtelin, and on the borders of Lake Como, differed in color from the common kind, and seemed to he more industrious. At the close of the war, he retired from the army, and returned to his ancestral castle, on the Rhaetian Alps, in Switzerland; and to occupy his leisure, had recourse to bee-culture, which had been his favorite hobby in earlier years. While studying the natural history, habits and instincts of these insects, he remembered what he had observed in Italy, and resolved to procure a colony from that country. Accordingly, he sent two men thither, who purchased one, and carried it over the mountain, to his residence, in September, 1843.

"In May, 1847, this colony, the queen of which had never failed to produce genuine Italian brood,[2] began to show signs of weakness, but suddenly recovered in the following month; and it was evident that it had supplied itself with a new queen, which had fortunately been impregnated by an Italian drone, as she produced genuine, or pure brood.[3] On the 15th of May, 1848, this queen issued with a swarm, and he hoped that, as be had placed the parent-hive in a rather isolated location, her successor would he impregnated by an Italian drone. But in this, he was doomed to disappointment; she produced a bastard progeny, while the emigrant queen produced genuine brood, as before. Similar disappointments awaited him from year to year; and in June, 1851, he possessed only one colony of the pure stock.

"Among the points which he considered as definitely established, by his observations on the Italian bee, are the following: 1. The queen, if healthy, retains her proper fertility at least three or four years. 2. The Italian bee is more industrious, and the queen more prolific, than the common kind; because, in a most unfavorable year, when other colonies pro-

---

from A. mellifera mellifera, the sub-species that Linneaus described from northern Europe.)

[2] *While the colony may have had only one queen during these 5+ years, it would seem unlikely. Thus, it probably had replaced its queen sometime previously, as well, which may say something about assortative matings. There are comments later in the text about the differences in drone flight between the two races, which may aid in the separation of the two sub-groups during mating.*

[3] *Beekeepers were convinced, at that time, that queens only mated once. By what we know today about multiple mating it is somewhat remarkable that they were able to get a "pure mating," but see footnote above. We also know that body color is determined by eight genes and thus a degree of blending could have occurred without much notice.*

duced few swarms and little honey, his Italian colony produced three swarms, which filled their hives with comb, and, together with the parent-stock, laid up ample stores for Winter; the latter yielding, besides, a box well filled with honey. The three young colonies were among the best in his Apiary. 3. The workers do not, at most, live longer than one year; for, though the bees and brood in the parent-hive, when the first swarm and old queen left, were of the Italian stock exclusively, few of this kind remained in the Fall, and none survived the Winter. 4. The young queen is impregnated soon after she is established in a colony, and continues fertile during life. Were this not so, the genuine queens would not have continued to produce pure brood during those seven successive years. 5. The queen leaves the hive to meet the drones. If not, it would scarcely have happened, that all the young queens bred in those seven years, with only one exception, were impregnated by common drones, and produced a bastard progeny. 6. The old queen regularly leaves with the first swarm, or the genuine Italian brood would not invariably have been the product of the swarm, but occasionally, at least, of the parent colony, which never happened in all that time.

"These observations and inferences impelled Dzierzon—who had previously ascertained that the cells of the Italian and common bees were of the same size—to make an effort to procure the Italian bee; and, by the aid of the Austrian Agricultural Society at Vienna,* he succeeded in obtaining, late in February, 1853, a colony from Mira, near Venice. On the following day, he transferred the combs and bees into one of his own hives, and, when the season opened, placed the hive on a stand in his Apiary, and screwed it fast, that it might not be stolen. He never moved it during the ensuing Summer, but took from it combs with worker and drone-brood, at regular intervals, supplying their place with empty comb. In this way, he succeeded in rearing nearly fifty young queens, about one-half of which were impregnated by Italian drones, and produced genuine brood. The other half produced a bastard progeny. He continued thus to multiply queens by the removal of brood, till the parent-stock, and several of his artificial colonies, suddenly killed off their drones, on

---

* Some of the Governments of Europe have recently taken great interest in disseminating among their people a knowledge of Dzierzon's system of Bee-Culture. Prussia furnishes annually a number of persons from different parts of the Kingdom, with the means of acquiring a practical knowledge of this system; while the Bavarian Government has prescribed instruction in Dzierzon's theory and practice of bee-culture, as part of the regular course of studies in its teachers' Seminaries.

the 25th of June. The bees of the original colony still labored very assiduously, but gradually became less diligent, till when the buck-wheat came into blossom, they were surpassed in industry by many colonies of the common bees. But, as young bees continued to make their appearance he felt satisfied that the colony was in a healthy condition. Later in the season, he unfastened the hive, preparatory to putting it into winter quarters; and on attempting to lift it, found he was scarcely able to move it. he now discovered why it had so greatly fallen behind the ether colonies in industry. Having early rid itself of drones (as probably is done instinctively in Italy), it had, in consequence of its extraordinary activity, filled all the cells with honey, in a very short time, and was thenceforward doomed to involuntary idleness. It had attained a weight which scarcely any of his colonies reached in the Summer of 1846, when pasturage was so superabundant; whereas, the Summer of 1853 was a very ordinary one in this respect.[*]

"'The general diffusion of this species of bee' says Dzierzon, 'will form as marked an era in the bee-culture of Germany, as did the introduction of my improved hives.[†] The profit derived by the farmer from feeding stock, depends not alone on due attention to the habits and wants

---

[*] "His experiments on this colony made it manifest, that frequent disturbance had not produced any injurious effect. Until Midsummer, he not only removed a brood-comb containing about 5,000 cells, every other day, but had, on numerous other occasions, taken out comb after comb, several times a day, to find the queen, and show her to bee-keeping friends, who visited him. When, in consequence of such interruptions, the queen retreated to the opposite end of the hive, he usually found her, half an hour thereafter, on the same comb she had occupied before, engaged in laying eggs. Such disturbances, if the combs be not broken, or materially damaged, he thinks, do no injury; but that, on the contrary they not unfrequently produce a certain excitement among the bees, which impels them to issue in greater numbers, and labor with increased assiduity." —S. WAGNER.

[†] After my application for a patent on the movable-frames was favorably decided upon, the Baron Von Berlepsch, of Seebach, Thuringia (see p. 126), invented frames of a somewhat similar character. Carl T. E. Von Siebold, Professor of Zoology and Comparative Anatomy, in the University of Munich, thus speaks of these frames: "As the lateral adhesions of the combs built down from the bars:" (see pp. 15, 16 of this Treatise), "frequently rendered their removal difficult, Berlepsch tried to avoid this inconvenience, in a very ingenious way, by suspending in his hives, instead of the bars, small quadrangular frames, the vacuity of which the bees fill up with their comb, by which the removal and suspension of the combs are greatly facilitated, and altogether such a convenient arrangement is given to the Dzierzon-hive, that nothing more remains to be desired."

of the animals, but mainly on the character of the breed itself. So also with the bee. We find marked differences in point of industry, even among our common bees; but the Italian bee surpasses these in every respect. A chief difficulty in the way of a more general attention to bee-culture, arises from the almost universal dread of the sting of this insect. Many fear even the momentary pain which it inflicts, though no other unpleasant consequences follow; but in some persons it causes severe and long protracted swelling and inflammation. This, especially, deters ladies from engaging in this pursuit. All this can be avoided by the introduction of the Italian bee, which is by no means an irascible insect.[*] It will sting only when it happens to he injured, when it is intentionally annoyed, or when it is attacked by robbing bees; then it will defend itself with undaunted courage and such are its extraordinary vigor and agility, that it is never overpowered, so long as the colony is in a normal condition. Colonies of common bees may speedily be converted into Italian stocks, by simply removing the queen from each, and after the lapse of two or three days, or as soon as the workers decidedly manifest consciousness of the deprivation, supplying them with an Italian queen. We are thereby also enabled to note the gradual disappearance of the old race, as it becomes supplanted by the new. Besides the increased profit thus derivable from bee-culture, this species also furnishes us with no small gratification, in studying the nature, habits, and economy of the insect to greater advantage, because, by means of it, the most interesting experiments, investigations, and observations may be instituted, and thus the remaining doubts and difficulties be cleared up.'

"He further says: 'It has been questioned, even by experienced and expert Apiarians, whether the Italian race can be preserved in its purity, in countries where the common kind prevail. There need be no uneasiness on this score. Their preservation could be accomplished, even if natural swarming had to be relied on, because they swarm earlier in the season than the common kind, and also more frequently. Captain Baldenstein's want of success was most probably the result of a deficiency of drone-comb [*] in his Italian hives, as a consequence of which, only few

---

[*] Spinola speaks of the more peaceable disposition of this bee; and Columella, 1800 years ago, had noticed the same peculiarity, describing it as "*mitior moribus.*" Both its superior industry and peaceableness have been noticed from the earliest ages.

[*] "Dzierzon guarded against this, by giving to a very large colony, which ordinarily produced drones in great numbers, a fertile queen very early in the season. Thousands of drones soon made their appearance, and he immediately formed an artificial colony

drones were produced.'

"The main thing to be attended to in any localities where common bees are found or kept, is to secure the production of drones in numbers overwhelmingly large; though Dzierzon is under the impression, that where both kinds of drones exist in about equal numbers, the Italian queens will usually encounter Italian drones, both queens and drones being more active and agile than the common kind. Besides, the wings of both queens and drones are finer and more delicate than those of the common kind, and the sounds produced in flying are clearer and higher toned. Hence, probably, they are readily able to distinguish each other when on the wing.†

"The Baron of [von] Berlepsech, one of the most enthusiastic and skillful Apiarians, on a large scale, in Germany, says he can, from his own experience confirm the statements of Dzierzon, in relation to the Italian bee, having found,

"1. That the Italian bees are less sensitive to cold than the common kind. 2. That their queens are more prolific. 3. That the colonies swarm earlier and more frequently, though of this he has less experience than Dzierzon. 4. That they are less apt to sting. Not only are they less apt, but scarcely are they inclined to sting, though they will do so if intentionally

---

by removing this queen, with a sufficient number of workers, adding worker-brood from other colonies. On the twelfth day following, he heard a young queen '*teeting*' in the parent hive and, to his surprise, a large swarm issued from it on the same day, though the weather was then cool and cloudy. This swarm came forth suddenly, without any previous indication of its intention, just as after-swarms usually do. On a similar day, Dzierzon says, he had never seen a first swarm of common bees leave. So cold was the weather, that some of the bees became chilled before the swarm was hived. As the swarm was unusually large, he divided it into two, as he was able to procure an additional queen from the parent hive. Both throve well, and each of the queens was impregnated by an Italian drone. From this occurrence, he judged that these bees have an instinctive proclivity to swarm early. Our common kind would have lingered long, rather than 'swarm in weather so cold and cloudy.'"—S. WAGNER.

† "If, at the time when young queens are emerging, the bees and drones be tempted to sally out earlier than usual in the day, hours before the common drones come forth, by feeding them with diluted honey, the perpetuation of the genuine breed will the more probably be secured. But this end will the most certainly be attained, if measures are taken to have Italian queens and drones bred early in the season, before the common drones make their appearance; and again late, after the latter have been 'killed off.' This may readily be done by the improved hive, and the application of certain known principles in bee-culture."—S. WAGNER.

annoyed or irritated. 5. That they are more industrious. Of this fact he had but one Summer's experience, but all the results and indications go to confirm Dzierzon's statements, and satisfy him of the superiority of this kind *in every point of view*. 6. That they are more disposed to rob than common bees, and more courageous and active in self-defense. They strive on all hands to force their way into colonies of common bees; but when strange bees attack their hives, they fight with great fierceness, and with an incredible adroitness.[*]

"From one Italian queen sent him by Dzierzon, Berlepsch succeeded in obtaining, in the ensuing season, one hundred and thirty-nine fertile young queens, of which number about fifty produced pure Italian progeny.[†]

"Busch (*Die Honig-biene*, Gotha, 1855) describes the Italian bee as follows : 'The workers are smooth and glossy, and the color of their abdominal rings is a medium between the pale yellow of straw and the deeper yellow of ochre. These rings have a narrow black edge or border, so that the yellow (which might be called leather-colored) constitutes the ground, and is seemingly barred over by these slight black edges, or bor-

---

[*] Spinola speaks of these bees as "*velociores motu*"—quicker in their motions than the common bees.

[†] "It is a remarkable fact that an Italian queen, impregnated by a common drone and a common queen impregnated by an Italian drone, do not produce workers of a uniform intermediate cast, or hybrids: but some of the workers bred from the eggs of each queen will be purely of the Italian, and others as purely of the common race, only a few of them, indeed, being apparently hybrids. Berlepsch also had several bastardized queens, which at first produced Italian workers exclusively, and afterwards common workers as exclusively. Some such queens produced fully three-fourths Italian workers; others, common workers in the same proportion. Nay, he states that he had one beautiful orange-yellow bastardized Italian queen which did not produce a single Italian worker, but only common workers, perhaps a shade lighter in color. The *drones*, however, produced by a bastardized *Italian* queen are uniformly of the Italian race, and this fact, besides demonstrating the truth of Dzierzon's theory, renders the preservation and perpetuation of the Italian race, in its purity, entirely feasible in any country where they may be introduced."—S. WAGNER.

(It is interesting that none of these "Apiarians" could make the intuitive jump to multiple mating even though much of the data pointed to that fact—"[These matings]...do not produce workers of a uniform intermediate cast, or hybrids: but some of the workers bred from the eggs of each queen will be purely of the Italian, and others as purely of the common race, only a few of them, indeed, being apparently hybrids." *It is clear to us now that there were "hybrids" produced, it's just that they were not prepared to accept the fact of multiple mating at that time.)*

ders. This is most distinctly perceptible when a brood-comb, on which bees are densely crowded, is taken out of a hive. The drones differ from the workers in having the upper half of their abdominal rings black, and the lower half an ochre-yellow, thus causing the abdomen, when viewed from above, to appear annulated. The queen differs from the common kind chiefly in the greater brightness and brilliancy of her colors.'

Otto Radlkofer, Jr., of Munich, in a communication to the *Bienenzeitung*, says that a colony of Italian bees, which he transferred in February, began to build new comb before the middle of March, while his common bees had not; at the date of his communication (the last of April), begun to build any new comb. 'Not only,' says Mr. Radlkofer, 'are the Italian bees distinguished by an earlier awakened impulse to activity and labor, but they are remarkable also for the sedulous use they make of every opening flower, visiting some on which common bees are seldom or never seen. They have also demonstrated their superior agility in self defense; nay, they would not tolerate the presence of other bees on comb that had been strewed with flour for their common use. In all these respects, the palm of superiority must be awarded to the Italian bee.'

"Considerable difficulty has been encountered, even by experienced Apiarians, in inducing a colony of common bees, deprived of its queen, to accept an Italian queen in its stead, and many failures have occurred, involving the loss of the offered queen, and causing grievous disappointment. The safest course appears to be, to remove the queen several days before the substitution is intended to be made, and to destroy all the royal cells and embryo queens the day before the Italian queen is introduced. At the time of her introduction, the combs should again be thoroughly examined, and, if any more royal cells have been started, they must likewise be destroyed. The Italian queen should be placed in a cage for her protection, and a small quantity of pure honey in open cells should be put in the cage. The conduct of the workers will speedily show whether and when they will receive her. Mr. Lange advises that the Italian queens be introduced immediately after the bees of a deprived colony manifest undoubted consciousness of the loss they have sustained and before they have started any royal cells, or made arrangements for doing so. —Yours truly,

Samuel Wagner."

"Rev. L. L. Langstroth."

The chief obstacle to the rapid diffusion of this valuable variety has been the difficulty experienced by the ablest German Apiarians in preserving the breed pure, even Berlepsch having failed entirely to do so. By means of my *non-swarmer*, however, this difficulty may be readily overcome.

Let the beekeeper who obtains an Italian queen in the Spring, give her, with proper precautions (p. 200), to a populous colony, whose hive is well furnished with drone-combs having first deprived it of its queen. When the drone cells are filled with sealed brood, let *nuclei* (p. 189) be formed from this stock, and replace the combs removed, with others containing workers ready to hatch. By thus keeping the parent-stock always populous, a large number of nuclei may be formed from it. Just before the young Italian queens mature, adjust the non-swarmer (Plates II., V., Figs. 5, 17) to all the hives containing common drones, so as to *shut them in*, while free egress is given to queens and workers. As only the drones bred by the Italian queen have their liberty, all the young females will be fertilized by them.[4] As fast as the queen of the nuclei become fertile, they may be given to the various stocks, and from these, in a short time, other nuclei which will raise Italian queens, maybe formed. In this way, an expert, who can be sure of having Italian drones until late in the season, might easily convert an Apiary of a thousand or more hives into stocks containing none but the new variety.

To secure the requisite number of drones, part of the Italian drone-brood should be given to some of the nuclei, so that, in case the parent-stock kill its drones, others may be on hand. If the Apiarian removes the queen from this colony before the drones are killed, the bees will tolerate their presence much longer. The same object may also be accomplished by liberal feeding as soon as natural forage falls (p. 224).

Dzierzon found that a queen which had been *refrigerated* for a long time, after being brought to life by warmth, laid only male eggs,

---

[4] *This idea was expanded Dr. John Hogg of Augusta, Michigan in a technique he called after-hours mating. He kept his virgins and selected drones closed up until after the normal mating time before he released them. Since they had been confined, they immediately took flight and he had successful matings. The advantage of this technique is that he avoided all other drones including feral ones, and so had more "pure" mating.*

whilst previously she had also laid female eggs. Berlepsch refrigerated three queens by placing them thirty-six hours in an ice house,[*] two of which never revived, and the third laid, as before, thousands of eggs, but *from all of them only males were evolved.* In two instances, Mr. Mahan has, at my suggestion, tried similar experiments, and with like results. It does not seem to have occurred to the German Apiarians *that by this refrigerating process we may secure as many Italian drones as we need.*[5] All that is necessary is to convert by it one or more of the queens of the nuclei into *drone-layers.* The reception of an Italian queen quite late in the season may thus be turned to good account.

If the Apiarian is in the vicinity of hives to which he cannot apply the non-swarmer, it will be necessary for him to seek some place where the common drones cannot interfere with his proceedings. Unless the breed is kept pure, the advantages proposed by its introduction cannot be secured.

Italian queens may be safely sent in my hives to any part of the county. A hive for this purpose should be made to hold only one comb, which ought to be old and very securely fastened. Into such a hive, suitably provisioned, an Italian queen may be introduced, with a few hundred bees to keep her company, and, if sufficient ventilation is given, with a little water daily, they will bear a journey of many days. If received at a season unsuitable for rearing new queens, she may be given to some strong colony and reserved for future operations.

It is hardly necessary to say, that a species of the honey-bee so much more productive than the common kind, and so mach less sensitive to cold, will be of very great value to all sections of our country.[*]

---

[*] A short exposure of a queen to pounded ice and salt, will answer every purpose. The spermatozoids are in some way rendered inoperative by severe cold.

[5] *Today we can accomplish the same trick with less risk by using carbon dioxide to anesthetize a virgin queen. By using at least two such CO_2 treatments a virgin queen will begin to lay unfertilized eggs, if she is not allowed to mate. It is the anesthesia with CO_2 that essentially allowed instrumental insemination of queen honey bees to be successful. First, because it allowed the queen to be held quietly while being inseminated, and secondly because the carbon dioxide stimulated the queen to begin laying. Without two carbon dioxide anesthesia treatments the queens would not commence oviposition sometimes for several weeks.*

[*] An attempt was made, in 1855, by Mr. Wagner, to import the Italian bees, but, unfortunately, the colonies perished on the voyage. Mr. Richard Colvin, of Baltimore, Mr.

Its superior docility would make it worthy of high regard, even if in other respects it had no peculiar merits. Its introduction into this country will, it is confidently believed, constitute a new era in bee-keeping, and impart an interest to its pursuit which will enable us, ere long, to vie with any part of the world in the production of honey.

Wagner, and myself, have made arrangements to have them brought to this country this Spring (1859).

(This would appear to be the beginning of the Italian bees in the United States. Frank Benton, of the U.S.D.A. subsequently traveled to Europe and shipped many queens to this country; in so doing developed the 3-hole Benton mailing cage which is still in use (probably copied from Langstroth). Benton, himself, later became established in Austria and obviously thought more of the Carniolian bees as he advertised and shipped Carniolian queens to the U.S. in the 1880's.

Today we think nothing of shipping a queen from all over the world via airplanes. However, in the 1860's it took weeks to ship a queen from Italy to the United States even with a fast clipper ship. Thus, it is remarkable that queens [or colonies earlier] made it to this country.)

CHAPTER XX.

SIZE, SHAPE, AND MATERIALS FOR HIVES — OBSERVING HIVES.

NOTWITHSTANDING the almost innumerable experiments which have been made to determine the best size, shape and materials for beehives, the ablest practical Apiarians are still at variance on these points. In most districts in this country, it is pretty generally agreed that hives holding less than a bushel, in the main apartment, are not profitable in the long run. As regards, however, the size, both of the main hive and the apartments for spare honey, so much depends on seasons and localities, and on whether the bees swarm or not, that no rule, applicable to all cases, can be given. Every bee-keeper must determine these questions by reference to the honey-resources of his own district. As the plan of my hives admits of their being enlarged and again contracted, without destruction or alteration of existing parts, the size, either of the main hive or surplus storage room, may be varied at pleasure.[1]

Being able to remove any surplus, I prefer to make the interior of my hives considerably larger than a bushel. Many hives cannot hold one-quarter of the bees, comb, and honey which, in a good season, may be found in my large hives; while their owners wonder that they obtain so little profit from their bees.[2] A good swarm of bees, put, in a good season, into a diminutive hive, may be compared to a powerful team of horses harnessed to a baby wagon, or a noble fall of water wasted in turning a petty water-wheel.

---

[1] *This has to be one of the most important aspects of the movable-comb hive as far as the development of a commercial beekeeping industry. However, I don't think Langstroth ever imagined that a beekeeper could operate thousands of colonies averaging 300 lbs./colony.*

[2] *Modern research has born out Langstroth, as bees will produce more honey if given more space. In economic terms the total amount of space has to be balanced with reason since it doesn't pay to have six or seven supers with 20 pounds in them as opposed to three with 35 pounds each.*

A hive *tall* in proportion to its other dimensions, has some obvious advantages; for, as bees are disposed to carry their stores as far as possible from the entrance, they will fill its upper part with honey, using the lower part mainly for brood, thus escaping the danger of being caught, in cold weather, among empty ranges of comb, while they still have honey unconsumed. If the top of this hive, like that of an old-fashioned churn, is made (on the Polish plan) considerably smaller than the bottom, it will be better adapted to a cold climate, besides being more secure against high winds. Such a hive is deficient in top-surface for the storing of honey in boxes, and it would be impossible to use frames [*] in it to any advantage; but, to those who prefer to keep bees on the old plan,[†] one of this shape, made to hold not less than a bushel and a half, is decidedly the best.

A hive long from *front to rear*, and moderately low and narrow, seems, on the whole, to unite the most advantages. Such a hive resembles a tall one, laid upon its side, and, while affording ample top-surface for surplus honey, it greatly facilitates the handling of the frames, besides diminishing their number and cost.[‡]

---

[*] The *deeper* the frames, the more difficult it is to make them hang *true* on the rabbets, and the greater the difficulty of handling them without crushing the bees or breaking the combs.

[†] It is instructive to see how the very first departure from the olden way proves the truth, in bee-culture at least, of the hackneyed quotation:

"A little knowledge is a dangerous thing."

Even so simple an improvement as that of top-boxes will, as used by many, eventually destroy their bees; for, while in favorable years such boxes may be safely removed, in others the surplus honey which they contain, is the life of the bees.

[‡] Mr. M. Quinby, of St. Johnsville, New York, in calling my attention to some stocks, which he had purchased in box hives of this shape, informed me that bees wintered in them about as well as in tall hives, the bees drawing back among their stores in cold weather, just as in tall hives they draw up among them. My hive, as at first constructed, was fourteen and one-eighth inches from front to rear, eighteen and one-eighth inches from side to side, and nine inches deep, holding twelve frames. After Mr. Quinby called my attention to the wintering of bees in his long box-hives, I constructed one that measured twenty-four inches from front to rear, twelve inches from side to side, and ten inches deep, holding eight frames. I have since preferred to make my hives eighteen and one-eighth inches from front to the rear, fourteen and one-eighth inches from side to side, and ten inches deep. Mr. Quinby prefers to make my movable frames longer and deeper.

(*Modern studies show that more brood will be produced in deep frames. Shallow frames may have an advantage in wintering. The shallow frames win with me simply*

The common Dzierzon hive * is long and flat, but, as the combs run from side to side, instead of from front to rear, the bees, unless the hive is uncommonly well protected, will suffer from cold in Winter. As the German Apiarian uses slats instead of frames, it would be inconvenient for him to remove any very long combs from his hive.

The variety of opinions respecting the best *materials* for hives, has been almost as great us on the subject of their proper size and shape. Columella and Virgil recommend the hollowed trunk of the *cork tree*, than which no material would he more admirable if it could only be cheaply procured. Straw-hives have been used for ages, and are warm in Winter and cool in Summer. The difficulty of making them take and retain the proper shape for improved bee-keeping, is an insuperable objection to their use. Hives made of wood are, at the present time, fast superseding all others. The *lighter* and more *spongy* the wood, the poorer will be its power of conducting heat, and the warmer the hive in Winter and the cooler in Summer.† Cedar, bass-wood, poplar, tulip-tree, and soft pine, afford excellent materials for bee-hives. The Apiarian must be governed, in his choice of lumber, by the cheapness with which any suitable kind can be obtained in his own immediate vicinity.

A serious disadvantage attaching to all kinds of wooden hives, is the ease with which they conduct heat, causing them to become cold and damp in Winter, and, if exposed to the sun, so hot in Summer as often to melt the combs. The Winter inconveniences are greatly increased if the hives are well painted, while, if this is neglected, they

---

because the supers are lighter. Today the more "perfect" size would be a 7-½ in. super as this would use an 8-in. board without waste in making the boxes, and also be slightly lighter in weight.)

* Dzierzon builds hives in structures for two, four, and even many more colonies. On Plate XXII., Fig. 71 (the Frontispiece to the first edition of my work), I have given a representation of a triple hive. The little that can be saved in the first cost of such hives, seems to me to be more than lost by the great inconvenience of handling them.

† Mr. Wagner informs me that Scholz, a German Apiarian, recommends hives made of *adobe*—in which frames or slats may be used—as cheaply constructed, and admirable for summer and Winter. Such structures, however, cannot be moved. But in many parts of our country, where both lumber and saw-mills are scarce, and where people are accustomed to build adobe houses, they might prove desirable. The material is plastic clay, mixed with cut straw, waste tow, &c.

cannot ordinarily be exposed to sun or weather without serious injury.*

To make the movable-comb hives to the best advantage, the frames at least should be cut out by a circular saw, driven by steam, water, or horse-power. In buildings where such saws are used, the frames may be made from the small pieces of lumber, seldom of any use, except for fuel, and may be packed almost solid in a box, or in a hive which will afterwards serve for a pattern.[3] One frame in such a box, properly nailed together, will serve as a guide for the rest. The other parts of the hive can easily and cheaply be made by any one who can handle tools, and can never be profitably manufactured to be sent far, unless made where lumber is cheap, and the parts closely packed, to be put together after reaching their destination.

MOVABLE-COMB OBSERVING HIVES

Each comb in these hives is attached to a movable frame, and, as both sides admit of inspection, all the wonders of the bee-hive may be exposed to the light of day, as well as that of (pp. 23, 116) lamps and gas.

In the common observing-hive, experiments are conducted only by cutting away parts of the comb; whereas, in this, they can be performed by the simple removal of a frame; and if a colony becomes reduced in

---

* The abundant ventilation now given to my hives, will enable the Apiarian to dispense with paint, except on the joints and roofs; and if the latter are, in Summer, covered with straw, battened to them so that the air can circulate under it, they may be safely placed in the sun, if not exposed to a close, suffocating heat.

[3] *It seems to me that Langstroth anticipated the development of the bee-supply industry, and the shipping of "knocked-down" beehives and frames—the beekeeper assembling the equipment after it was delivered to him. If a beekeeper is going to make their own equipment it is better to have a model to use. This is especially true of the frames. Most people that do make some of their equipment usually end up buying the frames because there are so many saw cuts to be made on each one.*

numbers, it may be recruited, in a few minutes, by giving it maturing brood from another hive.*

These observing-hives may be constructed to accommodate a full swarm. I do not, however, recommend such a hive for ordinary purposes, but one holding only a *single frame* (Pl. IV., Figs 14, 15), which, while it gratifies curiosity, admits of easy control, and requires only a few bees to be diverted from more profitable hives.

A parlor observing-hive of this form may be conveniently placed in any room in the house—the alighting-board being outside, and the whole arrangement such that the bees may be inspected at all hours, day, or night, without the slightest risk of their stinging. Two such hives may be placed before one window, and put up or taken down in a few minutes, without cutting or defacing the wood-work of the house. In one, the queen may always be shown, and in the other, the process of rearing young queens from worker-eggs. These miniature hives may he stocked in the same way that a nucleus is formed, or a small after-swarm may be hived in them.

An observing-hive will prove an unfailing source of pleasure and instruction; and those who live in crowded cities, may enjoy it to the full, even if condemned to the penance of what the poet has so feelingly described as an "endless meal of brick." The nimble wings of these agile gatherers will quickly waft them above and beyond "the smoky chimney-pots;" and they will bear back to their city homes the balmy spoils of many a rustic flower, "blushing unseen," in simple loveliness. Might not their pleasant murmurings awaken in some the memory of long-forgotten joys, when the happy country child listened to their soothing music, while intently watching them in the old homestead-garden, or roved with them amid pastures and hill-sides, to gather the flowers still rejoicing in their "meadow-sweet breath," or whispering of the precious perfumes of their forest home!

---

* A writer, in a description of the different hives exhibited at the World's Fair in London, laments that no method has yet been devised, to enable bees to cluster, in cold weather, in an observing-hive, so as to preserve them alive in Winter, even in the moderate climate of Great Britain. By the use of movable frames, this difficulty can be easily obviated, as, on the approach of cold weather, the frames, with the bees, may be put into a suitable hive, and returned in the Spring to their old abode.

"To me more dear, congenial to my heart,
 One native charm than all the gloss of art;
 Spontaneous joys, where nature has its play,
 The soul adopts and owns their first-born sway;
 Lightly they frolic o'er the vacant mind
 Unenvied, unmolested, unconfined.
 But the long pomp, the midnight masquerade,
 With all the freaks of wanton wealth array'd,
 In these, ere triflers half their wish obtain,
 The toilsome pleasure sickens into pain;
 And e'en while fashion's brightest arts decoy,
 The heart distrusting asks, IF THIS BE JOY."
                                    GOLDSMITH.

Preface to Chapter XXI

A large section of this chapter is mostly a translation of an article from the German bee journal **Bienenzeitung.** *The translation of the article was provided by Samuel Wagner. It is not clear, from the chapter, but I suspect that Langstroth had not ever used the system of a "clamp" to overwinter his bees. The system did, however, include most of the ideas that he considered good wintering practices, and this is the reason that he embraces the writing of the Rev. Scholtz. I suspect that our modern "clamp" is our air-conditioned wintering houses. In some ways accomplishing much of the same conditions that these beekeepers were trying to give to their bees—a moderately cool, steady temperature.*

Moving bees to a wintering house may be hard work but it is relatively easy compared to the effort in producing this clamp. However, prior to the movable-comb hive conditions inside the hive would have been mostly unknown, and, thus, this was a way to keep the hives alive until spring.

We have come a long way in our understanding of wintering. Yet, in his own way Langstroth departs a great amount of inf6rmation on how to winter successfully. Even when he is giving instructions on what we consider far too much work, there is much to learn here. Reading about the system will help a beekeeper to think about all the things that make for successful wintering of bees, and I think that in itself is helpful, especially to a new beekeeper.

CHAPTER XXI.

WINTERING BEES.

As soon as frosty weather arrives, bees cluster compactly together in their hives, to keep warm. They are never dormant, like wasps and hornets (p 110), and a thermometer pushed up among them will show a Summer temperature, even when, in the open air, it is many degrees below zero. When the cold becomes intense, they keep up an incessant tremulous motion, in order to develop more heat by active exercise; and, as those on the outside of the cluster become chilled, they are replaced by others.

As all muscular exertion requires food to supply the waste of the system, the more quiet bees can be kept, the less they will eat. It is, therefore, highly important to preserve them, as far as possible, in Winter, from every degree, either of heat or cold, which will arouse them to great activity.

The usual mode of allowing them to remain all Winter on their Summer stands, is, in cold climates, very objectionable.[1] In those parts of the country, however, where the cold is seldom so severe as to prevent them from flying, at frequent intervals, from their hives, perhaps no better way, all things considered, can be devised. In such favored regions, bees are but little removed from their native climate, and their wants may be easily supplied, without those injurious effects which commonly result from disturbing them when the weather is so cold as to confine them entirely to their hives.

If the stocks are to be wintered in the open air, they should all he made populous, and rich in stores, even if to do it requires the number of colonies to be reduced one half, or more.[*] The beekeeper who has

---

[1] *Langstroth was starting a "revolution" yet could not foresee the impact of the movable-comb hive would have on beekeeping. Today we might move bees to winter locations if the beekeeper is a migratory beekeeper and has his colonies on movable pallets. Most beekeepers do not move their apiaries from their permanent location for winter. The reason that they don't is that modern beekeepers have too many colonies to make that shift even if it might save more colonies..*

[*] Small colonies consume, proportionally much more food than large ones, and often perish from inability to maintain sufficient heat. Stocks should not, however, be made

ten strong stocks in the Spring, will, by judicious management with movable-comb hives, be able to close the season with a larger Apiary than one who begins it with thirty, or more, feeble colonies.

If two or more colonies, which are to be united in the Fall, are not close together, their hives must be gradually approximated (p. 280) [moved closer together], and the bees may then, with proper precautions (p. 203), be put into the same hive.

If the central combs of the hive are not well stored with honey, they should be exchanged for such as are, so that, when the cold compels the bees to recede from the outer combs, they may cluster among their stores. If the fullest honey-combs are not of worker size, the caps of their cells may be sliced off; and the combs put in the upper apartment, where the bees can remove the honey, and store it in the centre of the hive. In districts where bees gather but little honey in the Fall, such precautions, in cold climates, will be specially needed, as, often, after breeding is over, their central combs will he almost empty.[2]

As bees are natives of a warm climate, they do not instinctively place their honey where it will be most accessible to them in cold weather, but simply where it will least interfere with the raising of brood. Neither, if, while the weather is warm, they can easily communicate through the combs of the hive, can they be depended on to make such passages through them, as will allow them to pass readily, in cold weather from one to another.

The Apiarian, should, therefore, late in the Fall, cut, with a penknife, a hole, an inch in diameter, in the centre of each comb, about one-third from the top.[*]

---

over-populous, as their great internal heat would create restlessness, and engender dysentery, by leading to an inordinate consumption of food (p. 256).

[2] *Modern beekeeping practice would often have two brood chambers, with the top nearly full of honey, but with some space in the center combs. Bees do need honey within their cluster area but they also need empty cells in which to cluster and to raise brood. It is also good to have pollen within this winter cluster area. Yes, movable-comb hives are wonderful as you can determine this before winter begins.*

[*] If these holes are made before they feel the need of them, they will frequently close them. Mr. Wm. W. Cary (p. 204) has invented a process of making these holes without removing the combs. He makes a hole in the side of the hive, which, when not in use, is covered with a button or plug (Pl. V., Fig. 16), through which he slowly worms an instrument in the shape of a *flour or butter-taster* (sharpened at the end), until it strikes the opposite side of the hive. By this process of making the Winter passages, only a

Great care should be taken to shelter hives from the piercing winds, which in Winter so powerfully exhaust the animal heat of the bees; for, like human beings, if sheltered from the wind, they will endure a low temperature far better than a continuous current of very much warmer air.†

In some parts of the West, where bees suffer much from cold winds, their hives are protected, in Winter, by sheaves of straw, fastened so as to defend them from both cold and wet. With a little ingenuity, farmers might easily turn their waste straw to a valuable account in sheltering their bees.

If the colonies are wintered in the open air, the entrance to their hives must be large enough to allow the bees to fly at pleasure. Many, it is true, will be lost, but a large part of these are diseased;[3] and, even if they were not, it is better to lose some healthy bees than to incur the risk of losing, or greatly injuring, a whole colony by the excitement created by confining them when the weather is warm enough to entice them abroad.*

---

very few bees are hurt. As the queen always runs away from danger, she is not liable to be hurt. An application of a patent on this device is now pending. If the patent issues, the right to use it will be free to all owning the right to use the movable-comb hive.

I strongly advise every one using my hives to make Winter passages for their bees. As the frames touch neither the top, bottom, nor sides of the hives, the bees have such extraordinary facilities for intercommunication, that they cannot be depended on to leave any holes in their combs.

† The Winter of 1855-6 will long be remembered, not only for the uncommon degree and duration of its cold, but for the tremendous winds, which, often for days together, swept like a Polar tornado over the land. Apiaries standing in exposed situations were, in many instances, nearly ruined.

[3] *Langstroth was very insightful as most beekeepers dismissed any distress as dysentery. They did not recognize that much of the dysentery is caused by the protozoan parasite,* Nosema apis, *or now* Apis ceranaae. *Nosema disease affects most colonies but it is more pronounced in the Northern states where the bees are confined for a long period of time.*

* If the sun is warm and the ground covered with new-fallen snow, the light may so blind the bees, that they will fall into this fleecy snow, and quickly perish. At such times, it would probably be best to confine them to their hives. If the snow is hard enough to bear up a healthy bee, it is seldom lost, unless tempted to fly by the sun shining full upon its hive as it stands in a sheltered place.

(*Recent studies on the necessary <u>body</u> temperature for bees to continue to fly would indicate that the lower limit is probably above $30°$ C. [$86°$ F.]. Thus, if the bee's body is quickly chilled it will fall to the snow and die. The bee is able to fly out of the hive*

The best Apiarians are still at us to how much air should be given to bees in Winter, and whether hives should have *upward ventilation*, or not. If the hives have no upward ventilation, then I believe that they need as much, or even more, air, than in Summer. If upward ventilation is given, the smaller the *lower* openings the better, as it is not desirable that there should be a strong current of cold air passing through the hives.

In my hives, all the lower passages can easily be closed airtight, and the bees allowed to go in and out through the *Winter-entrance*, which is made at the top of the hive (Pl. I., Fig. 1; Pl. V., Fig. 17).[†]

If the hive has an upper box-cover, as in Pl. III., Fig. 9, the holes in the honey-board must be left open, or closed only with wire-cloth, that the dampness, which would otherwise condense or freeze on the combs and interior walls of the hive, may escape without injuring the bees.

If an upper hive, as in Plate V., Fig. 16, is placed on the top of the one in which the bees are wintered, its roof should be slightly elevated, to allow the escape of moisture. If a single hive, like that in Plate I, Fig. 1, or Plate V., Fig 17, is used, the same opening must be allowed for the escape of dampness.[*]

As facts observed have a value far above theories, I shall give the substance of numerous observations made by me, at Greenfield, Massachusetts, in the Winter of 1856-7, on wintering bees in the open air:

---

cluster since the temperature there may be above to the lower limit. By the same token, bees will fly out of the hive at lower temperatures as long as the sun is shining since they absorb radiant solar energy enough to keep their flight muscles functioning.)

[†] The lower entrance may be closed in the Fall, while the bees are still flying, and they will quickly accustom themselves to the upper one. Mr. Wheaton suggests making this Winter-entrance in the back of the hive, and in the Fall reversing the pile, stand and all. *This entrance is merely proposed for trial.*

[*] Small strips of wood, one-eighth of an inch thick, may be placed between the side of the hive and the under-surface of the roof, and when the roof is securely fastened, the dampness can escape from the front and rear of the hive, where the openings are sheltered by the clamps, from the snow and rain.

(I have taken this idea and incorporated the upper entrance into this moisture escape. Instead of one-eighth inch I use about three-eighths inch, tapered spacers [front to back] under the telescoping cover. The bees can exit from this gap and yet it is under the edge of the telescoping cover to prevent wind from blowing in. This upper entrance now provides both an escape of moisture from the winter cluster and an upper entrance.)

JAN. 9TH, 1857. —Examined a number of stocks with *Winter-passages* in their combs, and with all the holes in their honey-board uncovered. The previous month had been extremely cold, and, for three days before the examination, the thermometer had been one-half of the time below zero, and only once ten above, the wind blowing an almost continuous gale. In none of these hives could I find any frost or dampness, or any bees killed by being caught away from the main body of the colony. In a temperature below zero, they would rush up from their combs on the slightest jar of their hives, rapidly pouring through the Winter-passages, and showing their ability to reach any of their stores.[†] In a few colonies, to which no upward ventilation had been given, the interior walls of the hive, and many of the combs were coated with frost.

JAN. 14TH. —Carefully examined three hives. No. 1, made of boards seven-eighths of an inch thick, had stood with its honey-board removed, the same as would show by removing (f) in Plate III., Fig. 9. It had a good stock of bees, and, although the mercury in the morning was $10\frac{1}{2}°$ below zero, there was scarcely any frost in the hive. The bees were dry and lively, and the central combs contained eggs and unsealed brood. No. 2 contained an equally strong stock, in a thin hive holding eighteen frames, ten of which (five on each side) had no combs. This hive had no upward ventilation, and was very frosty. The central combs had eggs and unsealed brood. No. 3 was most thoroughly protected by double sides, filled in with charcoal, and all the holes in its honey-board were left open. It had a little frost, as No. 1. and its central combs contained eggs and some sealed brood. Although it had a better stock of bees than either of the others, it appeared to have begun to breed only a few days earlier.[4]

[†] On a cold November day, I have found bees, in a hive without any Winter-passages, separated from the main cluster, and so chilled as not to be able to move; while, with the thermometer many degrees below zero, I have repeatedly noticed, in other hives, at one of the holes made in the comb, a cluster, varying in size, ready to rush out at the slightest jar of their hive.

(We like our combs too neat. These communication holes are very beneficial.)

[4] *Years ago we installed 24 recording thermocouples in each of three colonies during winter. During December the central cluster temperature ranged from about 82° to 88° F. On January 4th the first colony recorded 95° F. as it began brood rearing. The next day the second colony also reached brood temperature and the third colony a couple of*

JAN. 30TH. —This month has been the coldest on record for more than fifty years. My hives have been exposed to a temperature of 30° below zero, and for forty-eight hours together the wind blew a strong gale, and the mercury rose only once to 6° below zero. No. 1 was again examined, and the bees found in good condition. The central comb was almost filled with sealed brood, nearly mature; all the combs were free from mould, and the interior of the hive was dry. In a hive as well protected as No. 3, but which *had no upward ventilation*, the vapor, or *breath of the bees*, which had frozen in it, having melted in consequence of a sudden thaw, both combs and bees were in a wretched condition.

As long as the vapor remains congealed, it can only injure the bees by keeping them from stores which they need; but, as soon as a thaw sets in, hives which have no upward ventilation are in danger of being ruined.*

Mr. E. T. Sturtevant, of East Cleveland, Ohio, so widely known as an experienced Apiarian, in a letter to me, thus gives his experience in wintering bees in the open air:

"No extremity of cold that we ever have in this climate, will injure bees; if their breath is allowed to pass off so that they are dry. I have never lost a good stock that was dry, and had plenty of honey.

"In the Winter of 1855-56, I had twenty stocks standing in a row, all but one of which would have been regarded as in a good condition for wintering—not too tight below, nor yet too open above. One was in a hive suspended twenty inches from the ground, *and without any bottom-board*. The chamber for surplus honey-boxes was open to the north; and had eight one-inch holes, *all uncovered.*

"I left home about the 12th of February, the weather being very cold, and the hives all banked up with drifted snow. Returning the last

---

days later. All during this period the outside temperature was near zero degrees F.. The trigger to cause this response is the increasing day-length.

* In March, 1856, I lost some of my best colonies, under the following circumstances: The Winter had been intensely cold, and the hives, having no upward ventilation, were filled with frost, and, in some instances, the ice on their glass sides was nearly a quarter of an inch thick. A few days of mild weather, in which the frost began to thaw, were followed by a temperature below zero, accompanied by furious winds, and in many of the hives, the bees, which were still wet from the thaw, were *frozen together in an almost solid mass.*

of the month, I examined the whole row, and found the nineteen thawed out, but in a sadly wet and miserable plight. If I could have taken them into a room, out of the reach of the frost, until they were dry, they might have been saved. The weather changed to severe freezing before the next morning, and all the nineteen swarms soon died; while the one that was apparently so neglected, came out strong and healthy. Before adopting upward ventilation, I had lost my best swarms in this way, until I became discouraged."

In the coldest parts of our country, *if upward ventilation is neglected*, no amount of protection that can be given to hives, in the open air, will prevent them from becoming damp and mouldy, *even if frost is excluded*. Often, the more they are protected, the greater the risk from dampness. A very thin hive unpainted, so that it may readily absorb the heat of the sun, will dry inside much sooner than one painted white,[5] and in every way most thoroughly protected against the cold. The first, like a *garret*, will suffer from dampness for a short time only; while the other, like a *cellar*, may be so long an drying, as to injure, if not destroy, the bees.

Much has been said in Germany, within the last few years, of the danger of bees that have upward ventilation perishing in Winter for want of water. Mr. Wagner has furnished me with a translation of an able article in the *Bienenzeitung*, by Von Berlepsch, and G. Eberhardt, the substance of which is as follows:

"The Creator has given the bee an instinct to store up honey and pollen, which are not always to be procured, but not water, which is always accessible in her native regions. In northern latitudes, when confined to the hive, often for months together, they can obtain the water they need only from the watery particles contained in the honey, the perspiration which condenses on the colder parts of the hive, or the humidity of the air which enters their hives.

Vital energy in the bee is at its lowest point in November and De-

---

[5] *Obviously the "tradition" of painting hives white was strong even in the 1850's. There are other treatments of wood, though most will still trap moisture from going into or out of the hive. The difficulty is to have a treatment that protects the wood from weathering, or rotting, yet allows the moisture to escape. The best solution may be one of the wood preservatives that are registered for hive (honey) use. A white hive may prove more beneficial in the South where temperatures can be very high in the summer.*

cember. If, at this time, an unusual degree of cold does not force her to resort to muscular action, she remains almost motionless, a death-like silence prevailing in the hive; and we know, by actual experiment, that much less food is consumed than at any other time. Breeding having ceased, the weather-bound bees have no demands made on their vital action and we have never known them at this time to suffer for want of water. As soon, however, as the queen begins to lay, which occurs in many colonies early in January, and in some by Christmas, the workers must eat more freely both honey and pollen, to supply jelly for the larvae and wax for sealing their cells. Much more water is needed for these purposes, than when they can procure the fresh nectar of flowers; and the want of it begins to be felt about the middle of January. *The unmistakable signs of the dearth of water in a colony, are found in the granules of candied honey lying on the bottom of the hive.* The suffering bees will now open cell after cell of the sealed honey, to obtain what remains uncandied, and when these supplies of moisture fail, will attack the unsealed larvae, and devour the eggs, if any are still laid. They now give way to despair, disperse through the hive, if the cold does not prevent, as though they had lost their queen, and perish amid stores of honey, unless milder weather permits them to go in search of water, or the Apiarian supplies it in their hive, when order will again be restored.

"After protracted and severe Winters, of every six bees that perish, five die for want of water, and not, as was hitherto supposed, from undue accumulation of fæces. Dysentery is one of the direct consequences of *water-dearth*, the bees, in dire need of water, consuming honey immoderately, and taking cold by roaming about the combs.

"On the 11th of February, we examined a number of colonies, on whose bottom-boards we noticed *particles of candied honey*, and found that in all of them, the sealed honey had been opened in various points, and that *breeding had entirely ceased*. The colonies that we had supplied with water on discovering that they needed it, *contained healthy brood, in every stage of development.*

"In March and April, the rapidly increasing amount of brood causes an increased demand for water; and when the thermometer is as low as 45°, bees maybe seen carrying it in at noon, even on windy days, although many are sure to perish from cold. In these months, in

1856, during a protracted period of unfavorable weather, we gave all our bees water, *and they remained at home in quiet*, whilst those of other Apiaries *were flying briskly in search of water*. At the beginning of May, our hives *were crowded with bees*; whilst the colonies of our neighbors *were mostly weak*.

"The consumption of water in March and April, in a populous colony, is very great, and in 1856; one hundred stocks required eleven Berlin quarts per week, *to keep on breeding uninterruptedly*. In Springs where the bees can fly safely almost every day, the want of water will not be felt.

"The loss of bees by *water-dearth*, is the result of climate, and no form of hive, or mode of wintering, can furnish an absolutely efficient security against it. The colonies may be put in yard long *lager-hives*, or in towering standards, in shapeless gums, in neat straw hives, or in well lined Dzierzons; in wool, or straw, or clay domiciles; or may dwell in hollow trees, or clefts of rocks; they may remain unshielded on their Summer stands; be protected by a covering of pine shatters or chaff; or be stored in dark chambers or vaults—still water-dearth may occur, here and there, earlier or later, and more or less injuriously; *because it is counter to the original instincts of the bee to dwell in Northern climates, confined to its habitation for months.*

"If water is regularly given to the bees, from the middle of January till the Spring fairly opens (unless the weather permits them to fly safely), they wilt not suffer. This water may be placed in a wet sponge in a feeding-box, directly over the bees, and protected by a cushion of moss. A hundred or more colonies may thus, without disturbance, be quickly supplied."

That bees cannot raise brood without water, has been known from the times of Aristotle. Buera, of Athens (Cotton, p.104), aged 80 years, said in 1797: "Bees daily supply the worms with water; should the state of the weather be such as to prevent the bees from fetching water for a few days, the worms would perish. These dead bees are removed out of the hive by the working-bees, if they are healthy and strong; otherwise, the stock perishes from their putrid exhalations." I have repeatedly known colonies to suffer severe losses, for want of water; and in my correspondence with bee-keepers, the last Winter

(1858-9),* have directed their attention to this point, and have had my estimate of the value of water to bees in Winter greatly increased. But as yet, I have had no satisfactory evidence that any colonies, *whose honey was not candied, have died from water-dearth.*[6]

The Baron Von Berlepsch says, that "death from this cause more rarely occurs in districts where there is late Fall bee-forage than in those like his own, where pasturage falls occasionally in July, and usually early in August. In such regions, the honey becomes very thick in Winter, and sometimes thoroughly candied * before Spring." It is fortunate that, in the coldest parts of our country, late forage is usually abundant.

Berlepsch and Eberhardt not only condemn upward ventilation, as depriving the bees of the moisture which they need, but insist that it often hastens the ruin of a stock, by causing an excess of dampness among the bees, although they are actually in want of water. Dzierzon thinks that these acute observers have here fallen into a great mistake; and, did my limits permit, I could show that their objections to upward ventilation do not accord with facts, as observed in this country. So far from its being true "that the hive in which perceptible condensation of moisture occurs needs water, and that in which it does not take place needs none"—moisture often condenses so as to wet *the*

---

* I am particularly indebted to Mr. William W. Cary, Mr. Richard Colvin, Rev. J. C. Bodwell, Mr. E. T. Sturtevant, and Rev. Levi Wheaton, for careful observations made— last winter, at my suggestion—on wintering bees.

[6] *It is important to keep the problem of not enough water for brood rearing separate from too much water in the winter cluster. Bees do need water to raise brood and will stop if none is available. During the coldest parts of winter there is an excess of water from the metabolism of the honey. This moisture needs to be vented off, as Langstroth indicates. In the accounts of not enough water from German writings, the bees were flying, thus the temperatures were quite warm (>55° F.) Under dry conditions, and with a lack of available water, the bees would stop brood rearing.*

* Madame Vicat, in some observations on bees, published in 1764—see Wildman, p. 231— speaks of finding, "on the 24th of March, when the weather was so cold that the bees of her other hives did not go abroad, much candied honey on the bottom of a hive, and bees which seemed to be expiring. A singular noise was made in the hive, at intervals, and at such times numbers of bees would fall into the candied honey, and perish. The bees not being able to swallow the candied honey emptied it out of their combs to get at such as they could swallow."

combs and the bees,[†] showing plainly that there is an excess of water instead of a deficiency. The following facts, which have been furnished to me by the Rev. J. C. Bodwell, of Framingham, Massachusetts, are highly important in this connection. His colonies were wintered in a very dry cellar:

"About the beginning of the year (1859), opened my single glass hive, and found the bees abundant, and apparently healthy, but no eggs nor brood.

"Feb. 2. Examined the same hive, and found sealed brood, and unsealed, but no eggs. A considerable part of the brood had perished, probably from lack of water.

"Opened another hive, not so full if bees, and found the same state of things, except that less of the brood had perished. Combs dry in both, and many honey-cells open. Gave water to all, to their evident joy, and closed up the glass hive at the top, for experiment as to dampness, leaving the rest with upward ventilation.

"Feb. 5.—Examined both hives. No eggs in glass hive. The bees had been busy expelling dead brood. In the other, found eggs in moderate quantity. Very small larvae in both.

"Feb. 11.—Opened glass hive, and found the cells mostly emptied of dead brood, and abundance of eggs, and larvae just hatched. Discovered an opening between the hive and top-hoard, permitting upward ventilation, and closed it.

"March 1.—Made a thorough examination of both hives. Eggs, larvae, and sealed brood in both. The glass hive *very wet*, water standing on the tops of the frames, and at least a gill on the bottom-board; combs mouldy, and whole aspect of things comfortless. The other, *quite dry*, both hive and combs. Examined two other glass hives, having top ventilation, and found them dry. All have been treated precisely alike, except that the closed-up hive has had less water, as the bees did not seem to want it—manifesting no pleasure at receiving it. This hive had not so many eggs as the other, though much the larger stock, and appeared in a less healthy condition generally."

In any of my hives which have an upper cover, the bees can be easily supplied with water, and in those which have none, it may be in-

---

[†] In very cold weather, ice and moisture may super-abound in a hive, but it may be so far from the cluster that they cannot obtain it, even when perishing for the want of it.

jected with a straw into the winter entrance, or poured through the roof by a small hole, stopped with a plug, care being taken not to give too much.*

If the colonies are strong in numbers and stores, have upward ventilation, easy communication from comb to comb, and water when needed—and the hive entrances are sheltered from piercing winds, they have all the conditions *essential* to wintering successfully in the open air.

[*Bold is mine, as this is the essence of wintering bees.* RH]

Great injury is often done by disturbing a colony of bees when the weather is so cold that they cannot fly. Many which are tempted to leave the cluster, perish before they can regain it, and every disturbance, by rousing them to needless activity, causes an increased consumption of food. About once in six weeks, however, it will be advisable to clean the bottom-boards of hives wintered in the open air, of dead bees, and other refuse. Where permanent bottom-boards are used, this may be done with a scraper (Plate XI, Fig. 30), made of a piece of iron-wire, about two feet long; this, when heated, is bent about four inches, and flattened to one-quarter of an inch wide, both edges being made sharp.*

Bees very rarely discharge their fæces in the hive, unless they are diseased or greatly disturbed. If the Winter has been uncommonly severe, and they have had no opportunity to fly, their abdomens, before

---

* Mr. Wheaton finds that they will easily supply themselves with water from a sponge put over a hole, and covered with a tumbler: "If the water is *sweetened*, they will always drain the sponge; if not, they pay little attention to it, unless prevented from going abroad."

Mr. Wagner suggests that a piece of roofing-slate, fastened to the underside of the bottom-board, will cause the water to condense over the bees, where they can easily get access to it. Mr. Cary, at my suggestion, has placed a pane of glass on the frames directly over the bees, and the water condensed on it has seemed to supply all their wants. It should be elevated, so that the bees can pass under it. It may be found that, by some such simple device, we can, without any supervision, supply all the moisture that a strong colony needs in the coldest weather, before breeding has begun very actively. There is little doubt that it would answer for bees that are not wintered in the open air.

* Where a ventilator is made on the back of the hive (Plate V., Fig. 16), any refuse may be blown out by a pair of bellows. A very little smoke should be used before cleaning the bottom-board. Palladius, who flourished nearly two thousand years ago, says that bees ought not to be disturbed in Winter, except for the purpose of cleaning their hives of dead bees, &c.

Spring, often become greatly distended,[7] and they are very liable to be lost in the snow, if the weather, on their first flight, is not unusually favorable. After they have once discharged their fæces, they will not venture from their hives, in unsuitable weather, if well supplied with water. Having given the necessary precautions for wintering bees out of doors, the methods for defending them against atmospheric changes, by placing them in special depositories, will be described.

In some parts of Europe, it is customary to winter all the stocks of a village in a common vault or cellar. Dzierzon says:

"A *dry* cellar is very well adapted for wintering bees, even though it be not wholly secure from frost; the temperature will be much milder, and more uniform than in the open air; the bees will be more secure from disturbance, and will he protected from the piercing cold winds, which cause more injury than the greatest degree of cold when the air is calm.

"Universal experience teaches that the more effectually bees are protected from disturbance and from the variations of temperature the hotter will they pass the Winter, the less will they consume of their stores, sad the more vigorous and numerous will they be in the Spring. I have, therefore, constructed a special Winter repository for my bees, near my Apiary. It is weather boarded both outside and within, and the intervening space is filled with hay or tan; &c.; the ground plat enclosed is dug out to the depth of three or four feet, so as to secure a more moderate and equable temperature. When my hives are placed in this depository, and the door locked, the darkness, uniform temperature, and entire repose the bees enjoy, enable them to pass the Whiter securely. I usually place here my weaker colonies, and those whose hives are not made of the warmest materials, and they always do well. If such a structure is to be partly under-ground, a very dry site must be selected for it."

Mr. Quinby, who has probably the largest Apiary in the United States, has for many years wintered his bees, with great success, in a

---

[7] *Honey bees have six very well developed rectal glands, similar to those found in desert insects, which preserve water by extracting it from their waste. In honey bees they need to concentrate their feces, since they can not leave the hive until it is warm enough to fly, so they remove as much water as possible. The excess water is vented out of the cluster through respiration. Even with the concentrated feces the bees end up winter with a very large swollen rectum that takes up a large portion of the abdomen.*

room specially adapted to the purpose.[8] To get rid of the dampness, he *inverts* the common hives, and removes the board that covers my frames.

Mr. Wagner has furnished me with the following translation of a very able article from the *Bienenzeitung*. The author, the Rev. Mr. Scholtz, of Lower Silesia, is widely known in Germany for his skill in bee-keeping:

"Farmers have long been in the habit of placing apples, potatoes, turnips, &c., *in clamps*, to preserve them during Winter. They are piled in a pyramidal form, on a bed of straw, and covered six or eight inches thick, with the same material, evenly spread, as in thatching; and the whole is covered, in a conical form, with a layer of earth twelve inches thick, taken from a trench which is dug around the clamp. The proper finish is given by beating this earth smooth and even, with the back of the spade. This mode of preservation, when well executed, is found to keep fruit, tuberous roots, &c., in better condition during cold weather, than can be effected in cellars or vaults.

"These facts suggested to me the idea of protecting bees during the Winter, in a similar manner. It was evident, however, that a *bee-clamp* would require various modifications, to secure proper ventilation, to prevent undue development of heat, and to obviate an accumulation of moisture; and an arrangement, also, for readily ascertaining, and effectually regulating the temperature. All this, too, without seriously disturbing the bees, after the hives have been deposited in the clamp.

"To attain these objects, a circular space, sufficiently large for the intended purpose, is to be marked off on the driest and most elevated part of a garden, or other suitable spot of ground. The surface-soil containing vegetable matter, liable to decay, is then to be removed,

---

[8] *Cellar wintering was quite widely practiced until the early part of the 20th century. The problem with cellars was that they probably were not always dry enough or had sufficient air exchange at times. The modern wintering houses have tried to duplicate some of the best of cellars. They keep a constant cold temperature such that the colony uses the minimum of food. The reduction in honey may be the greatest advantage of wintering house, and the saving in food helps pay for the building and maintenance of these houses. Most of these houses have back-up refrigeration units if the outside temperature rises too much, and they have specially designed air exchange systems to ensure that there is not an excessive build up of carbon dioxide.*

and in the central part of the plot, a pit, three feet square, and three feet deep (see Fig. 66), is to be dug, spreading the earth taken there from evenly around, and treading it down hard. This pit is designed to serve as an air-chamber, as will be fully explained hereafter.

"The area having been properly prepared, four trenches, one inch and a half wide and deep, are to be dug; one extending from the middle of each of the four sides of the pit, to the outer edge of the periphery of the plot (Pl. XXI., Fig. 66). Into each of these trenches, a lead pipe, one inch in diameter, is to be laid, so as to form a communication between the pit and the air outside of the clamp when finished (Pl. XXI., Fig. 66). When these pipes are covered with earth, and the ground again leveled, a narrow strip of board should be laid thereon, to designate the position of the tubes, that they may not be injured in subsequent operations.

"The area, including the air-chamber, is now to be covered with pieces of four-inch scantling,[9] placed radiating from the centre, as nearly as practicable at regular distances apart, to serve as a platform on which the lower tier of hives is to be placed. The scantling should be cut of unequal lengths, and placed end to end, four inches apart, so as to leave interstices for the free circulation of air; and where required, as the space widens towards the circumference, additional pieces are to be laid in, so that the hives may be set firm and level. On this platform, the hives are to be built up in tiers, so that the clamp, when completed, shall present the form of a pyramid. Thus, the lower tier may consist of four ranges, of four hives each; the second, of three ranges, of three hives each; and the third, of two ranges, of two hives each. The fourth, or apex; however, must be formed of two hives, instead of one, for reasons which will hereafter appear (Pl. XXI., Fig. 68). The whole will thus form a four-sided pyramid, consisting of thirty-one hives, which, if Dzierzon's double hives he used will contain sixty-two colonies, in a comparatively small space. The oblong clamp (Pl. XXI., Fig. 70), is constructed on similar principles, with the requisite variation in shape.

"These hives, which are placed on the platform directly over the pit, or air-chamber, must he set six inches apart so that it continuous

---

[9] *Scantling* was a term used for a timber of relatively small dimensions. This word was also used for left over pieces of wood from construction of buildings of the time."

funnel, or direct air-passage, may be formed from the centre of the air-chamber below, to the apex of the clamp; and on the opposite fronts of the two uppermost hives, is to be placed a kind of chimney (see p. 351), made of four pieces of board, eight inches broad, and thirty inches long, having a movable cap, with a suitable slope, to prevent the entrance of rain. Holes are to be made in the sides of the chimney, below the cap, to allow the upward passage of air from the interior of the clamp. The rest of the hives may he placed closer together, though it is advantageous that they should not *touch* each other, so as to obstruct circulation in the interior, as it is important that the proprietor

Plate XXI.

Fig. 66.

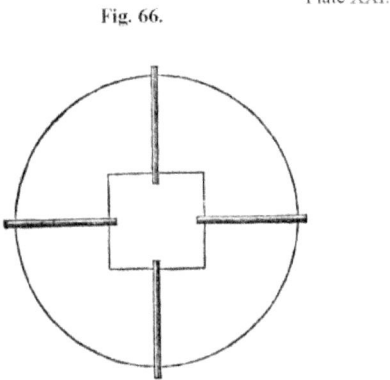

Fig. 67.   Fig. 68.

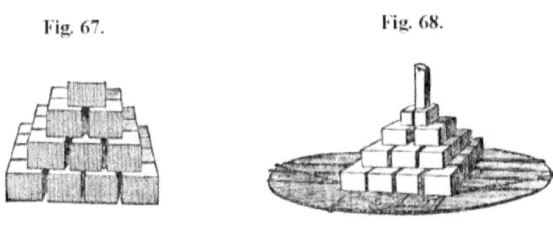

Fig. 70.   Fig. 69.

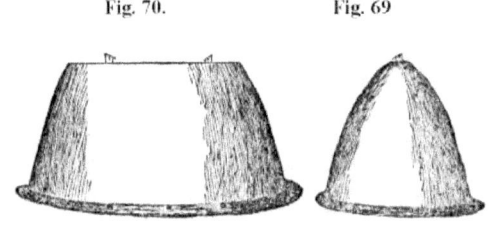

should be able to regulate the internal temperature uniformly. Very great exactness in arranging the hives, is, however, not requisite. It is essential only that they he set firm and level, so as to constitute a regular pyramid. Care must also be taken, not to commence by placing the hives too near the periphery of the area; because, between the outer edge of the lower tier of the hives, and the exterior mouths of the ventilating tubes, sufficient space must he reserved for the external covering, or mantle of the clamp (Pl. XXI, Fig.69).

"When the hives have been arranged in the manner described, and the chimney has been placed on the two upper ones, over the flue communicating with the pit, they are to be covered in with boards, cut to proper lengths, and placed vertically, side by side, around the sides of the pyramid. On and against these boards is to be laid a thick layer of rushes or old dry straw, forming a regular and dense coating, from base to apex. This coating is, in turn, to be covered with a layer of earth, five or six inches thick, spread as evenly as practicable, commencing below and proceeding upward to the chimney, so that the latter, having already been secured in its place by the boards and the straw or rushes, is now covered by the earth, to within six or seven inches of its top. The earth for covering, is taken directly from the base of the clamp, around which a trench six inches deep, and eighteen inches wide, is now to he dug, so as to expose the mouths of the ventilating tubes at the upper edge of the interior side of the trench. In digging the trench, care must be taken not to close or injure the mouths of the tubes, which should, moreover, be secured by a perforated tin cap, to exclude mice, and other vermin, and yet allow the free passage of air. The trench will serve to receive and carry off rain or snow-water, during the Winter; and to effect this more perfectly, several gutters or furrows should be drawn from it outwards. If sufficient earth be not obtained from the trench to cover in the straw or rushes completely, at least five inches thick, the deficiency must be supplied from other sources. The earth covering should be dressed smooth and oven with the back of a spade.

"In this state, the clamp should be allowed to remain till severe frosts occur, when an additional coat of leaves or pine shatters is to be given. This should be five or six inches thick, and applied as evenly as possible, from base to apex, leaving only about four inches of the chimney exposed. This material should be applied wet, as it will thus pack more closely, and afterwards better confine the heat. When finished, it should be well sprinkled with water from a watering-can, and allowed to freeze. A very compact structure will thus be formed (Figs. 69 and 70). The mouths of the ventilating tubes should next be protected, by placing a piece of board before each of them; and the trenches are then to be filled loosely with tangled straw.

"All this labor must be performed gently, so as to disturb the confined bees as little as practicable. The covering of leaves or pine-shatters should not be applied till after cold weather sets in; and it may be deferred till after the earlier snows have fallen and melted, and the severer weather of December or January makes additional protection desirable.

"If an extensive Apiary renders a clamp of larger dimensions necessary, two or three pits, or air-chambers, with their appurtenant ventilating tubes and chimneys (Pl. XXI., Fig. 70) may be introduced.

"On clear, mild days, the protecting boards maybe removed from the mouth of the ventilating tubes, that fresh air may freely enter the clamp, and carry off any dampness which may have formed within; and, as the entire interior is in direct communication with the air-chamber, a dry and healthy atmosphere will speedily he diffused throughout, by means of the draught of the chimney. Towards evening, the protecting-boards should be replaced. On the return of milder weather, or on the termination of severe and protracted frosts, the months of the ventilating tubes may be uncovered, and left open, day and night, to prevent the undue development of heat in the interior; but in clear weather, the direct rays of the sun should be excluded from the mouths of the tubes. If the holes in the sides of the chimney should at any time become closed with snow, the obstructions must be removed, by means of a rake or other convenient implement. When the exterior of the clamp is covered with snow, the mouth of one of the ventilating tubes should be kept open, even in cold weather, and of all of them when the weather is moderate, because the snow covering causes great internal warmth.

"To ascertain the interior temperature, a thermometer attached to a long rod may be introduced into the air-chamber, through the chimney, on removing the cap. This should be done frequently, to serve as a guide for opening or closing the mouths of the ventilating tubes. Ventilation seems, however, according to the numerous experiments which I have made, to be of less importance to the health of the bees than to preserve the combs and interior of the hives from dampness and mould; and it is in view of this fact, that I have adopted the peculiar arrangement of my clamps, which places it in the power of the Apiarian, at almost any time, to cause an adequate circulation of pure dry air within them.

"Apart from their cheapness, these clamps are far superior, for the purpose intended, to the best vaults or cellars ordinarily accessible. It might be objected to this mode of wintering bees, that the hives cannot be inspected during the Winter, however desirable such inspection might seem to be. That is so; but, in devising my clamps, I really had no reference whatever to that class of bee-keepers who are in the habit of operating among their colonies in Winter. Their case, in fact, seems to me to be a rather hopeless one at best, since colonies that are thus treated at that season, will scarcely ever enable their owner to found an Apiary worthy of the name. I prefer to let my bees remain undisturbed during cold weather, satisfied that if they were in good condition when enclosed in the Fall, they will pass the Winter uninjured, and be found with adequate supplies of honey even in April. Of this I am the more assured, since I have ascertained that bees preserved in clamps consume scarcely one-half of the quantity of honey required by such as are wintered in the open air, or in the Apiary.[10]

"To institute a comparison between different modes of wintering bees, I placed a portion of my colonies in a clamp of the fore going construction, on the 17th of November, 1856, and transferred the remainder into a well-protected dark chamber in my dwelling. house. Of some of the latter, I closed the entrances, but gave them air through a grate or ventilating-passage in the rear of their hives. Of the remainder, the entrances, as well as the ventilating-passages, were shut close. Several of these placed in the clamp were designedly selected as having only eight or ten pounds of honey each, that I might ascertain

---

[10] *Quantitative research has shown that the food consumption curve is U shaped. The lowest consumption occurring at 42° to 45° F., and increased consumption at higher and lower temperatures. This is one of the reasons that cellar wintering, or other wintering houses have been used. That is, constant 42-45° F. temperatures saves honey.*

whether they would survive with so small a supply of food. I placed therein, also, a late after-swarm, which had built only a few short combs, and contained not more than four or five pounds of honey. All the others had ample stores. I closed the entrance and ventilating. passage of one strong colony, and placed some pieces of empty comb in the rear of the hive, to test whether, if moisture were generated from want of ventilation, mould would form on those combs.

"From the 18th to the 23rd of November, the weather was very mild, and the ventilating-tubes were, therefore, all left open day and night. On the 24th, the clamp was covered with snow, and I closed three of the ventilating-tubes. On the 26th, a thaw commenced, and the weather continued to be very moderate to the end of the month, the thermometer standing at 33° in the open air. Two of the tubes were kept open. From the 1st to the 3rd of December, ten inches of snow fell, with the thermometer ranging from 20° to 22°; and I kept only one tube open. On the 6th, the weather moderated; from the 7th to the 12th, the thermometer stood at from 54° to 66°, and I again opened all the tubes, and kept them open till the end of the month, and to the 5th of January. On the 6th, the weather became cold and freezing, and I now added the outer mantle, or coating of leaves and pine shatters, closing all the tubes. The cold spell continued till the 17th of January. From the 18th till the end of the month, we had continuous fair, mild weather, and I opened all the ventilating-tubes. In February, the weather was particularly mild and fair, and, from the 18th to the 21st, the thermometer ranged from 76° to 78°. The bees belonging to some of my neighbors, and which were wintered in the open air, were now flying briskly every day, and most of the colonies in my chamber became so restless that I was constrained to remove them out of their Winter quarters. I did so with the less reluctance, as we had all the indications of an early Spring. The fair weather continuing, I deemed it wrong to keep my colonies longer confined in the clamp, and accordingly opened it on the 27th of February, to release them.

"Though the clamp had been exposed to the direct rays of the noonday sun, and the thermometer had daily ranged at from 76° to 78° for some time previous, yet, on removing the outer mantle, I found the earth-covering below it still frozen, so that it had to be removed with a hoe—satisfactory proof that the interior of the clamp could not have been affected by external variations of temperature. I now became exceedingly anxious to see whether rain or snow-water had penetrated to the straw covering, as I apprehended might be the case, having had no previous experience in such matters. To my surprise and gratification, however, I found it thoroughly dry-showing conclusively that the earth-covering had sufficed effectually to shed off the rain and snow-water, and that the ample and efficient internal ventilation had prevented the formation of moisture and mould. On removing the straw, I perceived no symptom of dampness on the boards; and when, finally, these latter were taken away, the hives presented themselves as clean and dry as when put there in the Fall.

"Anxious now to ascertain the condition of their inmates, I tapped against the hives, but, to my dismay, heard no response. I seized a stick, and, tapping harder and harder, finally proceeded to blows; still all remained mute within. An old man from the neighboring village, who chanced to be present, seemed vastly gratified at my chagrin and consternation, as he and his neighbors had kept bees for many years, but had no fancy for such novel contrivances and experiments as mine. I must admit that I was, for the moment, thoroughly disconcerted on finding, as I then supposed, all my anticipations and confident calculations thus suddenly and effectually nullified. But, resolved to know the worst, I removed the hives to the Apiary, where the sun shone bright and warm; and scarcely were the entrances opened, when the bees began to pour forth in masses, humming joyously, to my irrepressible delight, and to the utter discomfiture of the old villager. With special gratification did I notice that the bees came forth from their long imprisonment with bodies as attenuate and slender as they had in the preceding Autumn, whilst those which had been wintered in the dark chamber soiled their hives and all surrounding objects, by profuse discharges of fæcal matter. This led me to conjecture that these colonies had consumed comparatively little honey, which was found to be the fact on opening the hives and examining the condition of their stores. Those colonies which had only eight or ten pounds of honey in the Fall, had still a surplus remaining, and were healthy and strong; while the poor little after-swarm had not only well preserved its numbers, but had the greater portion of its small supply of honey still in reserve. Few dead bees were found, and those probably died of old age. The loss of bees was very much greater in the colonies which had been wintered in the house, and more than double the quantity of honey had been consumed by each of them; so that a very important saving can manifestly be effected by means of clamps, apart from the other important advantages which this mode of wintering bees possesses. The combs in all the colonies were clean and free from mould, and I could perceive no difference in this particular between the hives which had their entrances and ventilating passages closed, and those in which the latter had been left open, the pieces of old comb, even, having remained dry and free from mould. Satisfactory proof was thus furnished that, where the temperature is moderate and uniform throughout, condensation of moisture will not result from close confinement. Still, from various considerations, I would recommend ventilation in every hive; and previous experience has taught me that bees will remain more tranquil during the Winter in hives duly ventilated, than in such as are closed. A number of the colonies deposited in my dark room were purposely confined without ventilation. Three of these became very restless, consumed a disproportionate amount of their stores, and very many of the bees perished. Precisely these three colonies, though still strong and healthy in the Spring, were yet the weakest of the whole lot, though in as good condition as the others when removed from the Apiary in Autumn. Nothing similar occurred in the colonies which had even partial ventilation.

"Having thus, by these diversified experiments in wintering bees, arrived at certain and satisfactory results, I shall never hereafter winter my movable colonies otherwise than in clamps.

"Since the publication of my mode of wintering bees in clamps, some objections have been urged against it, which I shall briefly notice, before giving the results of my further experience in this matter.

"The expense of constructing: the clamps has been alleged as an objection to the use of them. In my case, the cost of labor was simply the hire, for one day, of two men, who assisted me in preparing the area, carrying the hives thither, and arranging and enclosing them. The materials used, with the exception of the scantling, cost literally nothing, as any old boards can be made to serve the purpose, and the rushes, or straw, leaves, &c., employed, are always worth their cost for litter.

"A second objection is, that rats and mice will be induced to collect and harbor in the clamps; if straw be used. I never use any but old straw, thoroughly divested of grain, and prefer using rushes when they can conveniently be procured. I have, however, thus far, not been annoyed by rats or mice.

"To show how very superior clamps are for wintering bees, in thin hives especially, I will state that one of my neighbors, whose hives are made of inch, boards, and who invariably lost many bees, and frequently entire colonies, when he left them to winter, as he usually did, in his open Apiary was induced by my success to place his hives in a clamp last Fall. They were put in on the 11th of November, 1857, and remained undisturbed till the 29th of March, 1858. When opened, all the colonies proved to be in excellent condition, strong, and entirely free from mould or moisture. Never, in any previous season, had he been equally successful, nor had his bees ever before required or received so little personal attention from him. He was 'a doubting Thomas,' when he saw me arranging my first clamp but is now a thorough convert to the system, and declares that he will, in future, use no other mode, as he cannot conceive that a better could be devised.

"My own colonies remained in the clamp from the 13th of November to the 29th of March, 1858, and were perfectly sound and healthy when I opened them. The earth under the outer mantle was still frozen, and had to be removed with a hoe, as in the previous year, thus showing that the bees were not affected by the prevalent mild weather. Long confinement had not injured them in the least degree, because, reposing in a low and equable temperature, they had consumed proportionally little honey, and remained without excitement or disturbance during the whole period. I am now fully convinced that bees may remain confined in this manner during the most protracted Winter, not only without injury, but with positive benefit, as they are altogether secure from the always detrimental, and frequently ruinous, effects of exposure to the vicissitudes of the weather in our variable climate.

"To simplify the construction of the clamps, I made my last one longer and lower than the one I prepared the previous Fall; and I was thus able to apply the successive covers, or mantles, more easily and conveniently. I also dispensed with the chimney, and could thus close the top more regularly and perfectly, laying over the apex, boards weighted down with stones to keep them in place. I found no disadvantage resulting from discarding the chimney, as the ventilating-tubes enabled me still to regulate the internal temperature, and give the bees a sufficient supply of fresh air. I also enlarged the air-chamber, making it three feet deep, as before, by only thirty inches broad, and lengthening it so as to extend the whole length of the *interior* diameter of the clamp. In every other respect, the construction remained the same."

When hives are wintered in a special repository, I should advise giving them upward ventilation. If they are in cellars or rooms, the upper cover may be entirely removed; and, if put in clamps, then it may be fastened, as advised on page 338, and some air be allowed to enter at the lower part of the hive.

In all the northern parts of this country, it is very obvious that those who mean to establish large Apiaries will have to so winter their bees, that they shall not be exposed to the usual atmospheric changes.[11] What way precisely is the best can only be determined by careful and long-continued experiments. These ought not to be conducted so as to hazard too much in one venture.

---

[11] *I like an abundance of honey within the hive during winter, if for no other reason than to buffer rapid changes in temperature. A large mass of honey will dampen the big shifts in temperature that occur. This large excess of honey needed for the actual winter is then the food that allows the colony to expand brood rearing in the Spring.*

Fig. 71. PLATE XXII.

So work the Honey bees,
Creatures, that by a rule in Nature, teach
The art of order to a peopled kingdom. — *Shakspeare*

Great loss is often incurred in replacing upon their Summer stands the stocks which have been kept in special depositories. Unless the day when they are put out is very favorable, many will be lost when they fly to discharge their faeces. In movable-comb hives, this risk can be greatly diminished, by removing the cover from the frames, and allowing the sun to shine directly upon the bees; this will warm them up so quickly, that they will all discharge their fæces in a very short time.*

After the stocks are placed on their Summer stands,† the precautions already described should be taken to strengthen feeble or impoverished colonies (p. 221).

---

* The following is an extract from my journal:

"Jan. 31st, 1857.—Removed the upper cover, exposing the bees to the full heat of the sun, the thermometer being 80° in the shade, and the atmosphere *calm*. The hive standing on the sunny side of the house, the bees quickly took wing and discharged their fæces. Very few were lost on the snow, and nearly all that alighted on it took wing without being chilled. More bees were lost from other hives which were not opened as few which left were able to return; while, in the one with the cover removed, the returning bees were able to alight at once among their warm companions."

† Dzierzon advises placing them on their former stands, as many bees still remember the old spot. Mr. Quinby uses this time for equalizing the colonies, as he finds that, "being all wintered in one room, their scent is so much alike that they mix together without contention."

(This seems to be a contradiction. First they "...remember the old spot." and then "...they mix together without contention." Most beekeepers experience that the bees do not remember their old location.)

CHAPTER XXII.

BEE-KEEPER'S CALENDAR — BEE-KEEPER'S AXIOMS.

THIS Chapter gives to the inexperienced bee-keeper brief directions for each month in the year,[*] and, by means of the full Alphabetical Index, all that is said on any topic can easily be referred to.

JANUARY.—In cold climates, bees are now usually in a state of repose. If the colonies have had proper attention in the Fall, nothing will ordinarily need to be done that will excite them to an injurious activity. In very cold climates, however, when a severe temperature is of long continuance, it will be necessary, unless the hives have thorough upward (p. 340) ventilation, to bring them into a warm room (p. 341), to thaw out the ice,[1] remove the dampness, and allow the bees to get access to their supplies. In January there are occasionally, even it very cold latitudes, days so pleasant that bees can fly out to discharge their faeces; do not confine them (p. 337), even if some are lost on the snow. In this month clean the bottom-boards (p. 347), but disturb the bees as little as possible. See, also, that they are properly supplied with water (p. 344), as healthy stocks have already begun to breed (p. 239).

FEBRUARY.—This month is sometimes colder than January, and then the directions given for the previous month must be followed. In mild seasons, however, and in warm regions, bees begin to fly quite lively in February, and in some locations they gather pollen. The bottom-board should be again attended to, as soon as the bees are actively on the wing, and, if any hives are suspiciously light, sugar-candy (p. 272) should be given them. Strong colonies will now begin to breed considerably, but nothing should be done to excite them to premature activity. See that the bees are supplied with water (p. 344).

MARCH.—In our Northern States, the inhospitable reign of Win-

---

[*] *Palladius*, who wrote on bees nearly 2,000 years ago, arranges his remarks in the form of a monthly calendar.

[1] *It is the change in temperature from the cluster to the outer frames or the hive body that causes the moisture from the bee's metabolism to condense. Insulation of the hive often allows this moisture to escape before it freezes, or condenses.*

ter still continues, and the directions given for the two previous months are applicable to this. If there should be a pleasant day, when bees are able to fly briskly, seize the opportunity to remove the covers (p. 361); carefully clean out the hives (p. 221), and learn the exact condition of every colony. See that your bees have water (p. 344), and are will supplied with rye-flour (p. 84).[2] In this month, weak stocks commonly begin to breed, while strong ones increase quite rapidly. If the weather is favorable, colonies which have been kept in a special Winter depository, may now be put upon their proper stands (p. 361). As soon as severe Winter weather is over, it will be necessary to shut off all upward ventilation.

APRIL.—Bees will ordinarily begin to gather much pollen in this month, and sometimes considerable honey. As brood is now very rapidly maturing, there is a largely increased demand for honey, and great care should be taken to prevent the bees from suffering for want of food. If the supplies are at all deficient, breeding will be checked, even if much of the brood does not perish, or the whole colony die of starvation.[3] If the weather is propitious, feeding to promote a more rapid increase of young (p. 268) may now be commenced. Feeble colonies must now be reinforced (p. 221), and should the weather continue cold for several days at a time, the bees ought to be supplied with water (p. 344) in their hives. In April, if not before, the larvae of the bee-moth will begin to make their appearance, and should be carefully destroyed (p. 248).

MAY.—As the weather becomes more genial, the increase of bees in the colonies is exceedingly rapid, and drones, if they have not previously made their appearance, begin to issue from the hives. In some

---

[2] *I like to get pollen supplement patties onto colonies either in late February (Here in the North), or early March. These are to supplement the pollen that may still be in the hive but outside the cluster area where the bees cannot access it. The patties are mixed with 2:1 sugar syrup into a bread dough consistency. These are put directly onto the top bars. The patties will stimulate brood rearing in greater amounts. The bees will stop feeding on these patties as soon as pollen is available in the fields.*

[3] *Bees will consume a great quantity of honey raising brood. The large amounts of honey (60+ lbs.) recommended for winter is mainly for this spring period when there still is not a nectar supply available. I think the threshold for the colony to stop brood rearing is about three frames of honey (~ 15 lbs.). Therefore always keep more than this in the hive.*

locations, the bees will now gather much honey, and it will often he advisable to give them access to the spare honey receptacles;* but in some seasons and locations, either from long and cold storms, or a deficiency of forage, stocks not well supplied with honey will exhaust their stores, and perish, unless they are fed. In favorable seasons, swarms may be expected in this month, even in the Northern States. These May swarms often issue near the close of the blossoming of fruit-trees, and just before the later supplies of forage, and if the weather becomes suddenly unfavorable, may starve, unless they are fed. Even if there is no danger of this, they will make so little progress in comb-building and breeding, when food is scarce, as to be surpassed by much later swarms. The Apiarian should have hives in readiness to receive new swarms, however early they may issue, or be formed. If new colonies are to be made by artificial processes, a seasonable supply of queens (p. 188) should be reared.[4]

JUNE.—This is the great swarming month in all our Northern and Middle States. As bees keep up a high temperature in their hives, they are by no means so dependent upon the weather for forwardness, as plants, and as most other insects necessarily are. I have had as early swarms in Northern Massachusetts, as in the vicinity of Philadelphia.

---

* If natural swarms are wanted, the bees should not be allowed to occupy too much surplus storage-room.

(In modern beekeeping we would normally want to prevent swarming and thus would provide more room. The general practice of having two brood chambers also allows the bees to put some honey into outer combs without adding extra supers.)

[4] *Queens can be safely reared at a fairly early time, but it is the mating that may be held up because of cool, Spring weather. Here in the North we can generally rear queens AND have them successfully mated after the 1st of May. The further south you go the earlier that you could advance the process. We normally would have drones by the first to the middle of April, so in theory we could mate queens any time after that. Some years yes and some years no. It just depends upon favorable mating weather.*

If the Apiary is not carefully watched, the bee-keeper, after a short absence, should examine the neighboring bushes and trees, on some of which he will often find a swarm clustered, preparatory to their departure for a new home.*

As fast as the surplus honey-receptacles are filled,† and the cells capped over, they should be removed, and empty ones put in their place. Careless bee-keepers often lose much, by neglecting to do this in season, thereby condemning their colonies to a very unwilling idleness. The Apiarian will bear in mind, that all small swarms which come off late in this month, should he either aided, doubled, or returned to the mother-stock. With my hives, the issue of such swarms may be prevented, by removing, in season, the supernumerary queen-cells. During all the swarming season, and, indeed, at all other times when young queens are being bred, the bee-keeper must ascertain seasonably, that the hives which contain them, succeed in securing a fertile mother (p. 218).

JULY.—In some seasons and districts, this is the great swarming month; while in others, bees issuing so late, are of small account. In Northern Massachusetts, I have known swarms coming after the Fourth of July, to fill their hives, and make large quantities of surplus honey besides. In this month, all the choicest spare honey should be removed from the hives, before the delicate whiteness of the combs becomes soiled by the travel of the bees, or the purity of the honey is impaired by an inferior article gathered later in the season.

The bees should have a liberal allowance of air during all extreme-

---

* "As it may often be important to know from which hive the swarm has issued, after it has been hived and removed to its new stand, let a cup-full of bees be taken from it, and thrown into the air, near the Apiary; they will soon return to the parent-stock, and may easily be recognized, by their standing at the entrance, and fanning, like ventilating bees." —DZIERZON. In my hives, it will be easy, from the back ventilator, to decide whether a stock is full enough to swarm, or has recently swarmed, even when there is no glass for observation.

† Mr. Quinby informs me, that he succeeds in making bees fill a double tier of small boxes, by placing one set on the hive first; when they have partially filled these, he put the second set *under* the first. By making a hole in the top, as well as in the bottom of the box (Pl. XI., Fig. 24), this can easily be effected.

(This practice of putting the newest super under is generally preferred in order to produce more honey. It often is not done in commercial apiaries simply because of the labor and time needed to accomplish the task)

ly hot weather, especially if they are in unpainted hives, or stand in the sun.

AUGUST.—In most regions, there is but little forage for bees during the latter part of July, and the first of August, and being, on this account, tempted to rob each other, the greatest precautions should be used in opening hives. In districts where buckwheat is extensively cultivated, bees will sometimes swarm when it comes into blossom, and in some seasons, extraordinary supplies are obtained from it. In 1856, I had a buckwheat swarm as late as the 16th of September! [5]

If any colonies are so fall of honey, that they have not room enough for raising brood, some of the combs should now be removed (p. 183). If the caps of the cells are carefully sliced off with a very sharp knife, and the combs laid over a vessel, in some moderately warm place; and turned once, *most* of the honey will drain out of them, and they may be returned to the bees, to be filled again.[6]

The beekeeper who has queenless stocks on hand in August, must expect, as the result of his ignorance or neglect, either to have them robbed by other colonies, or destroyed by the moth (p. 246).

SEPTEMBER.—This is often a very busy month with bees. The Fall flowers come into blossom, and in some seasons, colonies which have hitherto amassed but little honey, become heavy, and even yield a surplus to their owner. Bees are quite reluctant to work in boxes, so late in the season, even if supplies are very abundant; but if empty combs are inserted in the place of full ones removed, they will fill them with astonishing celerity. These full combs may afterwards be returned, if the bees have not a sufficient supply without them.

If no Fall supplies abound, and any stocks are too light to winter with safety, then, in the Northern States, the latter part of this month is

---

[5] *We may not know the all of the cues that cause swarming. Certainly the day-length would be the same in Spring and Fall. It also may be the dearth followed by a honey flow that helps turn on the swarming instinct. In nature all fall swarms would perish if they were not taken care of by a beekeeper. They just can not build a large enough supply of honey or bees to survive a cold winter.*

The biological window of a swarm is rather narrow. Too early and there isn't nectar available, and too late and they do not store enough surplus. The selection for the right time to swarm is rather intense.

[6] *Van Hruschka, of Austria, invented the extractor in 1865, or six or seven years after this book was written.*

the proper time for feeding them. I have already stated (p.274), that it is impossible to tell how much food a colony will require, to carry it safely through the Winter; it will be found, however, very unsafe to trust to a bare supply, for even if there is food enough, it may not always be readily accessible to the bees.[7] Great caution will still be necessary to guard against robbing; but if there are no feeble, queenless, or impoverished stocks, the bees unless tempted by improper management, will seldom rob each other.

OCTOBER.—Forage is now almost entirely exhausted in most localities, and colonies which are too light should either be fed, or have surplus honey from other stocks given to them, early this month. The exact condition of every stock should now be known, at the latest, and, if any are queenless, they should be broken up. Small colonies ought to be united, and all the hives put into proper condition for wintering. Some full honey-combs should be put in the centre of the hive, and holes, for easy intercommunication, made in the combs (p. 337); and, if the hives have a winter-passage, bees should now be accustomed to use it (p. 338). By the last of this month, the glass hives should be packed between their outer cases and the glass, with cotton waste, moss, or any warm material.

NOVEMBER.—I take for granted that all necessary preparations for Winter have, in our Northern States, been completed by the last of the previous month. If, however, the bee-keeper has been prevented from examining his stocks, he may, on warm days, in November, safely perform all necessary operations, the feeding with liquid honey excepted. The entrances to the hives must now be secured against mice, and it will be well to give the roofs a new coat of paint. If the hives are to be exposed to the sun, no color is so good as a pure white but, if they are set under the shade of trees (p. 280), a dark color will do them no harm, in the hottest weather, while early in the season, before the leaves are expanded, by absorbing instead of reflecting the

---

[7] *I think this is generally the most difficult problem with stores within a colony, which is that the honey is not accessible to them when they need it. Since the cluster must move upwards during cold weather, they may starve if they reach the top of the colony before they can spread out to reach honey in lateral combs. One of the easiest ways to make sure the honey is located properly is to feed the bees syrup in the Fall and let them place it where they want it.*

heat, it will prove highly advantageous to the bees.

By the latter part of November, in our Northern States, Winter usually sets in, and colonies which are to be kept in a special Winter depository, should be properly housed. The later in the season that the bees are able to fly out and discharge their fæces, the better. The beekeeper must regulate the time of housing his bees by the season and climate, being careful neither to take them in until cold weather appears to be fairly established, nor to leave them out too late. If colonies are carried in too early, and quite warm weather succeeds the first cold, it may be advisable to replace them on their Summer stands.*

As soon as freezing weather sets in, the colonies standing in the open air must have upward ventilation (p. 338).

DECEMBER.—In regions where it is advisable to house bees, the dreary reign of Winter is now fairly established, and the directions given for January are for the most part equally applicable to this month. It may be well, in hives out of doors, to remove the dead bees and other refuse from the bottom-boards; but, neither in this month nor at any other time should this be attempted with those removed to a dark and protected place. Such colonies must not, except under the pressure of some urgent necessity, be disturbed in the very least.

I recommend to the inexperienced beekeeper to read this synopsis of monthly management, again and again, and to be sure that he fully understands and punctually discharges the appropriate duties of each month, neglecting nothing, and procrastinating nothing to a more convenient season; for, while bees do not require a large amount of attention, in proportion to the profits yielded by them, they *must* have it at

---

* If bees are wintered on Mr. Scholtz's plan, it will neither be possible nor desirable to replace them on their Summer stands.

PLATE XXIII.

the *proper time* and in the *right way*. Those who complain of their unprofitableness, are often as much to blame as a farmer who neglects to take care of his stock, or to gather his crops, and then denounces his employment as yielding only a scanty return on a large investment of capital and labor.

BEE-KEEPER'S AXIOMS.

There are a few *first principles* in bee-keeping which ought to be as familiar to the Apiarian as the letters of his alphabet:

1st. Bees gorged with honey never volunteer an attack.

2nd. Bees may always be made peaceable by inducing them to accept of liquid sweets.

3rd. Bees, when frightened by smoke or by drumming on their hives, fill themselves with honey and lose all disposition to sting, unless they are hurt.

4th. Bees dislike any *quick* movements about their hives, especially any motion which jars their combs.

5th. Bees dislike the offensive odor of sweaty animals, and will not endure impure air from human lungs.

6th. The beekeeper will ordinarily derive all his profits from stocks, strong and healthy, in early Spring.

7th. In districts where forage is abundant only for a short period, the largest yield of honey will be secured by a *very* moderate increase of stocks.

8th. A moderate increase of colonies in any one season, will, in the long run, prove to be the easiest, safest, and cheapest mode of managing bees.

9th. Queenless colonies, unless supplied with a queen, will inevitably dwindle away, or be destroyed by the bee-moth, or by robber-bees.

10th. The formation of new colonies should ordinarily be confined to the season when bees are *accumulating* honey; and if this, or any other operation must be performed, when forage is scarce, the greatest precautions should be used to prevent robbing. The essence of all profitable bee-keeping is contained in Oettl's Golden Rule: KEEP

YOUR STOCKS STRONG (p. 303). If you cannot succeed in doing this, the more money you invest in bees, the heavier will be your losses; while, if your stocks are strong, you will show that you are a *bee-master*, as well as a bee-keeper, and may safely calculate on generous returns from your industrious subjects.

EXPLANATION OF PLATES OF HIVES.

DESCRIPTION OF WOOD-CUTS OF THE VARIOUS STYLES OF MOVABLE-COMB
HIVES, WITH BILLS OF STOCK FOR MAKING THEM.

ALL the engravings,* except those which are in perspective, are on the scale of $1\frac{1}{2}$ inches to the foot; so that every $\frac{1}{8}$ of an inch is an inch in a hive of full size. The thickness of stock used, is mostly $\frac{7}{8}$ ths of an inch– inch boards, when planed, being usually of that thickness – the measurements can be easily varied, to suit any required dimensions. In making a lot of hives (see p. 332), the small pieces, which otherwise would be refuse, should be used for the frames. Good stock will prove much the cheapest in the end.

These not accustomed to longitudinal and cross sections, will be greatly assisted by the perspective views. In the longitudinal sections, the hive is represented as sawed in two, from front to rear, and in the cross sections, from side to side. All the parts supposed to be cut by the saw, are marked by cross lines; the parts, which, though not cut, would be seen after the cutting, are also represented. Any measurement may be verified, by applying an accurate rule to the sections.

The reader will bear in mind, that those only who have purchased the patent right—Ministers of the Gospel excepted—can legally use these hives. For terms, see p. 391.

Figs. 1, 2, and 3, page 24, Hive No. 1.

Fig. 1 is a perspective view of a hive of the simplest form, the cover being removed, to show one of the frames. Fig. 2 is a vertical longitudinal section; and Fig. 3, a vertical cross section of the same.

---

* Since the publication of the second edition—for which most of these plates were engraved—some changes have been made in the construction of the hives, all of which are fully noted in the bills of stock, though not, in all cases, shown in the plates.

(*b*) Two pieces, front and rear of hive, $14\frac{1}{8}$" x $8\frac{7}{8}$" x $\frac{7}{8}$". (*c*) Two pieces, sides of hive, $19\frac{7}{8}$" x 10" x $\frac{7}{8}$", with outside lower edges beveled off—when a movable bottom-board is used—to avoid crushing bees, or giving lurking-places to moths or worms. When the bottom board is fixed in the hive, the sides- should be $19\frac{7}{8}$" x $10\frac{7}{8}$" x $\frac{7}{8}$", and the bottom board $25\frac{1}{2}$" x $14\frac{1}{8}$" x $\frac{7}{8}$", clamped on the under side. If another hive, of the same form, is put on the first, for surplus honey, as in Fig. 16 (p. 48), holes may be made through this bottom-board, as directed for Hives No. 2. (*d*) Two pieces, strips on upper part of hive, front and rear, forming rabbets for the frames to rest upon, $15\frac{7}{8}$" x $1\frac{1}{2}$" x $\frac{7}{8}$". (*f*) Movable cover, $25\frac{1}{2}$"x 18" x $\frac{7}{8}$". This should be tongued and grooved together, and may also be rain-grooved, as shown for the top of the hive in Fig. 23 (p. 96). The grain of the wood should run from front to rear. (*g*) Two pieces, clamps on under side of cover, 18" x 2" x $\frac{7}{8}$". The front and rear (*b*) of the hive should nailed between the sides (*c*), flush with their ends, but with the upper edges of (*b*) $\frac{5}{8}$" below the upper edges of (*c*). Some may prefer that the grain of the wood, both of the bottom-board and cover, should run from side to side, instead of from front to rear.

Movable Comb-Frames. Figs. 1, 4, and 22, pages 20, 24, 88.

(*t*) Two pieces, top, 19" 1" x $\frac{5}{16}$"; bottom, $17\frac{3}{8}$" x $\frac{7}{8}$" $\frac{1}{4}$". (*u*) Ends or vertical pieces,* two pieces, $8\frac{5}{8}$" x $\frac{7}{8}$" x $\frac{1}{2}$." (u) One piece, triangular-top comb-guide, $16\frac{3}{8}$" x $\frac{7}{8}$" x $\frac{7}{8}$" x $\frac{7}{8}$". This should he nailed to the top of the frame, centrally with regard to its width and length, and the frame may be stiffened by driving one nail through each end into it. If comb is used for guides (pp. 72, 130), or the other devices for securing straight comb succeed, those triangular guides may be dispensed with.

---

* The triangular pieces represented in many of the engravings, not answering the ends intended, I return to the shape originally used. The Winter passage (v) which was suggested for trial, is also discarded, Mr. Cary's method (p.337) being much better.

Double Movable Comb-Frames. Fig. 73, Plate X., page 96.

This frame is made up of the same parts as two single frames, differing from them only by having their end pieces in common, which are 8 $\frac{5}{8}$" x 2 $\frac{1}{4}$" x $\frac{1}{2}$". In putting this frame together, if the triangular guides are used, they are first to he nailed, as in the single frames, centrally to the top pieces; each top piece, when nailed to the end pieces, projects over their edges a sixteenth of an inch, and the bottom pieces come flush with the edges of the end pieces. As one side of a comb is usually a *fac simile* of the other, these double frames, which are *proposed for trial*, may answer a value-end (sic), in connection with the single ones. They rest very firmly on the rabbets, and are easily adjusted and handled.

All the parts of the movable frames should be cut out by circular saws (p. 332), and the measurements should be exact, so that the frames when nailed together may be square. If they are not *strong* and *perfectly square*, the proper working of the hive will be greatly interfered with. Ten single, or five double frames, equally distant from each other, are placed in the lower hive, and nine single frames, or four double frames and one single one, may be placed in the upper hive, for surplus honey.

Comb-Guides. Fig. 72, Plate VI., page 48.

This figure shows the form of a metallic stamp, invented by Mr. Wehring, of Bavaria, Germany, *for printing or stamping the foundations of the combs* upon the under side of the frames. After the outlines are made, he rubs melted wax over them, and scrapes off all that does not sink into the depressions. Mr. Wehring represents this device as enabling him to dispense with guide-combs, the bees appearing to be delighted to have their work thus accurately sketched out for them. In practice it is found to be inferior to the triangular comb guides. Mr. R. Colvin has invented a device* for securing the combs not merely *straight*, but of *uniform thickness*. It will be tested on a large scale, this season (1860), and the results given to the public. In those instances in which it has been tried, it has succeeded admirably.

Gage-Block for fastening the movable frames together. Figs. 6, 7, and 8, page 24.

Fig. 6 is a view of the front of this block; Fig. 8 a view of the back, and Fig. 7 is a cross-section.

(*a*) Foundation board, $21\frac{7}{8}$" x $9\frac{1}{8}$" x $\frac{7}{8}$". (*b b*) Guides; for sides (*u u*), of frames, fastened to (*a*), equally distant from its ends; and so as to leave $17\frac{3}{8}$" between (*b b*), and $\frac{1}{4}$" from upper edge of (*a*) to ends of (*b b*). (*c c*) Buttons for holding sides of frames (*u u*), against (*b b*), $6\frac{1}{2}$" x $1\frac{1}{2}$" x $\frac{7}{8}$". (*f f*) Guides in which the top triangular comb-guide is placed; in order to have the top strip (*t*) nailed thereto; each piece (*f*) is $21\frac{7}{8}$"x 2"x $\frac{3}{4}$" and they are beveled from one edge, back $\frac{7}{16}$", and are then fastened to (*a*), forming a triangular groove, each side of which is $\frac{7}{8}$". Two triangular pieces, $\frac{7}{8}$" x $\frac{7}{8}$" x $\frac{7}{8}$" x $2\frac{3}{4}$", are fastened (Fig. 6) at each end of the groove. (*g*) Guide-strip, $\frac{3}{4}$" x $\frac{5}{16}$" x $19\frac{1}{8}$, fastened to (*f*) $\frac{1}{8}$" from its beveled edge. (*h*) Guide-strip, $\frac{7}{8}$" x $\frac{5}{16}$" x $3\frac{3}{8}$", fixed on and across the pieces (*f f*), $\frac{1}{2}$" from their ends. To nail the frames together, put the triangular comb-guide (*u*) in the groove formed by the pieces (*f f*) place the piece (*t*) on the top of (*u*), and against the guides (*g*) and (*h*), and nail it to (*u*) with two brads each about 2" from the end. Proceed in this way until all the triangular guides are nailed to the top strips. Now turn over the gage-block and secure the vertical pieces (*u u*) against the guides (*b b*), by the buttons (*c c*), sad nail the bottom (*t*) to (*u u*), with two brads at each end. Turn the gage-block, and place the top of the frame (*t*), which has before been nailed to the guide (*u*), in its proper position and nail it to (*u u*) with two brads in each end.

---

* This device is substantially the same with the one alluded to on p. 208; Mr. Colvin's, however, was invented before mine.

Fig. 10, page 28, shows the arrangement of the circular saw to cut the triangular comb-guides.

The first piece cut is waste; as fast as a guide is sawed, the Piece from which it is cut must be turned over, end for end.[*]

Surplus Honey Box. Fig. 24, page 120.

Top and bottom, two pieces, $\frac{1}{4}$" x 6" x $5\frac{3}{4}$". Bore in the centre of the bottom, with $1\frac{3}{8}$" centre-bit, $\frac{1}{16}$" deep from the outside of the box; and then bore through with $1\frac{1}{4}$" bit. Sides, two pieces, $\frac{1}{4}$" x $5\frac{3}{4}$" wide x 5" high. Ends, glass, two pieces; 5" x 6", cut from glass 10" x 12". A block, $5\frac{1}{2}$"x 5"x $5\frac{3}{4}$", will be found very convenient to nail the boxes together upon.

Movable Stool for hives. Figs. 16 and 17, page 44.

Two pieces for uprights, or legs ; rear leg 7" wide, front leg 5" wide, both 20" x $\frac{7}{8}$". Take two pieces; 32" x $1\frac{3}{4}$"x $\frac{7}{8}$", and nail them to the top edge of the rear leg, flush with its ends; and projecting beyond it 4"; nail them also to the front leg in the same way, but let them project 9" . Then brace the legs and top strips, as shown in the figure. Hive No. 1, and any of the forms of Hive No. 2, will sit upon this stool, between the top strips ; cotton cloth (p. 279) is tacked to the alighting board, and to the longest ends of the top strips. Hive No. 5, also sits upon this stool, the top strips going between the clamps on the bottom of the hive. Hive No. 4 must be set *upon* the strips of this stool.

Movable Blocks for Entrance-Regulators,[†] Figs.11,16, 17, and 18; pages 28, 44; and 48.

Fig.11 is a right-angled triangle, $\frac{7}{8}$" thick x 4" x $5\frac{3}{4}$" x 7. In the bottom, grooves are cut $\frac{1}{8}$" deep x $\frac{1}{2}$" wide, as traps for the larvae of the bee-moth. Two of these blocks, made right and left; are used for a hive. By changing the position of these blocks on the alighting-board (see Fig. 18; page 48, in which some the positions are shown), the size of the

---

[*] To save beveling the first edge of the board by hand, the edge of the angular bed on the saw bench should be placed against the gage, with the saw passing through it, instead of against the saw, as represented in the figure.

[†] Figs. 12 and 19, pages 28 and 48, show the old arrangement for uniting the Non-swarmer with the entrance-blocks.

entrance to the hive be varied in a great many ways, and the bees always directed to it by the shape of the block, without any loss of time in searching for it.

Non-Swarmer. Figs. 5 and 17, pages 24 and 44.

Two pieces, $\frac{1}{4}$" thick x $4\frac{1}{2}$" long x $\frac{3}{4}$" wide; saw a slot through one of these, in the centre of its length and width, 2" long x $\frac{1}{16}$" wide; bevel the other piece to each edge, leaving a surface of $\frac{1}{8}$" in the middle of the width, the bevels being made for 2" only in the centre of the length of the piece; these pieces are to be fastened together with a piece between them at each end, $\frac{3}{8}$" thick x $1\frac{1}{4}$" long x $\frac{3}{4}$" wide, and the whole together then beveled off equally at each end, so as to make the length of one of the sides, where the passage appears, $2\frac{7}{16}$". A metallic slide, to be used in the slot; is 1" wide x $1\frac{15}{16}$" long, and is cut away on one edge to the exact depth of $\frac{5}{32}$", and on the other, $\frac{7}{32}$", leaving projections at each end, of $\frac{1}{4}$" each, which serve as feet, and rest on the plane surface left on the lower piece; sheet brass is the heat metal for the slide.[*] The Non-Swarmer may be varied from the above in length and bevel of the ends; so as to fit between the entrance-blocks in any of the positions shown in Fig. 18, page 48.

Movable Divider. No Figure

One piece, $18\frac{1}{8}$" x $9\frac{3}{8}$" x $\frac{7}{8}$", each end made $\frac{1}{4}$" beveling, for easy adjustment; the bevels should he parallel to each other. One piece, $\frac{5}{8}$" x $\frac{7}{8}$" x $19\frac{3}{4}$", nailed en the first piece, like the top piece (*t*) of the movable comb frames. By this divider the size of any hive may be diminished at will.

Temporary Movable Partition. No Figure.

$14\frac{1}{8}$" x $8\frac{7}{8}$" x $\frac{7}{8}$", from each and, cut to within 1" of the upper edge; $\frac{3}{8}$"; into the opposite, or short edge, drive two nails near the ends of the partition, letting them project $\frac{1}{2}$". These nails serve the purpose of feet to support the weight of the honey, which is stored in the short frames

---

[*] By making the slot wider, a wooden slide might be made to answer. These measurements may have to be slightly varied for the Italian bees. This Non-Swarmer is designed to prevent alterations by warping or swelling and to allow of adjustment without confusing the bees. It may also be used for excluding or confining the drones; see pp. 225, 326. It has not yet been fully tested.

resting by one end on this partition. The partition is further held, across the centre of the length of the hive, by two screws, one passing through each side of the hive into the partition, at the projections left upon the upper part of the ends. This partition is used only when a double set of small frames are put in a surplus honey-box of the same size as the lower hive.

Small Frames for Surplus Honey.

Top, $9\frac{5}{8}$" x $1\frac{1}{8}$" x $\frac{5}{16}$". Bottom, $7\frac{7}{8}$" x $\frac{5}{8}$" x $\frac{1}{4}$". Ends, or vertical pieces, two pieces, $8\frac{5}{8}$" x $\frac{7}{8}$" x $\frac{1}{2}$". Triangular comb-guide, (if used), 6 $\frac{7}{8}$" x $\frac{7}{8}$" x $\frac{7}{8}$" x $\frac{7}{8}$".

Hive No. 2, with Observing-glass at the back.

See perspective drawings (Figs. 16 and 17, p. 44), and the vertical longitudinal section (Fig. 9, p. 28), and the vertical cross-section (Fig. 13, p. 36), in which sectional drawings, and this bill of stock, and the two others immediately succeeding it, parts that are similar are marked with similar letters. This hive, in one of the three forms given, is recommended as the best for general use.

(*a*) Bottom-board, $24\frac{3}{4}$" x 15" x $\frac{7}{8}$", tongued and grooved together, with the grain of the wood running across the hive; the board to be rabbeted from one surface, at each edge, across the grain, $\frac{7}{16}$" x $\frac{7}{16}$", to fit into grooves formed in the sides (*c*); six holes are to be bored from the largest surface of this board, first with a $1\frac{3}{8}$" centre-bit, $\frac{1}{16}$" deep, and then through with a $1\frac{1}{4}$" bit.* The *centres* of these holes are to be in the intersections of lines gaged $3\frac{3}{8}$" from the centre of the width of the board, and $4\frac{3}{4}$", $10\frac{3}{4}$", and $16\frac{3}{4}$", from the rear of it. (*b*) Front of hive, $14\frac{1}{8}$"x $8\frac{7}{8}$" x $\frac{7}{8}$"; nail this between sides (*c*), $\frac{5}{8}$" below their upper edges, and 4" from their notched ends. (*c*) Sides of hive, two pieces, $24\frac{3}{4}$" x 10 $\frac{7}{8}$" x $\frac{7}{8}$"; notch out of one corner of each, to receive portico roof, 4" on the length of the pieces x $2\frac{1}{8}$" deep, and $\frac{7}{16}$" from the unnotched edge of each piece, make a groove to receive the bottom, $\frac{7}{16}$" square. Gage 4" on from the notched ends, and across the side pieces (*c*), for a line by which to set the outside of the front, which should come $\frac{5}{8}$" below the upper

---

* These holes, when not in use, are closed most conveniently by small covers cut out of refuse tin with a punch. They should be made only in the bottom-boards of those hives intended to be used one over another.

edges of the sides. (*d*) Ledges around sides and rear end of hive-body, nailed thereon $1\frac{1}{4}$" down from top edge; two pieces, $20\frac{3}{4}$" x $\frac{7}{8}$" x $\frac{7}{8}$", and one piece, 17" x $\frac{7}{8}$" x $\frac{7}{8}$". (*e*) Roof of portico, $17\frac{5}{8}$" x $4\frac{1}{2}$" x $\frac{7}{8}$", beveled off from $\frac{1}{2}$" thick at front edge, back $2\frac{1}{2}$" to full thick- ness, front edge rounded over from upper side only.[*] — One piece, $15\frac{7}{8}$" x $1\frac{1}{4}$" x $\frac{7}{8}$", nailed to the upper side of (*e*) flush with its rear edge, and in the centre of its length. — Cover for hive, $25\frac{3}{4}$" x 19" x $\frac{7}{8}$", tongued and grooved together, and rain-grooved, the grain of- the wood running front and rear of the hive. — Cleats for cover, two pieces, 19" x $1\frac{3}{4}$" x $\frac{7}{8}$", nailed on the under side of cover, flush with the ends. — Observing-glass at rear of hive, 14" x 5"; an outer glass of the same size can be used, if desired, for additional protection in Winter. — Shutter over glass, 14" x $\frac{7}{8}$" x $5\frac{1}{4}$" wide outside, and $5\frac{3}{4}$" wide inside, the bevel being made on the upper edge. — Clamps on this shutter, two pieces, $5\frac{1}{4}$" x 14" x $\frac{7}{8}$", nailed upon outside, each projecting $\frac{1}{4}$" over the end of the shutter, to cover the open joints. A piece, $14\frac{1}{8}$" x $2\frac{3}{8}$" x $\frac{7}{8}$", is nailed to a piece, $15\frac{7}{8}$" x $1\frac{7}{8}$" x $\frac{7}{8}$", centrally with regard to length, and so that one edge of both will he flush with each other. The ends of the longest piece are made dovetailing, to fit in the sides (*c*), as shown in Fig. 16, p. 44; the lower or flush edges of both picees coming $\frac{1}{2}$" above the bottom-board. The lower outer corner of this sash-rail, and the upper outer corner of the bottom-board, may be rabbeted a little to receive a covering of wire cloth, and the ventilator so formed may be furnished with a button slide arrangement, similar to those shown in the Fig., p. 13.[†] The upper sash rail is made up of a piece, $14\frac{1}{8}$" x $1\frac{1}{8}$" x $\frac{7}{8}$" nailed to a piece, $14\frac{1}{8}$" x $\frac{7}{8}$" x $1\frac{3}{4}$" wide on one side, and $2\frac{1}{4}$" on the other; gage $\frac{5}{8}$" from the square edge of the beveled piece, on its narrowest aide, for a mark to set the other piece to in nailing, and then nail the upper sash rail in place between the sides of the hive, the beveled piece being flush with the tops and ends of the sides. — Strips to bold the observing-glass, $\frac{1}{2}$" wide x $\frac{1}{4}$" thick, are nailed all around the place left to receive it, $\frac{8}{16}$" from the

---

[*] These parts marked with a (–), are not lettered in any of the figures.

[†] The ventilating passage may be closed by a strip of wood which nearly fills it; or it may be regulated by a slide as shown in the engraving on page 13. The objection to the strip is, that bees would be very apt to stick the strip fast with propolis within the ventilating passage. Mr. Wheaton uses no back ventilator, but depends upon a current of air from the front entrances of the lower and upper hive, the upper one being used for storing surplus honey on frames. The amount of ventilation needed will depend much upon climate and location.

interior of the hive. Two such hives, having one cover, are placed one on the top of the other (facing the same way), the upper one being designed to receive surplus honey, either in boxes placed over the holes in the bottom-board, or on frames.

Hive No. 2, without observing-glass.

This hive is similar to "Hive No. 2, with observing-glass," with the exception that these parts rendered necessary by the use of glass are omitted. The rear is 15" x $8\frac{7}{8}$" x $\frac{7}{8}$", and is halved into the sides (c) flush with their ends, and $\frac{5}{8}$" below their tops. The sides (c) are $23\frac{7}{8}$" long, but otherwise are the same as in the previous hive.

A strip, which forms the rear rabbet of the hive, in which the frames rest, is $15\frac{7}{8}$" x $2\frac{1}{8}$" x $\frac{7}{8}$"; this is nailed across the rear of the hive, to, and flush with, the tops of the sides (c). As the back ventilator will admit of all necessary inspection for general purposes (p. 365, note), a hive of this form will probably be best for those *largely* engaged in bee-culture.

Hive No. 2, with box-cover. Figs. 9[*] and 13, pages 28 and 36.

This hive may be made like either of the preceding hives, and has, in addition, a box-top, designed to cover small honey-boxes placed over the hive, or a large box, arranged to receive frames for the storage of surplus honey. The following comprises the additions referred to:

(f) Honey-board, $21\frac{1}{2}$" x $15\frac{3}{4}$" x $\frac{7}{8}$", tongued and grooved, and held together by cleats tongued and grooved to the ends of the board. Bore such holes through this board as are described in bottom-board (a), and at proper distances to receive the size of small honey-boxes used. (k) Honey-box cover; like (f), without the holes. (h) Front and rear of honey-box, front 14" x $9\frac{3}{8}$" x $\frac{7}{8}$" rear; two pieces, $14\frac{1}{8}$" x $1\frac{5}{8}$" x $\frac{7}{8}$"; nail the front between the sides; the lower edges flush; as is also one of the rear pieces, the other being $\frac{5}{8}$" below the top edges of the sides. (i) Two pieces, sides of honey-box, $19\frac{7}{8}$" x 10" x $\frac{7}{8}$". (j) Ledges at front and rear of honey-box, two pieces, $15\frac{7}{8}$" x $1\frac{1}{2}$" x $\frac{7}{8}$", nailed on flush with the top edges of the sides. (m) Observing-glass in rear of honey-box; 14" x 6". (n) Strips to hold observing-glass, $\frac{1}{2}$" x $\frac{1}{4}$", nailed all around the space

---

[*] Fig. 9 shows the construction, when neither observing-glass nor back ventilator are used, and when the front and rear of the hive are double thickness.

left for the glass; and within $\frac{8}{16}$" of the interior of the honey-box. (*o*) Top of box cover, tongued and grooved together, and rain-grooved, $26\frac{5}{8}$" x $19\frac{1}{8}$" x $\frac{7}{8}$". (*p*) Two pieces, front and rear of upper part of box cover, $17\frac{1}{2}$" x $8\frac{3}{4}$" x $\frac{7}{8}$"; these pieces are nailed between the sides. (*q*) Two pieces, sides of upper part of box cover, $24\frac{1}{8}$" x $8\frac{3}{4}$" x $\frac{7}{8}$". (*r*) Two pieces, front and rear of lower part of box cover; $17\frac{1}{2}$" x 5" x $\frac{7}{8}$". (*s*) Two pieces; sides of lower part of box cover, $24\frac{1}{8}$" x 5" x $\frac{7}{8}$". (*w*) Four pieces, 2" x 1" x $\frac{7}{8}$", buttons for holding the upper to the lower part of the cover, to which they are nailed; the upper inside part of the buttons is beveled off; to allow the upper part of the cover to set down readily on the lower part. The side pieces, (*q*) and (*s*), must be halved across the ends, to receive the front and rear; the upper and the lower parts of the box cover may be halved where they join, as shown in Hive No. 4, Fig. 23, p. 96.

A ventilator for the top cover should be made by boring a number of $\frac{3}{8}$" holes in the rear piece, as close as convenient to the roof; this ventilator may be opened and closed by means of the arrangement shown in the drawing opposite page 13.

Upper or Winter Entrance. Figs. 1, 2, 3, and 17, pages 20 and 44.

In all the Hives No. 2, a winter entrance for the bees may be made to open upon the portico roof for an alighting-board; gage from the upper side of the piece, forming the front rabbet, where the frames rest, 1" and $1\frac{1}{4}$", and then mortise a slot through, 3" long, in the centre of the length of the piece, between the gage marks, and slanting upwards, so that the lower side of the slot will come even with the top of the piece on which the frames rest. This entrance has been found on trial to be very important where bees are wintered in the open air. The lower entrance should be closed in winter.

Hive No. 3, Observing-Hive (p. 332). Figs. 14 and 15, page 36.

Fig. 14, is a side view, and Fig. 15; a vertical cross-section.

(*a*) Base-board, $24\frac{5}{8}$" x $4\frac{1}{4}$" x $\frac{7}{8}$". An entrance-hole $\frac{5}{8}$", is bored $3\frac{1}{2}$ inches deep into the end of (*a*), and two holes are bored in its centre, $\frac{7}{8}$" in diameter and $1\frac{1}{8}$" from centre to centre, the wood being cut out between them. (*b*) Bottom of hive, $2\frac{1}{4}$" x $18\frac{5}{8}$" x $\frac{7}{8}$"; make a rabbet at both upper corners, $\frac{3}{8}$" on x $\frac{7}{16}$" deep; start a $\frac{5}{8}$" hole, 1" from the end, and bore slanting, to meet entrance-hole in (*a*), and make a hole in the centre to match centre bole in (*a*); for a ventilator, and cover with wire-gauze on the inside. (*c*) Front and rear of hive, $1\frac{1}{4}$" x $2\frac{1}{4}$" x $10\frac{7}{8}$"; rabbet the inner corners, up and down, $\frac{1}{4}$" x $\frac{3}{8}$"; make a ventilator in each piece, like the one in (*a*); $\frac{5}{8}$" from the upper ends, cut in $\frac{7}{8}$"; and $\frac{7}{8}$" from the lower end, cut in $\frac{1}{4}$". (*d*) Side strips, $\frac{3}{8}$" x 1" x $20\frac{5}{8}$"; on one corner of each, rabbet on, $\frac{1}{4}$", and in, $\frac{1}{8}$" for the glass. (e) Movable cover, $21\frac{5}{8}$" x $4\frac{1}{4}$" x $\frac{1}{2}$"; holes may be made in this cover; as in Fig 21, over which glass receptacles for honey may be placed. (*f*) Glass, two panes, $9\frac{1}{2}$"x $18\frac{1}{2}$". (g) Alighting-board, 4"x $4\frac{1}{4}$" x $\frac{1}{2}$". (*h*) Clamps on base-board, $4\frac{1}{4}$" x 2" x $\frac{1}{2}$". (*i* and *j*) Clamps on cover, and ledges on hive, four pieces, $4\frac{1}{4}$" x $\frac{7}{8}$" x $\frac{1}{2}$".

Hive No. 4, Double-story Glass Hive. Figs. 19, 20, 21, 22, and 23, pages 48, 68, 88, and 96.

This and the following hive are not intended for general use in the Apiary, but for those who want one or more elegant hives.

Fig. 19 is a perspective view with the cover down. Fig. 20 is a perspective view with the cover elevated, so as to show the working of the bees, both in the main hive and the upper honey-box. Fig. 21 is a plan of the lower part of the hive, showing the surplus honey-board in place, and the holes made in it to allow the bees to pass up into the surplus honey-receptacles. On this board, receptacles of glass or wood, of any size or shape, may be set (see Glass hive opposite to the Frontispiece), instead of the upper box. Fig. 23 is a vertical longitudinal section, and Fig. 23 a vertical cross-section. This hive has glass on four sides, for purposes of general observation. A cornice under the projecting roof of the cover would improve its appearance.

(*a*) Main bottom of hive, tongued and grooved, 31" x $20\frac{3}{8}$" x $\frac{7}{8}$". (*b*) Outer* bottom of hive, $27\frac{7}{8}$" x $18\frac{1}{8}$" x $\frac{3}{8}$". (*c*) Rabbeted strips for outer bottom, two pieces, $29\frac{7}{8}$" x $1\frac{1}{2}$" x $\frac{7}{8}$", and two pieces, $17\frac{1}{8}$" x $1\frac{1}{2}$" x $\frac{7}{8}$". (*d*) Front and rear of lower outer case of hive, one rabbet in upper outer

---

* This outer bottom may be dispensed with, and clamps, $27\frac{7}{8}$ x 2 x $\frac{7}{8}$ inches, nailed lengthwise to the bottom of the hive, about 1 inch under and from its sides, for getting hold of the hive to lift it, and to prevent dampness.

corner of each, $\frac{7}{16}$" x $\frac{7}{16}$"; front, $11\frac{1}{4}$" x $20\frac{3}{8}$" x $\frac{7}{8}$"; cut out of the centre of the lower edge, $14\frac{1}{8}$" x $\frac{1}{2}$"; rear, $4\frac{1}{4}$" x $20\frac{3}{8}$" x $\frac{7}{8}$". (*e*) Sides of lower outer part, with rabbets the same as front and rear (for form of this, see Fig. 20), two pieces, $31\frac{7}{8}$" long x $\frac{7}{8}$" thick, $4\frac{1}{4}$" wide at one end, and $12\frac{1}{8}$" wide at $4\frac{7}{8}$" from the other end, where a notch is cut out, $1\frac{5}{16}$" deep x 4" long. (*f*) Roof of alighting-board, $23\frac{1}{8}$" x $4\frac{1}{2}$" x $\frac{7}{8}$" ; $\frac{7}{8}$ thick in rear, and $\frac{1}{2}$" thick in front. (*g*) Board under which bees pass into the hive, $14\frac{1}{8}$" x 4" x $\frac{1}{2}$" (*h*) Front posts of lower hive, two pieces, $9\frac{1}{8}$" long x 4" x $\frac{7}{8}$". (*i*) Rear posts of lower hive, two pieces, 10" long x $1\frac{3}{4}$" x $\frac{7}{8}$" with tenon, $\frac{7}{8}$" x $\frac{7}{8}$" x $\frac{7}{8}$", on one end. (*j*) Front and rear strips of lower hive, on which the frames hang, two pieces, $15\frac{7}{8}$" x $1\frac{3}{4}$" x $\frac{7}{8}$" with rabbet, $\frac{7}{8}$" x $\frac{5}{8}$", and notch, $\frac{5}{8}$" x $\frac{7}{8}$", cut at each end from upper side. (*k*) Side strips from post to post, in lower hive, $21\frac{5}{8}$" x $\frac{7}{8}$" x $\frac{7}{8}$", with notch, $\frac{1}{4}$" deep x $1\frac{3}{4}$", cut in the under side of each end. (*l*) Spare honey-board, $17\frac{7}{8}$" x $21\frac{5}{8}$" x $\frac{7}{8}$", nine holes bored $1\frac{3}{8}$" diameter x $\frac{1}{16}$" deep, and then bored through with a $1\frac{1}{4}$" bit; these holes when not in use are covered with pieces of tin, cut out with a punch; they may be bored plain, and covered with pieces of glass or wood. (*m*) Front and rear of lower part of cover, $6\frac{3}{4}$" x $20\frac{3}{8}$" x $\frac{7}{8}$", rabbets (Fig. 22) $\frac{7}{16}$" x $\frac{7}{16}$", on both upper and lower edges. (*n*) Sides of lower part of cover, two pieces, $27\frac{7}{8}$" from front to rear x $6\frac{3}{4}$" x $\frac{7}{8}$" with rabbets $\frac{7}{16}$" x $\frac{7}{16}$"; for shape of these pieces see fig. 20. (*o*) Front and rear of upper part of cover, one piece, $5\frac{5}{8}$" x $20\frac{3}{8}$" x $\frac{7}{8}$", and one piece, $13\frac{1}{2}$" x $20\frac{3}{8}$" x $\frac{7}{8}$". (*p*) Sides of upper part of cover, two pieces, each $5\frac{5}{8}$" and $13\frac{1}{2}$" x $27\frac{7}{8}$" x $\frac{7}{8}$", with rabbets, $\frac{7}{16}$" x $\frac{7}{16}$"; for shape, see Fig. 20. (*q*) Top of cover, tongued and grooved from front to rear, and rain-grooved on top (Figs. 19 and 23), $24\frac{5}{8}$"x $30\frac{3}{8}$" x $\frac{7}{8}$". (*r*) Honey-box cover, $21\frac{3}{8}$" x $19\frac{5}{8}$" x $\frac{7}{8}$". (*s*) Clamps for honey-box cover, two pieces, $21\frac{3}{8}$" x $\frac{7}{8}$" x $\frac{7}{8}$". (2.) Triangular checks to hold the cover when elevated, two pieces, $1\frac{3}{4}$" x $1\frac{3}{4}$" x $2\frac{1}{4}$" x $\frac{7}{8}$". (3.) Four buttons, $1\frac{1}{2}$" x 2" x $\frac{7}{8}$". (*w*) Posts of surplus honey-box, four pieces, $1\frac{3}{4}$" x $8\frac{1}{8}$" x $\frac{7}{8}$". (*x*) Front and rear bottom-strips of honey-box, two pieces, $1\frac{3}{4}$" x $15\frac{7}{8}$" x $\frac{5}{8}$". (*y*) Side-bottom strips of honey-box, two pieces, $21\frac{5}{8}$" x $\frac{5}{8}$" x $\frac{7}{8}$"; (*x*) and (*y*) are halved together at ends. (*z*) Front, rear, and side top pieces of honey-box, made up of two strips, $1\frac{3}{4}$" x $\frac{5}{8}$" x $17\frac{5}{8}$", two strips, $1\frac{3}{4}$" $\frac{5}{8}$" x $21\frac{5}{8}$", halved together at ends; and two strips, $17\frac{5}{8}$" x $\frac{7}{8}$" $\frac{5}{8}$ x ", two strips, $19\frac{7}{8}$" x $\frac{7}{8}$" x $\frac{5}{8}$". (4.) Clamps for spare honey-board, two pieces, $21\frac{7}{8}$" x $\frac{7}{8}$" x $\frac{5}{8}$". Glass, two pieces 14 x 9, four pieces 18 x 9, and two pieces 14 x 8, for the double glass of lower hive; two pieces 18 x 8, and two pieces 14 x 8, for the spare honey-box.

> Hive No. 5, Single-story Glass Hive as made by Mr. Colvin, see drawing on page 389 perspective on page 13; also the Figures referred to in Hive No. 4.

(*a*) *Bottom-board* $\frac{7}{8}$" thick x 25" lengthwise, and $36\frac{1}{2}$" across the grain of the wood, in *two* pieces *only*, tongued and grooved together; and rabbeted on under side of ends $\frac{5}{16}$" on, x $\frac{9}{16}$" deep, forming tongues on ends at top edge $\frac{5}{16}$" x $\frac{5}{16}$", which are let into sides. (*d*) *Front and rear ends of case*, bottom part; *front*, one piece, $25\frac{1}{4}$" x $9\frac{3}{4}$" x $\frac{7}{8}$"; cut out from centre of length on lower edge, $14\frac{1}{8}$" x $\frac{1}{2}$"; rabbet top outside of edge $\frac{7}{16}$" x $1\frac{5}{16}$"; *rear*, $25\frac{1}{4}$" x $3\frac{1}{2}$" x $\frac{7}{8}$", rabbet outside edge at top $\frac{7}{16}$" x $\frac{7}{16}$", and cut out from centre of length same as front. (e) *Sides of case*, lower part, two pieces, $36\frac{1}{2}$" x $11\frac{1}{2}$" wide, at $4\frac{15}{16}$" back from front and x $\frac{7}{8}$" thick and $3\frac{1}{2}$" wide at the other end; at the wide end, where slant terminates, cut out for roof of portico, $1\frac{15}{16}$" x $4\frac{7}{16}$", and rabbet the outside of slant edge, $\frac{7}{16}$" x $\frac{7}{16}$"; cut a groove $\frac{9}{16}$" up from bottom edge, inside, $\frac{5}{16}$" x $\frac{5}{16}$" the whole length of sides, to let in tongued ends of bottom; rabbet back end inside, from $\frac{7}{8}$" up from bottom edge to top edge, $\frac{7}{16}$" deep x $\frac{7}{8}$" on, to let in back end; $4\frac{1}{8}$" back from front end, $1\frac{3}{8}$" up from bottom edge, cut groove $\frac{7}{16}$" deep x $\frac{7}{8}$" wide to top edge, to let in front; for shape of (*e*), (*n*), and (*p*), see Fig. 20, p. 48. *Portico roof,* one piece, 27" x $5\frac{3}{4}$" x $\frac{7}{8}$", bevel from $\frac{5}{8}$", at front edge, back $\frac{5}{16}$" on top side, to full thickness, and round the front edge from the upper side. (*g*) *Cover of passage-way into hive*, one piece, $14\frac{5}{8}$" x 6" x $\frac{1}{2}$" let into front posts, $\frac{1}{4}$", full thickness, $\frac{1}{2}$" up from bottom ends; bore four holes, as directed in (*l*), in the centre of its width, the centre of the end holes being $3\frac{1}{4}$" from the ends; space the others equally between. (*h*) *Front posts*,

---

· The sides are toe-nailed to *bottom*-boards with four nails *only*, one on each side of tongue and groove in bottom-board, and about one inch apart, so that when the bottom-board swells and shrinks, the *joint* in it is kept *closed* and stationary, while swelling forces the edges of bottom-board out, front and back, and shrinking draws them in again. Sliding in the grooves in the sides, which prevent its warping.

two pieces, $\frac{7}{8}$" x $9\frac{1}{8}$" x 6". (*i*) *Rear posts*, two pieces, $\frac{7}{8}$" x $9\frac{1}{8}$", 6" wide at bottom and $1\frac{3}{4}$" at top, slope commencing $3\frac{1}{2}$" up from bottom ends of posts, and made "ogee" in form; these posts are fastened to the case by screws passing through the front and back end boards of it into their edges, and are *not* mortised into the bottom-board, but rest on it; in each post, $\frac{1}{2}$" up from bottom end, cut a groove $\frac{1}{4}$" deep, $\frac{1}{2}$" wide, entirely across their width (6"), to let in covers of "passage-way," and "back ventilator;" also mortise, in one edge, $\frac{1}{2}$" up from bottom end, $\frac{3}{8}$" wide x $\frac{3}{4}$" long x $\frac{3}{8}$" deep, for bottom rail of sides of "bee-chamber." (*j*) *Rear and front top rails* of "bee-chamber" two pieces $15\frac{7}{8}$" x $1\frac{3}{4}$" x $\frac{7}{8}$": rabbet one edge $\frac{7}{8}$" wide x $\frac{5}{8}$" deep, and cut from the top of ends to the depth of rabbet, $\frac{7}{8}$" on. (*k*) *Side top rails*, two pieces 20" x $\frac{7}{8}$" x $\frac{7}{8}$". *Bottom side rails*, two pieces 19" x $\frac{7}{8}$" x $\frac{3}{4}$"; tenon on ends $\frac{3}{8}$" long x $\frac{3}{8}$" x $\frac{3}{4}$" in centre. (*l*) *Surplus honey-board*, $\frac{7}{8}$" x $21\frac{3}{8}$" x $15\frac{7}{8}$", the grain of the wood to run crosswise of the board, which is to have clamps, tongued and grooved against the end of the grain, and form part of the above dimensions; 9 holes are to be bored for surplus honey-boxes; they are first bored $\frac{1}{16}$" deep, with $1\frac{3}{8}$" centre-bit, and then through with $1\frac{1}{4}$" bit; these holes are arranged in three rows one in the centre, and the others $2\frac{1}{2}$" from the side edges of the board, the front and back end holes of each row being $3\frac{3}{4}$" from the ends. (m) *Front and rear of case, middle part;* two pieces $9\frac{1}{2}$" x $25\frac{1}{4}$" x $\frac{7}{8}$"; rabbet out $\frac{7}{16}$" x $\frac{7}{16}$" on inside of lower edges, and same on outside of upper edges. (*n*) *Sides of case, middle part*, two pieces (for shape of these, see Fig. 20, p. 48), $32\frac{1}{2}$" long x $9\frac{1}{2}$" wide (measuring on a straight line from front to rear of case for length, and square across this section for width); rabbet out inside lower edges and outside upper edges, same as ends; also rabbet $\frac{7}{16}$" in x $\frac{7}{8}$" on, inside ends, to let in end pieces. (*o*) *Front and rear of case, upper part*, front, $25\frac{1}{4}$" x $7\frac{3}{4}$ x $\frac{7}{8}$"; rear, $25\frac{1}{4}$" x 14" x $\frac{7}{8}$"; rabbet out inside lower edges, $\frac{7}{16}$" x $\frac{7}{16}$". (*p*) *Sides of case upper part*, $32\frac{1}{2}$" long x 14" wide at back end, and $7\frac{3}{4}$" wide at front end x $\frac{7}{8}$"; rabbet inside lower edges $\frac{7}{16}$" x $\frac{7}{16}$", and inside at ends, $\frac{7}{16}$" in x $\frac{5}{8}$" on, to let in ends (for shape see Fig. 20, p. 48). (*q*) *Top of upper part of case*, five pieces, of equal width and length,

---

· The middle part need not be made, unless the hive is intended to be used with two stories, as in Hive No. 4.

forming together 30" x 36" x $\frac{7}{8}$", tongued and grooved together, and rain-grooved [†] (see Fig. 23, p. 72). *Collateral side honey-boards*, for surplus honey-glasses, two pieces, $30\frac{1}{4}$" x $4\frac{1}{4}$" x $\frac{1}{2}$"; bore six holes as directed in (*l*), in the centre of width, the end ones 2" from ends, and the rest equally spaced between.[‡] *Collateral rear honey-board*, for surplus honey-glasses, same as covered passage-way into hive, let into posts and perforated with holes as in (*l*). *Cleats* for under side of collateral side honey-boards, six pieces $4\frac{1}{4}$" x $\frac{1}{2}$" x $\frac{1}{2}$", one of which is nailed under each end, and one under the middle of each side honey-board. *Collateral front honey- board*, one piece, $15\frac{7}{8}$" x $6\frac{1}{4}$" x $\frac{7}{8}$", clamped across the ends, same as (*l*), and bore holes same as (*g*); frames may be hung in the space under this honey-board, and glasses on it, or the glasses may be placed instead of frames (as preferred) to receive the surplus honey; two pieces, 6" x $\frac{5}{8}$" x $\frac{1}{4}$", are nailed on outer edges of tops of front posts, (*h*), to form rabbets for frames. *Triangular checks to hold the case when elevated*, two pieces, 3" long x $1\frac{3}{4}$" x $\frac{7}{8}$" at one end, and $\frac{7}{8}$" x $\frac{1}{16}$" at the other. *Guides on outside of case* (see Fig. 19, p. 48), four pieces, $1\frac{1}{2}$" x 2" x $\frac{3}{4}$". *Cover to upper ventilator of case*, one piece, 24" x 1" x $\frac{3}{8}$" hung on buttons with screws; this ventilator is made by boring holes about $\frac{3}{4}$" in diameter in the rear of upper part of case, $\frac{1}{8}$" below (q).

No. 2, Box Hive, as made by Mr. Colvin, see drawing on page 390, with box-cover and observing glass in rear end.

Bottom (in two pieces only, plowed and grooved together, the grain of the wood running across the hive), $24\frac{3}{4}$" x $14\frac{3}{4}$" x $\frac{7}{8}$"; rabbet across under side of ends $\frac{5}{16}$" on, $\frac{9}{16}$" in, forming tongue $\frac{5}{16}$" x $\frac{5}{16}$" on upper side of ends which are let into groove in sides. *Sides*, two pieces, $24\frac{3}{4}$" x $10\frac{7}{8}$" x $\frac{7}{8}$", cut out from front end on top edge, 4" x $1\frac{1}{2}$" deep, for portico roof; on inside, $\frac{9}{16}$" up from bottom edge, cut a groove $\frac{5}{16}$" x $\frac{5}{16}$" the entire length, to receive tongue on bottom- board; 4" back from front

---

[†] By increasing the width and length of this top so as to project $4\frac{1}{2}$ in. over sides, and placing turned "drops" or other ornaments under the eave it may be, at small cost, made highly ornamental. See drawing on page 389
[‡] When it is desired to close the opening under side-rails of bee-chamber, turn the collateral side honey-boards upside down.

end, from $1\frac{3}{8}$" up from bottom edge, cut a groove $\frac{5}{16}$" x $\frac{7}{8}$" to within $\frac{5}{8}$" of the top edge to let in front; (these sides are nailed to *bottom* same as No. 5). *Portico roof*, one piece, $17\frac{1}{2}$" x $4\frac{5}{8}$" x $\frac{7}{8}$"; bevel on top side, to $\frac{1}{2}$" thick at front edge, back $2\frac{3}{4}$" to full thickness: front edge rounded from upper side. *Front*, one piece, $14\frac{3}{4}$" x $8\frac{7}{8}$" x $\frac{7}{8}$", let into sides $\frac{5}{16}$" at each end. *Observing-glass in rear*, 14" x 6"; strips to form rabbet for glass, $\frac{1}{2}$" x $\frac{1}{4}$", nailed all around the space left for the glass, and within $\frac{3}{16}$" of inside of hive. *Rear end*, two pieces, $14\frac{1}{8}$" x $1\frac{3}{8}$" x $\frac{7}{8}$", one of these nailed to a piece $15\frac{3}{4}$" x $1\frac{1}{8}$" x $\frac{7}{8}$", so that the bottom edges will be flush with each other, is to be dove-tailed into the ends of sides $\frac{1}{2}$" up from the top side of bottom-board, the other to be nailed to a piece $14\frac{1}{8}$" x $\frac{7}{8}$" x $2\frac{1}{8}$", on one side, and $1\frac{3}{4}$" on the other, file top edge of the inside piece $\frac{5}{8}$" below the outside piece; then nail these pieces between the sides of hive, so that the square edge and widest side come flush with the ends and tops of sides. *Cover for observing-glass*, one piece, 14" x $\frac{7}{8}$" x $6\frac{3}{4}$" inside x $6\frac{3}{8}$" outside, the bevel being made on the upper edge; clamps on this cover, two pieces, $6\frac{3}{8}$" x $1\frac{1}{4}$" x $\frac{7}{8}$" screwed in the middle and nailed at ends on the outside of cover, each projecting $\frac{1}{4}$" over its end to cover the joint. *Ledges around sides and ends*, to support the box-cover, &c., screwed on $\frac{5}{8}$" down from top edge of hive; two pieces (sides), $21\frac{1}{2}$" x $\frac{7}{8}$" x $\frac{7}{8}$"; one piece (back), $17\frac{1}{2}$" x $\frac{7}{8}$" x $\frac{7}{8}$"; one piece (front), $15\frac{3}{4}$" x $\frac{7}{8}$" x $\frac{5}{8}$", this last piece to be nailed on the top side of portico roof; notch out of centre of length, 3" long x $\frac{7}{16}$", for winter entrance. *Honey-board*, 21" x $15\frac{1}{2}$" x $\frac{7}{8}$" (in two pieces only), plowed and grooved together; clamps tongued and grooved against ends and forming part of its dimensions, and toe-nailed to clamps $\frac{1}{2}$" each side of the groove *only*; six holes are bored in this, same size as in Hive No. 5, in two rows from front to back, and three rows across at the intersections of lines gauged $3\frac{3}{4}$" from its sides, and $4\frac{3}{4}$" x $10\frac{3}{4}$" x $16\frac{3}{4}$" from either front or back ends. *Box-cover*, front and rear, two pieces, $16\frac{5}{8}$" x $8\frac{3}{4}$" x $\frac{7}{8}$", cut out of centre of bottom edge of front, 3" x $\frac{7}{16}$" for winter entrance. *Sides*, two pieces, $23\frac{1}{8}$" x $8\frac{3}{4}$" x $\frac{7}{8}$", rabbet at ends, $\frac{7}{8}$" on, $\frac{7}{16}$" in, to let in ends: bore five holes in rear end, for ventilation, with $\frac{3}{4}$" centre bit 2" from each end, and $3\frac{1}{8}$" from centre to centre, within $\frac{1}{8}$" of top edge. *Cover for ventilator*, one piece, 15" x 1" x $\frac{3}{8}$", held in its place by two buttons. T*op of box-cover*, four pieces, $26\frac{1}{2}$" x $\frac{7}{8}$" x $5\frac{1}{4}$"; when tongued and grooved together, rain-grooved on each side of joints. *Cover for back lower ventilator*, one piece, $14\frac{3}{4}$" x $\frac{3}{4}$" x $1\frac{1}{4}$" rabbeted on under side and at ends $\frac{1}{4}$" in x $\frac{7}{8}$" in; button for securing this and the cover of observing-glass, $1\frac{1}{2}$" x $\frac{3}{4}$" x $\frac{5}{8}$"; cut out $\frac{8}{8}$" x $\frac{1}{4}$" from the lower end.

EXPLANATION OF PLATES

Colvin No. 5.

Colvin Ornamental

Preface to the Index

Today we go about indexes in a different pattern than you will see in this index. Rather than try to make a whole new index I chose to incorporate my additions to the one prepared by Langstroth and the editors of the book. My parts are in italicized text. Often, throughout the book, I have added a parenthetical statement to Langstroth's footnote. In this case the italicized item will just say footnote. When the subject is in a numbered footnote it is referenced that way.

The way the index is constituted makes each main item become almost a summary of the topics within the book. The subtopics are not alphabetized but are almost always in linear order from first pages to last. Thus, they also read somewhat like a summary of what is in the book on that topic.

INDEX.

A.

Absconding 116 (note 10), swarms of Africanized bees 123 (note 19); by swarms not liking new hive 138 (note 35); lack of food causing, 147 (note 4).

Achoria grisella (lesser wax moth), 228 (and note 1).

Adobe, for hives, 331 (note 2).

Advantages required in complete hives, 95-108.

Adventure, amusing, in search of honey, 254.

Adulteration, of honey with water, 277 (note).

Africanized bees, absconding 116 (note 10); distance swarm travels 123 (note 19); response to alarm pheromone 312 (note 2).

After swarming, 120; causes and indications of, 121, *caused by orientation flight of virgin queen* 122 (note 18); easily prevented in mov. comb hives, 124, 140; *small size* 124 (note 20); evils of, 140; author's mode of obviating evils of, before invention of mov. comb hive, 140 (note); excessive; exposes stock to bee-moth, 243.

After-swarms, easily strengthened in mov. comb hives, 140; when to expect, 122; often issue in bad weather, 122; often have more than one queen, 122; seriously reduce strength of parent-stocks, 124,140; wise arrangement concerning, 124; *virgin queen in* 124 (note 20); easily prevented in mov. comb hive, 124; weak, of little value, 140, 141; returning of, to parent stock, or doubling, unprofitable, 140; make few drone-cells the first season, 184 (note).

Age, of bees, 58; queen-bee, 49; of workers, proved from Italian bee, 59 (note); signs of old, 59; of colonies, 59; of queens, designated by the clippings of their wings, 223.

Air, necessary for bees 88; bees need in Winter, 89, 338; pure, necessary for eggs, brood, and bees, 89; pure, necessary for health of man, 91; abundance of, supplied by mov. comb hive, 94; new swarms require more than old, 281; cold, alarms bees, 311 (note); how to give in Winter, to mov. comb hives, 338.

Air-tight stoves, deficient in ventilation, 92.

Alarm pheromone and stinging, 312 (note 2), Africanized bees response more pronounced, 312 (note 2); chemicals of alarm pheromone, 313 (note 2).

Alighting-board, should shelter from wind and wet, 103; improved by attaching muslin, 279 (note): Pl. V., Figs. 16,17.

Alsike, *(Trifolium hybridum)* or Swedish white clover, 294; value of, for bees and stock, 295.

American women, their sufferings from bad ventilation, 92.

American foulbrood disease 257 (note); spores persisting for years, 257 (note 10); antibiotics 259 (note 13); shaking method for curing, 259 (note 13).

Analysis of royal jelly, 64.

Anaphylactic shock from sting, 315 (note 4).

Anger of bees, 308-314; difficult to repress, when once aroused, 170; excited by the human breath, quick motions, or jarring, 170; and sometimes by smoke, 168 (note 5); should not be violently repelled, 170; occasioned by disease, 256 (note); never necessary to provoke a colony to, 309; when provoked to, terribly vindictive, 310; of dyspeptic bees, troublesome, 310; bee-hat, a protection from, 310; Butler's directions bow to prevent the rising of, 311; warm breath provokes, 311 (note 2); when excited, how to act, 311; never excited away from home, 312; excited by disagreeable odors, and uncleanly persons, 313; aroused by a smell or the bee-poison, 314; and by rough and hairy substances, 317.

Ants, pests of bees 255 (note).

Ants, white, their fecundity, 32; sometimes injure bees, 255; small, harmless, 256 (note); extravagantly fond of honey, 287.

Aphides, singular mode of propagation of, 42; description of, 285; cause of honey-dew, 285.

Apiarians, see Bee-keepers.

Apiaries, must be closely watched in swarming-season, 143; Large, rendered difficult by natural swarming, 145; danger of crowded, 214; stocking, &c., 279-284; in establishing, a knowledge of the honey resources of the locality important, 279 (and note 1); should be protected from high winds, and from cattle, and sweaty horses, 279 (note 2); should be in sight of occupied rooms, 279; proper exposure for, 279; covered, objectionable, 280; shaded, agreeable to bees; 280; location or, how to change, 280; procuring bees for, 280; to secure bees in their hives, for removal to, 281; precautions to be observed in moving hives to, 281; transferring bees from common to mov. comb hive, for, 282; large, in Europe, 300; *current and former apiary size, 301 (note 11);* should be fenced against cattle and horses, 313.

Apis dorsata, single comb on limb 63 (note 35).

Apis florea, 63 (note 35).

Apple-tree, yields much honey, 292.

Apricot-tree, honey-yielding, 292.

Aristotle, noticed similarity of drone and worker-eggs, 42; observed that bees collect pollen from one kind of flower at a time, 83; observation of, concerning the flight and feeding of drones, 224 (note); on the difficulties which perplex the Apiarian, 276 (note); described the Italian bee, 318.

Artificial honey, recipe for, 276 (note).

Artificial rearing of queens, 188; the process to be performed late in the day, 188; honey and water to be supplied to bees in, 189; when to confine bees in, 189.

Artificial swarming, 143, 211; not performed by Columella 147 (note) ill success of ancient method of, 148; Huber's plan of, objectionable. 148; by dividing hives, unsatisfactory, 149; by removing full hives and substituting empty ones, worse, 150, 151; by self-colonizing hives, ineffectual, 151; causes of failure of, 152; has received great attention from author, 153; mode of, adapted to common hives, 154; cautious handling of combs in, needful, 155 (and note); how to prevent bees in, from returning to old stand, 156, 157; hot to be performed till drones appear, 158; tokens of the absence or presence of the queen in, 158; how to proceed if the queen is absent, 159; if done in morning or late in afternoon how to proceed to secure bees for the old stock, 160; proportion of bees necessary for old stocks in, 160; new and decoy-hive should resemble that of parent stock, or adjoining hives be covered, 160; mode or, by exchanging hives, 160; by juxtaposition, 161; by confining bees in parent stock, 161; preferable plan when to be done on a large scale, 162; rapidity of this plan, 162 (note); its advantages 163; Dr. Dönhoff's method of, 163; tow to attach bees to new places, in, 163 (note); difficult for persons ignorant of the laws which control the breeding of bees, 164; easily performed with mov. comb hive, 164: mode of performing it, 166; queen to be sought for, 166; supply of sealed queens provided for, 166; great care necessary in transferring sealed queens, 167; should not be attempted in cool weather, or when dark, 167; early morning best time for, 167; little danger attending, 167, 168; perfectly safe even at mid-day, 168; sugar-water often better than smoke, useful in, 168; honey-water objectionable, 169 (note); caution in, enjoined, 170; how to apply sugar-water in, 170; how to remove frames in, 170; rapidly performed, 173; *Hoffman top-bar frames, 170 (note 23);* best mode of, 180, 181; supply of queens to mother-stocks, in, 182; obviates the risk of after-swarming, 184; capable of safe expansion, 185; how to double stocks by, 185; Dzierzon's mode of, 186; author's mode of, for single apiaries, 186; mode or, resembling natural swarming, 186; mode of, by reversing position of hives, 187; how to provide a full supply of queens for, 188; nucleus for rearing queens for, 189; rapid increase of stocks by, 190; how to induce bees, in, to rear queens on convenient parts of the comb, 191; how to secure adhering bees for the nuclei in, 192 (and note 2); queens, in, made to supply several stocks with eggs, 193; mother-stocks, in, should be kept strong, 199; most successful when forage is abundant, 199; hazardous in a crowded apiary, 200; how to supply stocks, in, with stranger-queens, 200; queen-cage for, 201; union or bees of different stocks in, 203; practiced in ancient times, 210.

Artificial swarms, where should be put, 158; how to know whether they have a queen, 158; will accept a strange queen, 159 (note); cautions to be observed in locating, 159; how to make, by slightly changing position of parent stock, 161; how to form several with one natural swarm, 163; quickly made in mov. comb hive, 164, 173: when to force, in cases of retarded swarming, 174; cannot be formed by merely transferring combs and bees into an empty hive, 175; caution against too rapid multiplication of, 175 (note); the piling mode of forming, its advantages, 188; not to be increased so as to reduce the strength of the mother stock, 199; attempts at rapid increase of, in vicinity of sugar-houses, &c., 199; difficult to form when forage is scarce, 199.

Asters, furnish valuable pasturage for bees, 298; *large amounts of pollen, 299 (note 11).*

Attica, its yield of wax and honey,

304;
Austria, value of its honey crop, 304
Axioms, bee-keeper's, 369.

B.

Baldenstein, Capt., on Italian bee, 318; ill-success of, in propagating pure breed, 319.
Bar-hives, ancient, 210 (note), author's experiments with, 14.
Basket, used as a hiver, 133.
Bass-wood, see Linden.
Bears, destroyers of bees, 254, *(and note 11).*
Bee-bob, to attract swarms, 132.
Bee-bread, see Pollen.
Bee-dress, use of, recommended, 132, 209, 316.
Bee Eaters, (birds of Eastern Europe) Merops sp. 252 (note 10).
Bee-glue, see Propolis.
Bee-hat, author's, how made, 316 (Pl. XI., Fig. 25.)
Bee-journal, much needed in this country, 22.
Bee-keeping, depressed condition of, in America, 13, 145; *apiarian vs. apiculture, 33 (note 4);* a fascinating pursuit, 144, 146; estimate of profit of, 146 (note); better understood by the ancients than the moderns, 147 (note); with feeble stocks, unprofitable, 177; no "royal road" to, 211; demands care and experience, 211; in Spain, extensive, 222 (note 2); on a large scale, unprofitable to beginners, 282.
Bee-moth, *Langstroth patent, 12 (note 2); eggs laid in cracks of supers, 78 (note 3);* permanent bottom-boards, a security against, 97; *spinning cocoons 97 (note 6);* easily dislodged from mov. comb hive, 141; has more sins to bear than she commits, 216, 246; *); scientific name now* <u>Galleria mellonella</u>, *228 (note 1);* habits, &c., of, described, 228-252; mentioned by ancient authors, 228; pest of modern apiaries, 228, 251; when a moth-proof hive will be obtained, 228; Dr. Harris's account of, 228; to distinguish female of, from male, 229; cut of female and male, 230; nocturnal 230; interesting experiment with female, 230 (note 2); agility of, 230 (and note 3); eggs of, laid in the cracks of the hive, &c., 231 *(and note), 235, 248 (note 8);* cut of gallery of, 232; cocoons of, in empty combs, 233 (and Pl. XIX., Fig. 56); female will deposit eggs on pressure, 234 (note 2); condition of a hive destroyed by, 235 (and Pl. XX., Fig. 57); did not appear simultaneously in this country with the bee, 236; multiplied by the use of patent hives, 237, 241; movable frames a remedy for the evils of, 239, 241; first appearance noted, 240; rapid spread of, in Ohio, 241; commonly infest old stocks, 251 (note); eggs of, deposited on uncovered combs in weak stocks, 242; signs of presence of, in hives, 242; not developed in low temperature, 243; sulphur fumes will kill the eggs and larvae of, in combs, 243; will certainly destroy queenless stocks, 244 (and note); *far more trouble in warm or tropical climates, 244 (note);* fertility of, 244; instinct of, in discovering queenless stocks, 245; *find colonies by odor, 245 (note 6);* easily conquer stocks suffering from hunger, 246 (and note); mission of, 247 (and note) keeping stocks strong the surest defense against, 247; insecurity of other

contrivances, 247; placing hives so as not to endanger the loss of their queens, an important protection against, 248; adaptation of mov. comb hive to protect stocks from, 249; facilities of destroying, of no use to careless bee-keepers, 250; protection from, by an upper entrance, 250 (note); *protection against moths is not essential 250 (note);* caught by sweets and sour milk, 251; destroyed by fire, 251 (note 2).

Bee-moth, larva of (with cuts 229); *entering hives 97 (note 6), 231 (note), 248 (note 8);* how it secures itself from the attacks of the bees, 231; representation of its gallery, 232; *crochets or prolegs of larva 232 (note 3);* food of, 233, *pollen as food (note 233)*; appearance of their cocoons in empty combs, 233 (and Pl. XIX., Fig. 56); activity of, 233; transformation of, to the winged form, and effect of cold on, 234 (and note), 243 *(note 5);* movable frames a remedy against 239, 241; signs of presence of, in hives, 242; sulphur fumes fatal to, 243; *chemical control 243 (note 5);* should be destroyed early in the season, 248; extent of their ravages 249 (and note); how to entrap them, 249; traps for, of no use to the careless, 250.

Bee-palaces, objections to, 61, 242.

Bees, honey, will work in the light, 16; 23, 332; may be tamed, 24, 28, 308; intended for man's comfort, 24; never attack when gorged with honey, 25, 132, 169; when swarming, peaceable, 25, 132; always accept of offered sweets, 25, 168, 169, 170; sometimes attracted from other hives by sprinkling sugar-water, 7; gorge themselves when frightened, 27, 154, 169; subdued by smoke or drumming on the hive, 27, 154; and chloroform or ether, 210; the most timid may manage, 28; can flourish only in colonies, 29; how affected by loss of queen, 31; intelligence of, 48; breed in Winter, 48, 339; number of, in a colony, 54; honey-bag of, 56 (Pl. XVII., Fig. 54); pollen-basket, 56; proboscis of, 56 (Pl. XVI., Fig. 51, Pl. XIII, Fig. 63); sting, 56 (Pl. XVII., Fig. 53); loss of sting fatal, 57; age of, 58; industry of, instructive, 59; number of, in a colony, why limited 61; advantages of their being able to Winter in a colony state, 62; despair of, when without queen or brood-comb, 67, 245; work night and day 73; sagacity of, in the structure of their cells, 74; superstitions connected with, 80; not injurious to fruit, 85; need little air in Winter, if comfortable, 89; when disturbed or confined, require much air, 90; become diseased in impure air, 90; annoyed by thin hives in hot weather, 90; superior to man in ventilation, 91; why they do not cluster on sealed honey in hot weather, 91; averse to jarring, 96; not torpid in Winter, 110, 335; chilled by cold, 110; must live in communities, 110; conduct of, when queen is lost in swarming, 113; sometimes abandon hives to avoid starvation, *hearing 114 (note),* 116; why they do not select new homes before abandoning the old, 116; intercommunicate quickly on the wing, 117; send scouts to seek new abodes, 117; sight of, for distant objects, acute, 117; commotion of, during absence of queen for impregnation, 125, 217; native of hot climate, 128 (note); detest smell of fresh paint, 129; often perspire while swarming, and reluctant to enter heated, hives, 130; pleased to find comb in hive 131; modes of securing swarms in difficult places 135; acute of hearing, 138; refusing to swarm, should have plenty of storage-room, 139; may be advantageously kept in cities, 144; often refuse to swarm, 145; seldom colonize unless blossoms abound in honey, 147; ability of, to rear queens from worker-brood, when discovered, 148; without mature queens, build combs with large cells, 149, 150 (and note); diminish rapidly in number after swarming, 151 (and note); will not form independent colonies in intercommunicating hives 152; work better in new swarms than in old colonies, 153; laden with stores, welcomed by strange swarms, 155; without stores, expelled, 155; frightened by rappings on the hive, 155; disposition of, when moved, to return to old location 156; effect on, of temporary loss of home, 157; how to make adhere to old home, wherever put; 157; losing their queens, will accept of others, 159, (note); more irascible at night, 167; confounded by sudden introduction of light into their hives, 168, 169; difficult to subdue when once thoroughly excited, 170; use all available space for honey, 142 (note 2); tenacious adherence of, to their combs, 172; losing their queen when swarming, return to parent stock,, 174; their mode of .communication, 174 (note 1); storing surplus honey to be unmolested, 180 (and note 1); amusing conduct of, on finding a strange hive where their own should be, 181 (note 1); emboldened to self-defense by presence of queen, 182; judicious renewal of, for swarms, not injurious to mother-stock, 183; their instinct to become over-rich, 183 (note 2); their passion for forage, 186 (note when destitute of queen, will rear young ones, if they have brood-comb, 188; need water when confined, 189 (note); how encouraged to work in an upper hive, 189; do not always cluster on brood comb in nuclei, 192 (note); sometimes start queen cells that fail, 193; young do inside, and old, outside work, 194; young are wax-workers, 196; their occasional refusal to make royal cells explained, 197 (note) a worthy trait of, 197; their treatment of strange queens, 200; to cause, to receive strange queens, 200; to cause, to receive strange queens kindly, 201; of different colonies may be united, 203; distinguish their hive companions by smell and actions, 203; conduct of, when frightened, 203; when disturbed and scented, will readily mingle, 203 (and note); in too large hives, become dispirited, 208; in large apiaries, if the hives are alike, liable to mistake them, 214; effect on, of loss of queen, 217; enemies of, 228-255; vigilance of, against the moth 231; not a native of the New World, 235; a harbinger of civilization, 236 (note); can learn to defend themselves against new enemies, 240; destroyed by mice and by birds, 252; by toads and bears, 254; diseases of, 255-260; propensities of, to rob, and appearance of thieving bees, 261 habitual robbers become black, 262 (and note);

sometimes rob the humble bee, 262; grand battles of, 263; of. conquered colonies, incorporate themselves with the victors, 263; frantic fury of robbers, when deprived of their spoil, 265; how to cool them into temporary honesty, 265; feeding of, 267-278; are fond of salt, 272; infatuation of, for confectionery, 277.; compared to intemperate men, , 278; the avaricious, folly. of 278; fond of shade, 280; procuring for an apiary, 280; transferring from common to mov. comb hives, 282; get supplies from honey-dews, 287; flight of, its extent, 305; pacific temper of, 308; incident, illustrating good nature of, while, swarming, 308; readily taught by ill treatment to be vindictive, 310; human breath offensive to, 311; at a distance from their hives, never sting unless hurt, 312; kindness of; at home, a lesson for man, 312; their treatment of the sick, 312; their sense of smell, 313; dead, medicinal qualities of, 315 (note); will more surely sting hairy than bare parts, 317; maintain a high temperature in Winter, 335; eat less in Winter when kept quiet, 335, 355. wintering of, 335-361; uniting small colonies of, for wintering, 336; do not store honey so as always to be accessible in Winter, 336; cannot be relied on to make Winter passages in combs, 336; should be protected from Winter winds; 337; 348; if out of doors in Winter should be allowed to fly, 337; sometimes perish in snow, 338 (note 1); experiments on wintering, by author, 339; need water in cold weather, 342-346; need water to eat granulated honey, 342-344; injured by being disturbed in Winter, 347 355; seldom discharge their faeces in the hive, 347; on wintering in dry cellars, 348; in special depositories 349-360; eat less and fewer die in *clamps* than in other special Winter depositories, 355, 358.

Bee-keepers, common hives do not teach the laws of bee-breeding, 154; if timid, should use bee-dress, 209; ignorance of, the greatest obstacle to speedy introduction of mov. comb hive, 209; often captivated by shallow devices, 211; skepticism of many, in regard to the wonders of the beehive, 211; often mistake the cause of the loss of their queens, 216; careless, will he unsuccessful, 226, 250; should not encourage the destruction of birds, 253; specimen or, opposed to improvements, 357.

Bee-quack's secret, 238 (note).

Bees, queen of, see Queen Bees.

Beginners, should be cautious in experimenting, 179, 307.

Behavior, instinct 247 (note).

Berg, Rev. Dr., first informed author of Dzierzon's discoveries, 16.

Berlepsch, Baron of, his stacks injured by scientific experiments, 179 (note); uses frames similar to the author's, 321 (note 2); experiments on impregnation of queens, 126 (note); Italian bee, 323; his experiments on the effect or cold on queens, 327: shows that bees need water in winter, 242.

Bevan, on eggs, and larvae of bees 41-47; on "driving," or forced swarming (note), 154; an experiment of, in removing a queen, 218 (note); feeds salt to bees, 272; his description or honeydew, 286.

Birds, bee devouring, 252; why they should not be destroyed, 253 (and note).

Blocks, entrance regulating (Plate III., Figs. 11,12); useful to prevent swarming, 174 (and note); security against mice, 175, 252; against robber-bees, 264.

Bodwell, S.C., experiments of, in wintering bees 345.

Boerhave's account of Swammerdam's labors, 65 (note).

Bohemia, its production of honey, 304.

Boiling honey improves it, 287.

Borage, *(Borago officinalis)* valuable for bees, 298.

Bottom-boards should be permanently fixed to hive, 97; should slant towards entrance, 97; cleaning of, 98; dangers of movable, from the moth, 231; Spring cleaning of, 243; Winter cleaning of, 347.

Boxes for spare honey, 289, 290.

Braum, Mr. A., his experiment to ascertain the increase of honey in a hive, 303.

Breath, human, *useful in moving bees on comb, 28 (note 10);* offensive to bees, 170, 311.

Breeding "in-and-in," injurious, 54; early, encouraged by spring-feeding, 268.

Brood, temperature necessary for its development, *44 (note 15), 46, note 17) 48; production without pollen, 81 (note 1); needs water during development 189 (note);* attended to by young bees, 197; production of, checked by over-feeding, 268; found in hives in Winter, 48, 339; food reserve necessary for maintaining, 363 (note 3).

Brood-comb, see Comb.

Broodnest, space within preventing swarming, 152 (note 8).

Brown, Hon. Simon, his description of a combat between two queens, 205.

Buckwheat, valuable for late bee-pasture, 290; its yield, and quality of honey variable, 296 (and notes 1 and 2) *and (note 6);* its cultivation recommended, 296 (and note 3); *allelopathic properties, 296 (note);* blossoming of, may cause swarming, 366.

Buera, on the need of water for bees, 344.

Bumble bees 110 (note 2), mate in fall 110 (note 3); [see Humble bees].

Burnens, great merits of, as an observer, 33; laborious experiment of, 33 (note); Huber's tribute to, 194 (note).

Busch, his description of the Italian bee, 324.

Butler's description of the drone, 224; his drone-pot, 225; his anecdote of a honey-bunting swain, 254; his directions for procuring the favor of bees, 311, 317.

C.

Cage, see Queen Cage.

Calendar bee-keeper's, 362-370.

Candied honey, bees need water to dissolve, 342-344.

Candy, sugar, recommended for bee-feed, 272; recipe for making, 272 (note).

Carbon dioxide anesthesia, 327 (note 3).

Cary, Wm. W., his mode of uniting colonies, 204; of fastening comb in frames, 283 (note); his mode of making winter passages in combs, 337 (note); on wintering bees, 346 (note 2).

Casts, see After-swarms.

Catalogue of bee-plants, 298.

Cellars, dry, good for wintering bees,

345,348.

Cells, or bees their contents, 29; covers of, 44; for breeding, become too small, 60; wood-cuts of, Plates XIII., XIV., and XV.; royal 62, 218; thinness of their sides 71 (note); sizes of, 74, Pl. XV., Fig. 48; *foundation prevents odd sizes 149 (note 6); bee's eyes determine shape 74 (note 10);* demonstrate the existence of God, 75.

Cherry-tree yields honey, 292..

Chickens, curious use of, 248.

Children of the rich, compared to pampered bees, 268; may learn from bees how to treat their mothers, 312,

Chloride of lime, useful as a disinfectant of foul hives, 257.

Chloroform, subdues bees by stupefaction, 210.

Clamps, for wintering bees, 348-360.

Clipping queen's wings 223 .

Clover, white, *found in city lawns, 144 (note 2);* most important source of honey, 294; *white clover honey region, 294 (note 3);* Mr. Holbrook, on the value of, for stock, 294; Swedish, 294.

Clustering of swarms, 113, 116.

Cocoon, complete one, spun by drone and worker-larvae, 46; imperfect one, by queen-larvae, 46; of larvae, never removed from cells, 60; of the moth, 231, (Pl. XIX.)

Colonies, of bees (see also Stocks of bees; rapid increase of, in Australia 51 (note); age or, 59; new, composed of young and old bees; 119; impossible to multiply rapidly, by natural swarming, 147; folly of attempting to multiply, by dividing-hives, 149; to remove, from old locations, 156, 157; artificial, not to be formed till drones appear, 158; artificial, time necessary to form, 173; cautions against too rapid increase of, 175 (note), 176-178; weak, easily strengthened by use of mov. comb hive, 178; possible extent of multiplication of, 178; most profitable rate of increase, 179; to form one new colony from two old ones, 180; mother, easily supplied with young fertile queens, in mov. comb hive, 182; sometimes overstored with honey, 183 (notes 1 and 2); table illustrating rapid increase of, 185; new, must remain where first put, 185; many bees may be removed from, when the queens are fertile, 186; new, formed by reversing position of hives 187; piling mode of forming, 188; should, when moved, be supplied with water, 189 (note); to supply queens for rapid increase of, 190-193 bow they may be safely mingled, 203, 336; if small, should be confined by movable partition, to suitable limits, 208; endangered by loss of queen, 217, 246; having young queens should be watched, 218, 222; signs that, have no queen 219; Spring care of; 221; queenless in October, to be united with other colonies, 223; old,. more liable than young, to the ravages of worms, 233, 251 (note); queen less, will be destroyed by the moth, 244 (and note); when hopelessly queenless, their destruction certain, 246; how to be treated when infected with dysentery 256; how, when attacked with foul brood, 257-260; suspected, used by Dzierzon to rear surplus queens for artificial stocks, 260; strong, can, in a season; supply materials for four swarms, 260; feeding of, 267-278; should be strong when honey harvest closes, 269; weak, in the Fall, should be added to other stocks, 270, 336; location of, how to change, 280; removal of, to new apiaries, 281; weak ill-success of; has led to the belief that we are over-stocked, 299; only strong, profitable, 299, 303 (and note); itinerating, 305 (note 2); when broken up for their honey, the queens should be removed beforehand, 206 (note); of common bees, readily converted into Italian 322.

Color, aids in recognizing their hive, 214, 216.

Columella, notice of his Treatise on Bee-Keeping, 147 (note); his remedy against the over-storing of hives, 183 (note 2); advice of, concerning Spring examination of stocks, 221 (note 1); recommended that weak stocks be strengthened from strong ones, 221 (note 2); his suggestion as to the proper time to remove surplus. honey, 224 (note); his mode of feeding bees, 271 (note 1); his directions how to gain the favor of bees, 311.

Colvin his method of securing straight comb, 373; manner or making the mov. comb hive, 383.

Comb, *capping color and age of comb, 45 (note 16); foundation press, invention 51 (note 24), Mehring, inventor of foundation, 72 (note 9); drone comb, amount of, 52 (note 25)* 69-76; too old, can be easily removed in mov. comb hives, 60, 209; *size affected by age of, 60 (note 33);* materials of, 69; woodcuts of, representing various kinds of cells, Plates XIII., XIV., and XV.; empty, great value of; to beekeeper, 71; should not be melted into wax 71; rapidly refilled by bees, 71; easily supplied to bees in mov. comb hive, 71; how attached to frames , 72 283 (and note); drone-comb, not to be put in breeding apartments, 72, 130; artificial, suggestion concerning, 72; author's experiments to induce bees to make it from old wax, 72; building of, carried on most actively by night, 72; comb-building and honey-gathering simultaneous, 73; danger to, in hot weather 91; caution respecting, in artificial swarming from common hives, 155 (and note); generally built somewhat waving, 171, *Africanized bees build straight comb 150 (note 7);* bow to examine; when in mov. comb hive, 172; brood, used for nuclei, 189; worker, used to rear queens, 191; building of; by young bees, 196; worker, should never be destroyed, 207 (and note 2); preferable to artificial comb-guides, 207, 208; *replacement of comb needed, 207 (note);* control of, essential to a system of management, adapted to the wants of all bee-keepers, 208; safely taken from hive when bees are filled with honey or sugar-water, 210; old, most liable to be infested with worms, 233, 251 (note); empty, should sometimes be removed from feeble stocks, 243; *when to replace 251 (note);* new, unsafe to move in warm weather, 281; containing bee-bread, has inferior honey, 288; very old brood, not worth rendering into wax, 288; to make Winter bee-passages in, 337 (and note 1).

Comb foundation, size of cells in, 149 (note 6); 171 (note 24); 184 (note); makes straight combs 208 (note 43).

Commercial beekeeping and mov.

comb hive 145 (note 3).

Composition for corners of hives, to secure them from moths, 78.

Cooling during flight or overheating 129 (note 24).

Confectioners, how they may prevent annoyance from bees, 277; of comb, essential to a true system of bee-culture, 208.

Corsica, ancient, yield of honey of, 304.

Cut-comb honey, 289 (note).

D.

Dampness, injurious to bees, 90, 95, 338-342, 345, 348; produces dysentery, 256.

Dances, communication of pollen, 83 (note 6; communication of new nest site 117 (note 12).

Dandelion, furnishes honey and pollen, 292.

Dangers of too rapidly multiplying stocks, 176-178; of using hives of uniform size, shape, and-color; 214.

Daylight, needed for operations on bees, 167.

Denmark, its honey-produce, 304.

Desertion of hives by swarms, indications and prevention of, 115; *see also absconding.*

Diseases of bees, 255-260.

Dishonesty, as poor policy in bees as in men, 262.

Dissection of queen bees 34, 213 (note).

Disturbing bees in cold weather, injurious, 256, 335,347; 355.

Dividing hives, worthless for artificial, 149, 150; *need for excess bees in new part 160 (note 15); queen attracts bees in, 166 (note 18)*

Division of labor in workers 193 (note).

Dönhoff, Dr., on artificial impregnation of a drone-egg, 41; n on thickness of sides of cells, 71 (note); his mode of forced swarming, 163; his experiment indicating a division of labor among bees according to age, 194; on food of bee-moth larvae, 233 (note); on eggs of bee-moth, 234 (note 2).

Double-stocks, produce a large yield of honey, 135.

Doubling stocks yearly, 185.

Draining combs of honey, 288.

Drawings, explanation of, for making mov. comb hive, 371.

Drifting, of queen during mating flight 125 (note 21); foragers between colonies 203 (note 39); caused by wind 186 (note); caused by hives all painted alike, 214 (note 2); to end of row 215 (note 3); preventing by placement of hives, 216 (note).

Drone-comb, wood-cut of, Pl. XV., Fig. 48; the cause of excess of, 51; excess of, should be removed from breeding apartments, 51, 225; if new, advantageous in boxes for surplus honey, 130.

Drone-eggs, not impregnated, 37; attempt of bees to rear a queen from, 39; artificial impregnation of, 41; *young queens laying, 41 (note 11);* laid by superannuated queens, 49.

Drone-laying queens, 38, 40, 213 (note); use to be made of, 214 (note), 327.

Drones, or male-bees, produced by retarded impregnation of queens, 36; always by unfecundated eggs, 37; often by unfecundated queens, 37, 127 (note); their development from egg to insect, 46; description and wood-cuts of, 49; Pl. XII., Figs. 33, 34 (natural and magnified size); office of, to impregnate young queens, 49; time of their appearance, 50; often very numerous, 50; how to prevent excessive multiplication of, 51; why destroyed by workers, 52, 224; wisdom displayed in providing so many, 53; *congregation areas (DCA's) 54 (note 26); production by laying-workers, 55 (note 30);* length of life, 58; perish in impregnation of queen, 125, 126 (note); never molest queens in hive, 127 (note); on leaving the hive, are filled with hone?, but on returning are empty, 224; Butler's description of, 224; destroyed by ancient bee-keepers, 51, 225; easily destroyed by use or mov. comb hive, 225; *over concern with number of drones, 225 (note 9);* their anxiety when excluded from the hive, 225; their odor, 226 (note 1); how to prevent common, from impregnating Italian queens, 326; refrigerated queens produce only, 327.

Drought, failure occasioned by, 178 (note);

Drumming, *moving bees from box hive to modern hive, 28 (note 9);* on hive subdues bees, 210 (note).

Dunbar, his description of how queen lays, 43.

Dysentery from bad ventilation, 90; *usual cause is Nosema apis 99 (note 8);* from dampness and sour honey, 256; how prevented, 256; makes bees cross, 310; caused by want of water in Winter, 343.

Dzierzon, *identified parthenogenesis, 15 (note); leaf hive, 15 (note);* facts connected with the invention of his hive 19; rise of his system, 19; his apiary nearly destroyed by "foul brood," 19; committee of apiarian convention report favorably on his system, 20; it creates a revolution in German bee-keeping, 20; profits of his apiary, 21; discovered that unfecundated eggs produce males, 37; thinks some brood may be raised without pollen, 81; discovered ryemeal to be a good substitute for pollen, 84; supposes sound of queen's wings excites drones, 127 (note); his mode of forcing swarming 186; his estimate of the value of a queen, 192 (note); his treatment of foul brood. 257; recommends the cultivation or buckwheat, 296; on the difficulty of estimating profits of bee-culture, 306 (note); his experiments with the Italian bee, 320; thinks bees not injured by the opening of their hives, 321 (note); his mode of wintering bees, 348.

E.

Eggs, *number laid per day, 32 (note 3);* of bees how fecundated, 35; fecundated produce females, unfecundated, males, 37; sex of, determined by queen, 38; what is necessary to their impregnation, 41; no difference in size between drone and worker eggs, 42; process of laying, 43; description of, 44; Pl. XIII., Fig. 39; degree of heat necessary to hatch them, 46; power of queens over their development, 47; laid ten months in the year, 48, 339; *egg laying and day-length, 48 (note 20);* supernumerary, how disposed of, 48; ventilation necessary for hatching, 89; of workers transferred to royal cells, 219 *(note 5);* of beemoth, 234 (note 2).

Ehrenfels, profits or his large apiary, 300.

Enemies of bees, 228-255; moth, 228-

252 mice 252; birds, 252, toads, 254; bears 254; ants 255; wasps, spiders, &c., 255; all agreed in fondness for honey, 255.
Energy of bees, instructive, 197.
Engravings, see wood-cuts.
Entrance of hives, should not ordinarily be above the level of the bottom-board, 98; should be readily varied without perplexing the bees, 98; a small upper one, uses of, *96 (note 3)*, 250, 388 (and note); should be nearly closed when colony is threatened by robbers, 264; how to regulate in Winter, 338.
Epitaph on bees killed by sulphur, 239.
Equalizing colonies, by exchanging position 161 (note 15); by adding brood, 269 (note 2).
Ether used for stupefying bees, 210.
European foulbrood disease 257 (note).
Evans, Dr., quotations from poem or, on bees, 50, 60, 69, 76, 77, 78, 79, 109, 267, 292.
Experiments, an interesting one, 67; of Huber, showing the use of pollen, 80; author's to the same effect, 81; numerous, of author, 179; cautions concerning, to beginners, 179; bee-keepers invited to make, 180; of Huber, showing two kinds of workers, 193 (note) difficulty of demonstration by, 193 (note); Dr. Dönhoff's showing that young bees are nurses and old bees honey-gatherers, 194; of author, in wintering bees, 339; of E. T. Sturtevant, 340; of Berlepsch and Eberhardt, 342; of J. C. Bodwell, 345; of Mr. Scholtz, 348; further, needed, in wintering bees, 360.
Examination of combs and bees in hive, importance of, in Spring, 221.

Experience renders bee-keeping profitable, 282.
Extractor, invention by Van Hruscha, 72 (note 8).

F.

Facts, however wonderful, should be received, 42.
Faeces. appearance or, in young and old bees, different, 197; healthy bees do not discharge, in hive, 347; how to make bees in mov. comb hives, safely discharge, 361 (and note).
Faint-heartedness, rebuked, 198.
Famine causes bees to abandon hives, 116.
Fear, effect of, in taming bees, 27; in uniting swarms, 204.
Feeble stocks unprofitable, 141,177, 269, 336.
Feeder, convenience of, in mov. comb hive, 270 construction of, 271; PL XL., Fig. 26.
Feeding bees, 267-278; few things more important in practical bee-keeping, 267; Spring feeding specially necessary, 267 (and note), *178 (note)*; caution in, required, 268; over-feeding, like pampering children, 268; to be submitted to only in extremities, 268; how done, in common hives, 269; difficult to build up small colonies by, 269; equitable division of resources, in, 270; when it should be done for Winter, 270; what should be used in, 270; unprofitable in late Fall stocks, 270 (note); mode of, by means of a feeder, 271; *feeding on top of combs better, 271 (note);* water should be supplied, 271, 342; importance of salt, in, 272; sugar-candy a good and cheap article for, 272 (and note), and 273 (note);

Kleine's mode of using candy, 273, 274; value of grape-sugar for, 273; Sholz' sugar-honey for, 274; granulated sugar for, 274 (and note); quantity of honey needed for, to Winter bees, 274; weight of hives, unsafe standard to determine amount of honey for, 275 (note); caution to be observed in, 277; should not be too early in the Fall, 298; cheap honey, to sell again, unprofitable in, 275.

Feral (wild) colonies, number, 51 (note 22) 143 (note 1); exposed nests 118 (note 13).

Fertility of queens, 32; diminishes with age, 141, 223; diminished by hunger and cold, 223 (note 1).

Fishback, Judge, his precautions to prevent loss of young queens, 216; his experience with the bee-moth, 240 (note).

Flight of bees, its extent, 305; its rapidity, 305 (note 2); *bee's body temperature needed for flight, 338 (note)..*

Flowers for bees, Nutt's catalogue of, 298; garden, furnish little bee-pasture, 297.

Follower boards 96 (note 4).

Foul-brood, *American and European foulbrood separated, 18 (note 14); resistance to, 18 (note 15);* its malignity, 19, 256; dry and moist, 256; remedy, 257, 258; a disease exclusively of the larvae, 259; supposed cause, 256 (note), 259; liable to appear the second time, 259.

Food sharing, "corporate stomach" of bees 178 (note).

Foraging, distance and moving swarms, 157 (note 12); bee, age when starting to forage 195 (note 32).

Forcing-box, its size and use, 154,165.

Foundation (see comb foundation).

Frames, movable, invented by author, 15; *Hoffman self-spacing frame 131 (note 25), 170 (note 23);* how they must be made to be lifted out of hive, 150, 171, 209 (note); process of removing from the hive 171; 370 **(Pl. XXIV.);** *wire for comb strength 171 (note 24);* with comb used for patterns, 208; effect on bee-culture, 211 (note); a protection against the ravages *or* the moth, 239, 241; render the cleaning of hive easy, 243; used by Berlepsch, 321 (note 2); *size and brood rearing, 331 (note);* approved of by Siebold, 321 (note 2), not well adapted to tall hives, 330.

Friesland, East, its productiveness in honey, 304.

Fructose corn syrup, 274 (note 8).

Fruit, honey-bees beneficial to, 85-87; wasps and hornets injurious to, 86.

Fruit-trees; blossoms of, yield honey, 292.

Fumigation of hives with puff-ball, objectionable, 210.

G.

Galleria mellonella (wax moth) see bee moth.

Gardeners might manage their employers' bees, in mov. comb hive, 226.

Garden plants insufficient to furnish bee-pasture, 297.

Gentleness of bees, selection, 13 (note 7)

Glass, vessels of, for spare honey, should hive guide-combs, 290; objections to, 290 (note);

Gloves, India-rubber, to protect the hands, 317 (Pl. XI., Fig. 27); wool-

en, objectionable, 317.
Glucose=dextrose=grape sugar, 273 (note 7).
Goldsmith, on spontaneous and fashionable joys, 334.
" Good old way" of corn-raising, 237.
Golden-rod, some varieties of, furnish food for bees, 298.
Governments, of Europe, interest of some in disseminating knowledge or bee culture, 320 (note).
Greek slatted-top hive, 14 (note 8).
Grafting queen larvae 192 (note 31).
Grape-sugar, as food for 'bees, 273 (=glucose or dextrose)..
Guide for combs, artificial, secure regularity in building comb, 130, 207; cannot be invariably relied on 208; German invention of, (Pl. VI., Fig. 72).
Gundelach, on the necessity of pollen for rearing brood, 81.
Gutta-percha, plastic combs, 72 (note 7).

H.

Hairy objects, why offensive to bees, 317.
"Hanging out" bees on front of hive 139 (note 37).
Haploid, drones, 42 (note 13).
Harris, Dr., his account of the bee-moth, 228.
Hartshorn, spirits of, remedy for bee-stings, 316.
Health, bad ventilation of houses impairs, 92.
Hearing, in bees, acute, 138.
Heat, degree required to hatch the eggs of bees and develop the pupa, 46; great, attendant on comb-building, 71.
Hens, too much crowded, mistake their nests, 215; not good tenders of moth-traps, 248;
Heyne, on over-stocking, 301.
Hiver, basket for, 133.
Hives (see Mov. Comb Hive), Huber's, author's experiments with, 14; made with slats, 15, 210 (note); should be made of sound lumber, 78; mixture for sealing corners of, 78; thin, annoying to bees in hot weather, 90; sixty-one requisites for complete, 95-108; size of, should admit of variation, 96; "improved," often bad, 107; qualities of best, 107; paint on, should be very dry before Living, 129; heated in the sun, should not be used for new swarms, 129; should incline forward, but stand level from side to side 130; if clean, need no washing or rubbing with herbs, 131; five stocks in one, 137; should be placed where it is to stand, as soon as swarm is secured, 138; if not ready to swarm, bow to proceed, 139; difficult to rid of bee-moth, 141; common, difficult to remove infertile queen from, 141; Huber's, 148; "dividing," and objections to, 149; self-colonizing, ineffectual, 151; thorough inspection or, necessary for success, 152; non-swarming, likely to exterminate the bee, if generally used, 153; decoy when to be used 155; for surplus honey should be undisturbed, 180, (and note); like Dzierzon's, even with movable frames, give inadequate control of bees, 187 (note); should be opened before or after sunlight, when forage is scarce 199; royal combat witnessed in author's observing, 205; with poor arrangements, educate bees to regard their keeper as an enemy, 210 (note); wonders of, unknown by many bee-

406 INDEX.

keepers, 211; in crowded apiary, 214-216; condition of, should be ascertained, 221; patent, evil results of, 237, 241; should be cleaned in early Spring, 243; common, furnish no reliable remedy for loss of queen, 245; infected with foul-brood, to disinfect, 257; common, how prepared for removal when occupied by stocks, 281; to transfer bees from common to mov. comb, 282; size, shape, and materials for, 329-332; size of author's can be varied at pleasure, 329; tall, advantages and disadvantages of, 329; most advantageous form of, 330; Dzierzon's, disadvantages of, 331; double and triple, 331 (note); proper materials for, 331; suggestions as to making mov. comb. 332.

Hives, mov. comb, see movable Comb Hives.

Hives, patent, see Patent Hives.

Hiving bees, directions for, 129; expertness in, makes pleasant, 129; should be conducted in shade, 130; should be attended to soon after swarm settles, 132; process of, 133; basket for, 133; sheet for, how arranged, 133; how to expedite, 133; process of, must be repeated when queen not secured, 134; when settled out of reach, how to secure the swarm, 134; when swarm alights in difficult place, or two swarms cluster together, 135; how to secure the queen 136; old-fashioned way of, bad, 136; so as to prevent swarms uniting, 138; when done, remove swarms to proper stands, 138; danger of delaying 138; what to do if no hive is ready, 139.

Holbrook, Hon. F., on cultivation of white clover, 294.

Home, should be made attractive, 220.

Honey, *movement of moisture through capping, 45 (note 16), 270 (note 3);* 285-299; its elements, 70; quantity consumed in secreting war, 71, 176; gathered by day, 72; sometimes gathered by moonlight, 73 (note); honey gathering and comb building simultaneous, 73; surplus, incompatible with rapid increase of colonies, 176; how to secure the largest yield of, 180; *yield reduced by manipulation of hive 183 (note);* more abundant fifty years ago than now 236; reasons assigned for the deficiency, 237; foreign, supposed cause of foul brood, 256, 258; from foul-brood colonies, infectious, 256 (note 2); infected, how purified, 257; West India, used for bee-feed, 256 (note), 270; and sugar (Scholz' composition), 274; quantity of, necessary for wintering stocks, 274; poor, not convertible into good, 275; not a secretion of the bee, 275 (and note 2); retains the flavor of the blossoms from whence it is taken, 275; evaporation produces the principal changes in, 276 (and note 1) " making over" honey not profitable, 276; recipe for artificial, 276 (note); a vegetable product, 285; quantities of vary, 287; hurtful qualities cured by boiling, 287 (and note); should not be exposed to low temperature, 287; old, more wholesome than new, 287 virtues ascribed to it by old writers, 287 (note); to drain from the comb, 288, 366; to make liquid when candied, 288; caution as to West India, 288 (note); of Hymettus, 293 (note); yield of, affected by soil, 294 (note) from the raspberry, delicious 296; yield of, by plants uncertain, 296

(note 2); large amount gathered in a day, 303; on the hands, protects them against bee-stings, 317; bees eat less in Winter, when kept quiet 335, 348, 358; how to get in centre of hive for Winter, 336; candied by bees need water to dissolve, 342-344.
Honey-bag, worker's, 56 (Pl. XVII., Fig. 54).
Honey-bees, see Bees.
Honey-board, spare, holes in, left open in Winter, 338, sometimes strongly glued by bees 172 (note); care in placing necessary, 173.
Honey-dews, 285; of California, 285 (note); *why and how produced by Homoptera, 285 (note 1);* when most abundant and where found, 286; *German Black Forest honey, 287 (note 2).*
Honey-hornets, Mexican, 58 (note) 87.
Honey-resources, how to increase, 293
Honey-suckle, juice of, a remedy for bee stings, 315
Honey, *19th century yield per colony, 18 (note 13); yield in large cities 144 (note 2);* surplus, in much, incompatible with rapid multiplication of stocks, 176, 178 best yield of, from undisturbed stocks 180; receptacles for, when to admit bees to, 288, 364; *no longer boil honey, 288 (note 3);* how secured, 289; quantity from one stock, 289 (note 2); large boxes more profitable than small, for, 289 (and note 2), 290 (note 1), glass vessels and small boxes, for, 290, air-tight boxes, to preserve, 290 (note 2); receptacles of, how and when to remove them, 291, 365; boxes for, bees reluctant to fill, late in the season, 366.

Honey section boxes, Half-comb cassettes, 289 (note).
Honey-water, objectionable for subduing bees, 169 (note).
Hornets, fecundation of, 35; Mexican, honey 58 (note), 87; injure fruit, 86; should be destroyed in Spring, 87; torpid in Winter, 109.
Horses sweaty, very offensive to bees, 279, 313.
Horticulturists, honey-bees their friends, 85, 87.
Houses, ventilation of, neglected, 91.
Huber, Francis, *scientific apiculture, 13 (note 5)* tribute to, 32-34; discovered how queens are impregnated, 34; that unfecunded queens produce only drones, 36; experiments of, to test the secretion of wax, 69; to show the use of pollen, 80; his discovery of ventilation by bees; 88; his supposition as to development in queen of male eggs, 123 (note); big plan for artificial swarming and its objections, 148; effect of his leaf hive in pacifying bees, 168; his mistake as to the cause, 169; an inconvenience of his hive, 171 (note); his description of workers, 192 (note 2); his curious experiments showing a distinction among them, 193 (note); his tribute to Burnens, 194 (note); his account of the treatment by bees of strange queens, 200; his trial of two queens in a hive, 207 (note); splendid discoveries of, formerly ridiculed, 211.
Humble-bee robbed by honey-bees, 262.
Humidity needed by larvae during queen grafting, 283 (note 6).
Hunger impairs fertility or queen-bee, 223 (note 1).
Hunt, Rev. T. P., his mode of securing swarms, 132.

Hunter, Dr., discovers pollen in the stomach of bees, 80.
Hurting bees, important to avoid, 95.
Hyginus, on feeding bees, 267 (note).

I.

Immunoglobulins and sting reaction, 316 (note).
Impregnation, of queen-bees, 34-43; retarded, effect of, 36; remarkable law of, in aphides, 42; takes place in the air, 50, 320; act of, fatal to drone, 125, 126 (note); Shrimplin's experiment illustrative of, 127.
Insectivorous birds, 252 (note 10).
Inspecting colonies, reduces yield, 24 (note 1)
Instinct, versus learning, 247 (note)..
Italian honey-bees, 41; *(Apis mellifera ligustica)* singular result of crossing with common drones, 41, 324 (note 2); used to show a division of labor among bees, 194; account of 318-328; described by Aristotle and Virgil, 318; Mr. Wagner's letter on 318; their modern introduction to notice, 318; value of, in the study of the physiology of the honey-bee, 319; cells of, the same size as those of the common bee, 320; Dzierzon's experiments with, 320; frequent disturbances abate nothing from the industry of, 321 (note); general diffusion of, desirable, 321; superior to common bee, 322, 324,325; peaceable disposition of 322; may readily be introduced into hives of common bees, 322; furnishes now means of studying the habits of bees, 322 the purity of, can be preserved, 322; character of, as tested by Berlepsch, 324; number of queens obtained in one season, from one queen; 324; remarkable fact in relation to hybrids, 324 (note); description of, by Busch, 324 Radlkoffer's account of, 325; how to introduce an Italian queen to a stock of common bees, 325; advantages of author's non-swarmer in preserving the Italian bee pure, 326; how to produce abundance of drones of, 327; precaution suggested when non-swarmer cannot be used, 327; queens or, safely moved in mov. comb hive, 327; introduction of, into this country, important, 328; arrangements to that end, 328 (note).
Itinerating colonies, 305 (note 2).
Ignorance, the occasion of the invention of costly and useless hives, 209 (and note).
Increase of colonies, rapid, impracticable, by natural. swarming, 147; or by dividing hives, 149; rapid, cautions against, 175-178; rapid, incompatible with large yield of surplus honey, 176; a tenfold, *possible,* in mov. comb hive, 178; sure, not rapid, to be aimed at, 179; forming one new from two old colonies best, and how effected, 180; rapid, requires liberal feeding, 184; *limits based on wintering the splits* 199 (note 35).
Inexperienced persons should not begin bee-keeping on a large scale, 282.
Indian name for honey-bee, 236.
Industry taught by the bee, 59.
Intemperate men compared to infatuated bees, 278.
Intercommunication of bees in hives, important, 103, 336, 337 (and note), 339 (and note).
Irving, Washington, his account of the abundance of bees at the West, 236 (note).

J.

Jansha, on impregnation of queen, 36.
Japanese, veneration for birds, 253 (note).
Jarring, disliked by bees, 96, 170, 309.
Jelly, *worker, 44 (note 14);* royal the food of immature queen, 63; a secretion of the bees, 64; analysis of, 64; *chemical analysis, 64 (note 36);* effect of, in developing larvae, 64, 191; pollen necessary for its production, 197.
Johnson, M. T., the first American observer of the fact that queenless stocks are soon destroyed by the moth, 244 (note).

K.

Kaden, Mr., on over-stocking, 301.
Killing bees for honey; an invention of the dark ages, 239 (note); more humane than to starve them, 238; not necessary, 239.
Kindness of bees at home, a lesson for man, 312.
King-bird, eats bees, 252.
Kirby and Spence on ants and aphides, 285.
Kirtland, Dr. J. P., his letter on the introduction of the bee-moth, 240; on benefits of transferring stocks into mov. comb hive, 284.
Knight on honey-dews, 286.
Kleine, Rev. Mr., on making bees rear queens in selected cells, 191; his method of preventing robberies among bees, 265 (note); on feeding bees, 273; on over-stocking, 301; on accustoming the human system to the poison of bees, 316 (note).

L.

Langstroth, L. L., theology training, 25 (note 2)
Larvae of honey-bee, development of, 44 (Pl. XIII., Figs. 40, 41, 42; royal, 64; perish without ventilation, 89; of bee-moth, see bee-moth, larva of; of honey bee, disease of, 259.
Leidy, Dr. Joseph, his dissection of fertile and drone-laying queens, 34, 39, 213 (note); of a queen just impregnated, 126 (note).
Lemon Balm for attracting swarms, 131 (note 26).
Life span, 59 (note 32), 151 (note).
Light, bees will work when exposed to, 16, 205, 332; its sudden admission, effect of on bees, 168, 169; of day, needed for operations about the hive, 167.
Ligurian, or Italian, bee, 318 (note).
Linden, or bass-wood tree, yields much honey, 293 (and note); *also known as lime trees in Europe.*
Liriodendron, yields much honey, 292.
Locust, valuable for bees, 293.
Lombard, his interesting anecdote of swarming, 308.
Longfellow, H. W., his Indian warrior's description of the bee, 236.
Loss of queen, 213-227; frequent, though the queen is usually the last to perish in any casualty, 213; when by old age, bees prepare for her successor, 213; occurs oftenest when queen leaves hive for impregnation, 213, 214; how occasioned by queens mistaking their hives, 214, 215; bees, like hens, in this respect, 215; Judge Fishback's preventive of, 216; author's preventive, 217; effect of; on stocks, 217; sometimes not discovered by bees

for some time, 218 (and note); excitement in hive when discovered, 218; will not cause bees to abandon the hive if they are supplied with brood-comb, 218; nucleus system will remedy it, 219; indications of, 219; the most common cause of destruction or stocks by bee-moth, 219.

Lunenburg, number of colonies of bees in, 302; bees of, more than pay all the taxes, 302.

M.

Mahan, P. J., on causing bees to adhere to new locations, 163 (note); interesting observations of, 219 (note); his discovery that drones leave their hives with honey and return without any, 224; on the odor or the queen, 226 (note 2).

Maple-tree a source of honey, 292

Maraldi, anecdote from, of bees and a snail, 78.

Marking queens, color code 223 (note 7).

Materials for hives 331.

Mating, number of times queens mate, 50 (note 20), 319 (note 2).

Mating flight, queen finding hive on return, 31 (note 2), 125 (note 21); height of, 54 (note 27); queen pheromone attracts drones 127 (note); limited by poor spring weather 158 (note 13), 213 (note).

Mating sign after nuptial flights 125 (note 22).

Meal, a substitute or pollen, 84, 219.

Medicine, poison of bee used for, 315 (note).

Mice, ravages of, and protection against, 252.

Miller, see Bee-moth.

Miller, C. C., queen rearing with edges of comb, 192 (note).

Mills, John on marking hives with different colors, 216 (note).

Mixing of bees, of different colonies, 203; precautions concerning, 203.

Months of the year, direction for treating bees in, 362-369.

Moonlight bees sometimes gather honey by, 73 (note).

More, Sir J., on the sovereign virtues of honey, 287 (note).

Movable-comb hive, definition, 14 (note 10); management with, 14 (note 9); re-queen,, 14 (note 9); research on, 16 (note 12); similar development, 16 (note 12), 149 (note 5); patent rights, 20 (note 16); comparison to observation hive, 22 (note 18).

Moving colonies, distance needed 157 (note 12); 187 (note 28).

Moth, see Bee-moth.

Moth, death-head, 240 (note).

Moth, large honey-eating, from Ohio, 241 (note).

Mothers, unkind treatment of, reproved by bees, 312.

Mother-stock, in forced swarming, easily supplied with fertile queen, 182; exposed to perish without a prompt supply of queen, and by over swarming, if left to supply itself, 182; also to be robbed, 182; advantage of supplying with fertile queen, 183.

Moth-proof hives a delusion, 228, 238, 247; *250 (note).*

Moths, honey-eating, ravages of, 240 (and note).

Motions, in operating on hives should be deliberate, 170.

Mouse guards 252 (note).

Movable-comb hive, invention of 13-23, *148 (note 5);* superiority to Dzierzon's, 16, 18; enables each bee-keeper to observe for himself, 23, 164; admits of easy removal of old comb, 60; bees in it easily supplied with empty comb, 71; its facilities for ventilation 94, 276; *standardization of modern hives 95 (note 1);* size of, adjustable to the wants of colony, 96, 329; facilities of, for securing surplus honey, 100, 289, 329; advantages of, for preventing after-swarming, 124, 140; enables one person to superintend various colonies, 102, 226; not easily blown down, 103; *interchangeable combs a major asset 103 (note 15);* may be made secure against mice, 103, 252, and thieves, 104; durability of, 104; *size of boards 104 (note 16);* cheapness and simplicity of, 105; some desirables it does not possess, 105; invention of, result of experience, 105; perfection disclaimed for, 105; merits of, submitted to experienced bee-keepers, 108; desertion of, by swarms, easily prevented, 115; by use of, can employ all good worker comb, 130; furnishes storage-room for non-swarming bees, 139; importance of, in supplying extra queens, 141, 188; easily cleared of the bee-moth, 246; best for non-swarming plan, 153; enables the apiarian to learn the laws regulating the internal economy of bees, 164; enables artificial swarming to be quickly performed, 164; advantages of movable top of 168; affords facilities for supply of fertile queens to mother stocks, in forced swarming, 182, 192.; danger of being stung, diminished by use of, 209; the greatest obstacle to its speedy introduction, 209; the author sanguine of its extensive use by skillful bee-keepers, 211; should be thoroughly examined in Spring, 221; durable and cheap, if properly taken care of, 221; *cold Spring temperature should not stop examination, 221 (note);* advantages of, readily perceived by intelligent bee-keepers, 226; adaptation of, to protect stocks from the moth, 249; enables the apiarian to know the amount of honey stocks contain, 275 (note); how prepared for transporting bees, 281; to transfer into, from common hive, 283; designed to economize the labor of bees, 305; experiments concerning the size of, 330 (note 3); suggestions as to making, 332; observing, 332; how to get honey in centre of, for Winter, 336; how to make Winter passages in combs of, 337 (and note); how to ventilate, in Winter, 338; bills of stock, for making, 371.

Movable entrance blocks, see Blocks, entrance regulating.

Movable bottom-boards, dangerous, 231.

Movable stands for hives, 279.

Moving stocks, 281.

Munn, W. A., his "bar and frame hive," 209 (note).

Musk, used to stop robbing, 265 (note).

N.

Narcotics, in managing bees, worse than needless, 211.

Nasonov, gland 31 (note 1); pheromone 31 (note 1), 117 (note 11); Nasonov pheromone and orientation during swarming 133 (note 28); pheromone dominant over queen pheromone during swarming 159 (note).

Natural swarming and hiving of

swarms, 109-142; guards against extinction of bees, 109; not unnatural, 111; time of, 111; seldom occurs in northern climates, when hives are not well filled with comb, 111 (note); signs of, 111; only in fair weather, 112; time of day of, 112; preparation of bees for, 112; queen often lost in, 113; ringing of bells and tanging, useless, 113; how to stop a fugitive swarm, 114; after, ventilation should be regulated, 124; hiving should be done in shade, or hive be covered, 130; should be promptly attended to after swarm settles, 132; process of, 133; basket for, 133; sheet for, 133; how arranged, 133; how to expedite, if bees are dilatory, 133, 134; must be repeated if queen not secured 134; small limbs cut with pruning shears in, 134; when swarm out of reach, how to secure, 134; when in difficult places, or two swarms cluster together, 135; how to secure queen, 136; old fashioned way, objectionable, 136; more than one swarm in a hive, 137; to prevent swarms uniting while hiving, 138; swarms as soon as hived, should be removed to their stands, 138; an expedient if no hive be ready, 139; suggestions for making more profitable, 139-142; excessive, prevented by use of mov. comb hive, 140; affords no facilities for strengthening late and feeble stocks, 140; objections to, 139-147; uncertainty of, 147; why some stocks refuse to swarm, 147; *broodnest space above or below? preventing swarming 152 (note 8).*

Nectar sugar types, 276 (note 10).

"New England Farmer," extract from, describing a combat of queens, 205;

Newspaper method of uniting a swarm with another colony, 135 (note 30).

Night-work, on bees, hazardous, 167.

Non-swarmer, author's, prevents swarming, 174; excludes drones 228; facilities it offers to preserve pure the Italian bee, 326; wood-cut of, Pl. II., Fig. 5.

Non-swarming colonies, may lose their queens, or queens become unfertile, in common hive, 153; queens may be supplied to, in mov. comb hive, 153.

Non-swarming hive, advocated by many, 154; objections to, 153; mov. comb hive best for, 153.

Nosema (Nosema apis) [see also dysentery] 99 (note 8); 256 (note).

Number of colonies managed (102 (note 13).

Number of bees in colony, 54 (note 28).

Nuclei, *Nucs or nucleus hives, 190 (note);* what they are, and how to form them, 189; to obtain adhering bees for, 192 (and note); must not be allowed to get too much reduced, 197; always furnish plenty of queens, 219.

Nutt, his list of bee flowers, 298.

Nymph, bee, see Pupa.

O.

Objections to natural swarming, 143-147.

Observing-hive, mov. comb, 332-334; Hon. S; Brown's experiment with, 205; its facilities for observing the internal operations of the bees, 332; for wintering, 332 (note); those with single frames recommended, 333; adapted for the parlor, 333; how to stock with bees, 333; source of pleasure and instruction, 333; may be kept in cities, 333.

Odor, of Queens, *(See Pheromones),* 226,266; of drones, 226 (note 1); of workers, 203.

Odors, *detection of, 67 (note 37);* unpleasant, offensive to bees, 313; used to prevent robberies, 265 (note); excite bees to anger, 313.

Oettl, remarks of, on over-stocking, 303; his golden rule in bee-keeping, 303; his statistics of bee culture, 303.

Old age, signs of in bees, 59;

Oliver, H. K., observations of, on bee-moth, 251.

Onions, blossoms of, yield much honey, 293.

Ovaries of queen-bee, 35, (Pl. XVIII.); of workers, are undeveloped, 29, 54.

Over-stocking, 299-307; no danger of 299; Wagner's letter on, 300; Oettl and Braun's statistics on, 303.

Oviposition (egg laying), number of eggs per/day, 32 (note 3); competition between queen and honey stored 183 (note).

Ovum, what necessary to impregnate it, 41.

P.

Package bees, installing with syrup and direct release of queen 119 (note 14); package shippers remove 8,000 bees/week 183 (note).

Paint, smell of fresh, detested by bees, 129; if fresh be used, it should contain no white lead, and be made to dry quickly, 129; recipe for, preferable to oil paint, 129; color of, for hives, 368, *129 (note 23), 280 (note 1); all hives painted alike and drifting 214 (note 2), 280 (note 1).*

Parthenogenesis, 37 (note 7) 42 (note 13).

Pasturage for bees, 292; effect of, on removal of colonies, 157; honey-yielding trees and plants 292-299; gardens too limited for, 297; catalogue of bee plants, 298; range of, 305.

Patent hives, deceptions in vending, 61 (note) 106 146 (note); have greatly multiplied the bee-moth, 237; and done more harm than good, 237,241.

Peach-tree, yields honey, 292.

Pear-tree yields honey, 292*; pear nectar low in sugar, 292 (note 1).*

Peppermint, use of in uniting colonies, 203.

Perfection, folly of claiming for hives, 106.

Perfumes, disagreeable to bees, 313 (note).

Perseverance or bees, worthy of imitation by man, 197.

Perspiring by bees, 130 (note 24).

Persons attacked by bees, directions for, 312, 314.

Pesticide poisoning prevented by sprinkling syrup on bees, 281 (note 2).

Peters, Randolph, interesting experiment of, 219 (note).

Pheromones, 31 (note 1); alarm 95 (note 2); Nasonov (aggregation) 117 (note 11); Nasonov during swarming 133 (note 28); Nasonov, during hiving of swarm 133 (note 28); scenting and fanning of Nasonov with queen loss 218 (note); drone 226 (note).

Pillage of hives, secret, cause and remedy of, 266.

Piping of queens, an indication of after-swarming, 121, distinctive sound 121 (note 17).

Plantain, a remedy for bee-stings, 315.

Plum-tree, a source of honey, 292.
Poison of bees, smell of, strong and irritating to bees, 314; effect of, on the eye, 314 (note); remedies for, 314-317; effect of, when taken into the month, 315; cold water the best remedy for, 315; a homeopathic remedy, 315 (note); the human system can be inured to, 316 (note).
Poisonous honey, and how to remove its injurious qualities, 287.
Pollen, or bee-bread, 80-87; found in stomachs of wax-makers, 80; may aid in secretion of wax, 80; whence obtained, 80; food of immature bees, as shown by Huber's experiments, 80; author's, to the same effect, 81; Gundelach's opinion of, 81; useful in secretion of wax, 82; bees prefer fresh to old, 82; in mov. comb hives, excess of, in old stocks, can be given to others, 82; *need in winter cluster, 82 (note 2);* how gathered and stored by bees, 83; bees gathering, aid in impregnating plants, 83; bees collect, only from one kind of flower at a time, 83, *(note 7);* wheat and rye meal a substitute for, 84; necessary for the production of wax and jelly, 197; *needed for brood food but NOT for wax 197 (note 33);* the gathering of, by bees, indicates a fertile queen in the hive, 219 (and note).
Pollen-basket, on leg of bee, 56, *83 (note 5).*
Pollen substitutes, 84 (note 9); patties for spring feeding, 363(note 2).
Pollination of fruit, time needed, 86 (note 10).
Poppy, white, a remedy for bee-stings, 315.
Population of hive, 54 (note 28).
Posel, discovery of, on use of spermatheca, 36 (note).

Proboscis of a worker, 56; wood-cuts of, Plates XIII., XVI., Figs. 63, 51.
Profits of bee-keeping, Dzierzon's experience in 21; Sydserff's calculation of, 146 (note); dependent on strong stocks, 176; difficulty of estimating, 306 (note); safe estimate of, 306.
Propolis, 76-80; whence obtained, 76; curious sources of, in Mexico, 77; *named by Aristotle, 77 (note 1);* its uses, 77; bee-moth lays her eggs in, 78; curious anecdotes, illustrating its uses, 78.
Prussia, bee-keeping encouraged by government of, 320 (note).
Pupa, or bee-nymph, 45; heat required for its development, 46.
Punk, smoke of, subdues bees, 27, 154.
Push-in cage for queen introduction 200 (note 36).

Q.

Queen-bee, wood-cut of (natural and magnified size), Pl. XII., Figs 31, 32; wood-cut of ovaries and spermatheca of, 35, Pl. XVIII.; description of, 30; the mother of the whole colony, 30; affectionate treatment of, by the other bees, 31; effect of her loss on the colony, 31; her fertility, 32; how her eggs are fecundated, 34-41; Huber discovers impregnation of, to take place out of hive, 34. dissection of, by Dr. Leidy, 34, 126 (note), 213 (note); effect of retarded impregnation on, 36; she determines the sex of the egg, 38; Dr. Leidy's dissection of a drone laying, 38, 126 (note), 213 (note); attempt of bees to rear, from a drone-egg, 39; account of a drone laying, afterwards laying worker eggs, 40; a drone laying, with shriveled wings, 40; Italian, impregnated by common drones, produce Italian drones, while the females are a. cross, 41, 324 (note 2); becomes incapable of impregnation, 42; process of laying, 43; development of, in pupa state, 46; enmity of, to each other, 46, 120, 205-207; can regulate development of eggs in her ovaries, 47; *Age at first flight, 47 (note 18); piping, 47 (note 19);* disposition by, of supernumerary eggs, 48; fertility of, decreases with age, 49, 223; longevity of, 49, 58; when superannuated, lays only drone-eggs, 49; why impregnated in the air, 53; office of no sinecure, 58; Italian, use of, to show how long workers live, 59; *length of life, 59 (note 32);* manner of rearing, 62; larvae of, effects of royal jelly on, 63; process of rearing in special emergency, 66; development of, an argument against infidelity, 68; old, leads first swarm, 111; often lost in swarming, 112; loss of, in swarming, causes bees to return to parent stock, 113; how to prevent, from deserting new hive, 115; influence of, in causing bees to cluster, 117; prevented by bees from killing inmates of royal cells, 121; *piping of, 114 (note),* 121 *(note 17)*; several sometimes accompany after-swarms, 122; emerges from her cell mature, 122; young more active on wing than old, 123; young often reluctant to leave hive, 123; telling *age of queen in a swarm 123 (note 19);* young, does not leave for impregnation till established as sole head, 51, 125; her precautions to regain her hive, 125; never molested by drones in hive, 127 (note);

begins laying two days after impregnation, 128; lays mostly worker-eggs the first year, 128; never stings, except in combat with other queens, 136, 204; alacrity of, in entering hive for new swarm, 136; young, often lost after swarming, 141; her loss easily remedied by. mov. comb hive, 141; unfertile, difficult to remove in common hives, 141; when immature, bees do not build worker-comb, 149; seldom enters side-apartments, 152; signs indicating her presence or absence in forced swarms, 158; *attraction of swarm or division to, 166 (note 18);* supply of sealed, for forced swarming, how to secure, 166; how to cut sealed ones from comb, 166; *knowing the age of queen cell by grafting 167 (note 19);* fertile, deprived of wings to prevent swarming, 173; *clipping does not cause supersedure 174 (note);* may be confined to prevent swarming, 174; unfertile, should not be confined, 175; fertile, easily supplied to destitute mother stocks, 182; young, in after-swarms, lay few drone-eggs, 184 (note); to raise, for artificial-swarming, 188; when to be given to newly-forced swarms, 189; to induce bees to raise, on what part of the comb you please, 191; her value, 192 (note); can she be developed from *any* worker-larvae? 192 note 2; made to supply several stocks with eggs, 193; will lay eggs while under inspection, 196 (note); *introduction via push-in cage 200 (note 36);* caution needed in giving, to strange stocks, 200; stranger, how to induce stocks to receive, 201; protected by queen-cage, 201; *mailing cage 201 (note 38);* care to be used in catching, 202; never stings, but sometimes bites, 202, 204; may be lost if allowed to fly, 202; her great appetite, 202; her life indispensable to the safety of the colony, 204; loss of, see "Loss of Queen;" young, dangers besetting, 213; should be given to queenless stocks in Spring, 221; when unimpregnated, colony should be watched, 222; when unimpregnated hides, 222; wings of, may be clipped for artificial swarming, 222; how to mark the age or, 223; fertility of, diminished by hunger and cold, 223 (note 1); should be removed in their third year, and new one given, 223; regular and systematic, best, 223 (note 2); odor of 226; removal of, a remedy for foul-brood, 258; surplus, reared by Dzierzon, in suspected hives, 260; deserted by her subjects when they have been conquered by stronger stocks, 263 (and. note); should be removed before smothering the bees, when stocks are broken up for their honey, 306 (note); *age of queens, 319 (note 1);* Italian, how to propagate, 326; after being chilled, lay only drone-eggs, 327.

Queen bees, why, when two fight, both are not killed, 205; combat of, as witnessed in one of author's observing hives, 205

Queen-cage, use and construction of, 201, 325; *candy for, 274 (note 9).*

Queen cells, see Royal cells.

Queen excluders, use during hiving of swarm 134 (note 29).

Queenless stocks, signs of, 219 245 to be supplied with queens, 221; in October, should be united with other stocks, 223; a sure prey to the moth, if not protected in time, 244 (and note).

Queen substance (see pheromone), 31 (note 1)

Quinby, M., author or a very valuable work on bee-keeping, 249 (note); on the ravages of the larvae of bee-moth, 249 (note) on shape of mov. comb hives, 330 (note 3); on wintering bees, 348; on equalizing colonies when removed from Winter repository, 361. (note 2); on making bees work in a double tier (if surplus honey-boxes, 365 (note).

R.

Radikofer, Doctor, on over-stocking, 300; on the Italian bee, 325.

Rapping on hives, its effect on bees, 27, 155, 204.

Raspberry, one of the best bee-plants and very abundant in hill towns of New England, 296; *high nectar sugar, 297 (note 7).*

Reaumur, his account of a snail covered with propolis, by bees, 78; his error as to the treatment of strange queens by bees, 201; thought there were two species of bee-moth, 228.

Reid, Dr., on the shape of honey-cells, 75.

Religion, revealed, appeal to those who reject, 52.

Remedies for bee-stings, 324-327.

Riem, the first to notice fertile workers, 55.

Ringing bells, in swarming time, useless, 113.

Requisites of a complete hive, 95-108.

Robbers, highway, bees sometimes act the part of 262; *robbing bees' flight pattern, 262 (note 2).*

Robbing, by bees; *nectar availability and, 27 (note 7);* frequent, when forage is scarce, and caution against, 199, 261, 263; *induced by syrup feeding, 261 (note 1);* how prevented, 261-266; committed chiefly on feeble or queenless colonies, 261; signs indicating a bee engaged in, 261 *(and note),* 265; begets a disrelish for honest pursuits, 262, 264 (and note); movable entrance blocks protect bees against, 264; infatuation produced by, on bees, 264; caution needed in checking, when a hive is vigorously attacked, 265; how to stop bees engaged in, 265; secret, its remedy, 266.

Robbing screens 98 (note 7), 199 (note 35), 252 (note), 264 (note 4), used for moving bees, 282 (note 5).

Royal cells, described, 62; wood-cuts or, Plates XIII., XIV., and XV.; attention paid to, by workers, 62; why they open downwards, 63; number of, in a hive, 63; how supplied with eggs, 63; *reason for location on comb, 63 (note 35);* description of, 66 , when built, 111; queen prevented from destroying, 121; remains of, indicate number of queens hatched, 121; may be removed in mov. comb hives, to prevent after swarming, 124; bow to decide whether inmate of has been hatched or killed, 121; *cell cups always in hive 121 (note 16);* how to cut out of combs, 166; sign that the queens in, are nearly mature, 167; *grafting helps determine age of, 167 (note 19);* how to make bees rear, in convenient places on the comb, 191; t& be given to colonies second day after removal of queen,. 223.

Royal Jelly, see Jelly, royal.

Rural life and beekeeping, 12 (note 1)

Rye-meal, see meal.

S.

Sagacity of bees, 47, 48.

Salt, fondness of bees for, 272.

Scent, see smell and odor.

Schirach, on artificial rearing of queens, 148.

Scouts scent out by swarms to find a new home, 117; necessity or, 118.

Scraper for cleaning the bottom-board or mov. comb hive, 347.

Screening hives for moving, 282 (note 4).

Scudamore, Dr., on many swarms clustering together, 137.

Secret recipe for keeping stocks strong, sham vendor of, 238.

Scholtz, Mr., on wintering bees in clamps, 348-360.

Sex alleles, determination of drones; and inbreeding, 50 (note 21).

Sex of bees, determined by queen, 38; *determined by sex alleles, 42 (note 13).*

Shakespeare's description of the Hive, 268.

Shrimplin, experiment of, showing impregnation to take place in the air, 127.

Sick persons, the care of; beneficial to man, 313.

Siebold, Professor, extracts from his Parthenogenesis, 126 (note); his dissection of spermatheca, 127 (note); found spermatozoa in worker, but not in drone eggs, 41; on bee life, 144 (note); recommends movable frames, 321 (note 2).

Sight of bees, acute, for distant objects, 117.

Signs of swarming, 111; of queenless colonies, 219, 224; of presence of moths in hive, 242.

Size of hives, 329-332.

Smell, of hives, in gathering season

177 (note); strange bees distinguished by, 203; the same, to be given in uniting colonies, 203; sense of, in bees, acute, 313; or their own poison, irritates bees, 314.

Smoke, importance of, in subduing bees, 27, 154; its use in forced swarming, 165, 168, 169; its use of, very ancient, 210; drives clustered bees inside of hive, 281; useful in removing surplus honey, 289.

Smokers, not invented, 27 (note)

Smothering bees, cautions for preventing, 281.

Snails, sometimes covered by bees with propolis, 78.

Snodgrass, R. E. Honey bee morphologist, 34 (note 5).

Snow, bees perish on, when carrying out their dead, 98; sometimes fatal to bees, 338 (note 1); often harmless to bees, 361 (note 1).

Solidago, see Golden Rod.

Sontag, F., on meal as a substitute for pollen, 84.

Spare honey, see Honey, surplus.

Spermatheca, of the queen bee, woodcut and description of, 35; Pl. XVIII., Fig. 55; dissection of, 34, *finding by removing tip of abdomen, 36 (note 6);* 126 (note), 213 (note).

Spermatozoa, found in spermatheca of queen-bee, 34, *checking for sperm, 36 (note 6);* 126 (note).

Sphinx Atropos, see Moth, Deathhead.

Spinola described the Italian bee, 318 (note).

Spring, importance of sun-heat in, to hives, 101; feeble stocks, in, unprofitable, 177; examination of bees, in, important, 221; colonies should be fed, in, 267, 268.

Sprinkling bees, should not be done to excess, 170; cools their robbing frenzy, 203.

Starving of bees, often happens when there is honey in the hive, 336, 342.

Sting, Bevan's description of, 56; Pl. XVII., Fig. 53; microscopic appearance of, 57; *derived from ovipositor, 58 (note 31);* loss of, fatal to bees, 57; loss of, in stinging, a benefit to man, 58; of queen, 65; woolcut of queen's, Pl. XVIII; *ice treatment for, 316 (note 5).*

Sting, poison of, dangerous to some, 313; *hypersensitivity percentage in population, 313 (note 3);* remedies for, 314-317; smell of poison of, irritating to bees, 314; instant extraction of, important 314; rubbing the wound made by, should be avoided, 314; Mr. Wagner's remedy for, 315; different remedies answer for different persons, 315; human system may be inured to, 316 (note); amusing remedy for, 316 (note).

Stinging, *habituation to prevent, 24 (note 1);* bees when gorged disinclined to, 25, 169, 308; little risk of, unless bees are irritated, 28, 168, 170; *not caused by terror 167 (note 20); by virgin queen 135 (note 31);* risk of, diminished by use of mov. comb hive, 209; diseased bees inclined to, 310; risk of, not increased by proximity to the hive, 211 (note); not to be feared from a bee away from its hive, 312; effect of, sometimes dangerous, 312; Italian bee less inclined to than common bee, 322, 324.

Stocks, of bees (see also colonies of bees), enfeebled by "in-and-in breeding," 54; strong, will rapidly fill empty comb, 71; often lose young queens after swarming, 141; fewer in this country than there were years ago, 145; often refuse to swarm, 139, 145; 147; new, work better than old, 153; if weak in Spring, usually unprofitable, and sometimes require to be fed, 177; the less disturbed, the better for surplus honey, 180; best mode for rapid increase of; 184; doubling, trebling, &c., 185; subject to great loss of bees in storms, 186; rapid increase of, hopeless in vicinity of sugar-houses, &c., 199; hostility of, to strange queens, 200; when united, the bees should be gorged with honey, 204; will adhere to the hive when the queen is lost, if supplied with brood-comb, 218; queenless, should be broken up, if not supplied with a queen or brood-comb, 218; Spring-care of, 221; healthy, destroy the drones when forage is scarce, 224; weak, with uncovered comb, infested by moths, 242; suffering from hunger, are an easy prey to the moth, 246 (and note).

Stocks, union of, see Union of colonies

Stomach of worker, wood-cut of, Pl. XVII., Fig. 54.

Stoves, air-tight, deficient in ventilation, 92; Franklin, a good kind of, 22 (note).

Straw use of, for protecting hives, 337.

Stupefaction of bees, by smoke, chloroform, and ether, 210.

Sturtevant, E. T., on wintering bees, 340.

Suffocation of bees, symptoms, 90.

Sugar, *sweetener instead of honey, 25 (note 3);* its elements 70.

Sugar-candy, see Candy:

Sugar-water, use of to pacify bees, 26, 154, 168-170; how to apply it, 170; *introducing queen using, 201 (note 37);* used in mingling stocks, 203.

Sulphur, *killing bees to rob honey, 12 (note 3)* use of, in killing eggs and worms of bee moth, 243.

Sun, heat of, important to bees in Spring, 101, 368.

Supers, space induces more honey, 329 (note 3).

Superstitions about bees, 79

Surplus honey, see Honey, surplus.

Swallow, address of Grecian poet, to a bee-eating, 253.

Swammerdam, his drawing of queen's ovaries described, 35; great merits of, as an observer, 65 (note); his drawing or queen's ovaries; Pl. XVIII.; how he learned the internal economy of the hive, and his reverence in studying the works of Nature, 164 (note); spoke of two species of bee-moth, 228.

Swarms, *Africanized bees and docility, 25 (note 4); old swarms and stinging, 26 (note 5); queen flying and egg laying, 39 (note 10);* new, often construct drone-comb to store honey, 51; *after-swarms, 51 (note 23);* number of bees in a good one, 54; first ones led *by* old queens, 111; no sure indications at first, 111; will settle without ringing of bells, &c., 113; more inclined to elope, if bees are neglected, 114; how to arrest a fugitive, 114; how to prevent, from deserting a new hive, 115, *with comb and brood 115 (note 9);* indications of intended desertion, 115; clustering of, before departure, of special benefit to man, 116; send out scouts, 117; *communication within 117 (note 11 & 12);* sometimes build comb of fence-rails, &c., 118; how parent hive is re-populated, after departure of, 119; composed of young and old bees, 119, *composed of young bees*

119 (note 14); none of the bees of new, return to parent hive, 120; *age of queen in swarm estimated by height of, 123 (note 19); reduction of honey yield from 124 (note 20);* signs and time of second, 122; sometimes settle in several clusters, 122; singular instance of plurality of queens (in Mexico),122; signs and time of third, 123; *distance traveled from parent hive 123 (note 19); determining age of queen in swarm 124 (note 21);* first, sometimes swarms again, 128; new, reluctant, to enter heated hives, 130; often take possession of deserted hives stored with comb, but seldom of empty hives, 131*; chemical for attracting 131 (note 26);* trees convenient for clustering of, 131; can be made to alight on a selected spot, 131, *131 (note 27); traps for catching 132 (note 27);* hiving of, should not be delayed, 132; *Swarm traps, 132 (note 27); caution using an excluder to keep swarm from leaving 134 (note 29); uniting with weak colony using newspaper 135 (note 30)* several, clustering together, 137; may be separated by hiving in large hive, 137; hissing sound of bees while swarming, causes other stocks to swarm, 137; *large Africanized bee swarm 137 (note 33);* how to prevent their mingling, 138; should be placed where intended to stand, as soon as hived, 138; *Scout bees selecting a new home site, 138 (note 35);* how to proceed when hive is not ready to receive, 139; feeble after-swarms, of little value, 140, 141; strong, tempted to evil courses, 141; many, annually lost, 143; danger of losing, in swarming season, 144; *need for some nectar available before swarm leaves, 147 (note 4);* decrease of in bees, after swarming, 151 (and note); *reason for mated queen in swarm 151 (note);* new, have greater energy than old 153, *swarm vigor 153 (note 10);* forced, 154; *orientation of bees to new home 156 (note 11);* will enter hives without the queen, 159 (note); *proportion of bees in swarm 160 (note 14);* when forced how to induce to adhere to new locations, 163 (and note); to avoid risk of losing, in swarming-time, 173; too rapid multiplication of, unprofitable, 176; second, usually valueless, unless early, and season good; 177; weak, may be strengthened by use of mov. comb hive, 178; one new, made from two old ones, 181 (note 3); *swarms and honey production 182 (note 27);* artificial, rapid increase of with mov. comb hive, 183; dangers attending, in large apiaries where the hives are uniform in appearance, and near together, 216; how to avoid the danger, 217; Washington Irving's account of, in the West, 236 (note); new, need more air than old, 281; precautions in moving, 281; a late one, 366.

Swarming, *reproduction of the honey bee species 109 (note 1); window of opportunity narrow 110 (note 5); lack of queen pheromone triggers 111 (note 6); time of day 112 (note 7); queen flight allowed by reduced egg laying 112 (note 8);* signs of, 111; indisposes bees to return to parent hive, 120; unseasonable, often caused by famine, 116; causes bees to mark the place of their new abode, 120; incident in, in Mexico, 123; after, care needed to preserve young brood in parent hive, 124; *and subsequent honey yield in parent colony 125 (note 20);* in tropical climates; at all seasons, 128; season of, 128; inconveniences of, 139-147; *instinct restricts foraging 140 (note 38);* artificial, mode of for common hives, 154; best prevented by use of authors hive, 153; for the season, can be accomplished in few days with author's hive, 173; time of natural, easily determined in author's hive, 173 (note); prevented by clipping wings of queen, 173, 223; *behavior needs to prevented 173 (note);* prevented by contracting the entrance of hive, 174; last plan not thoroughly tested, 174 (note 3); frequent, unprofitable, 176; best mode of artificial, 181; how to obtain extra queens in natural, 190 (note); *preventing by moving brood to weaker colonies, 269 (note 2);* interesting anecdote of, 308.

Swarming, artificial, see Artificial Swarming.

Swarming, natural, see Natural Swarming.

Swarming season, commencement and duration or, 111,128; *available nectar necessary for, 147 (note 4); daylength cue for, 366 (note 5).*

Sweaty horses, detested and often killed by bees, 313

Synserff's calculation of profits of bee culture, 146 (note).

T.

Table, illustrating the increase of stocks by artificial swarming, 185; of forming nuclei, 191.

"Taking up bees," facilitated by mov. comb hive, 209; suggestions as to time of, 306 (note).

Taylor, Dr. O. R. very large Africanized bee swarm, 137 (note 33).

Temperature of hive, rises at time of swarming, 130.

Temperature tolerated by eggs and larvae, 283 (note 6).

Theories often fail, when put to a practical test, 175 (note).

Thistle, Canada, a good bee-plant, 296.

Thompson, poetical extract from, upon killing bees, 239; on bees in linden trees, 293

Thorley, John, first stupefied bees by puff-ball smoke, 210.

Tidd, M. M., his experiment on a female moth, 230 (note 2); notices the difference between tongue of the male and female moth, 230.

Time of bees, economized in mov. comb hive, 95, 96; importance of saving, 305.

Timid persons may safely remove surplus honey, 289-291; should use bee-dress while hiving bees, 132,154; often stung while other persons seldom are, 168; some should not attempt to rear bees, 209.

Toad, eats bees, 254.

Tobacco, should not be used for subduing bees, 169.

Top-bar hives, 14 (note 8)

Top-boxes, for surplus honey, should be used with caution, 330 (note).

Transferring bees from common to mov. comb hive, 282-284; mode of, 282; best time for, 283; results of, 284,

Transportation of bees, easy in mov. comb hive, 281.

Traps for moths, usually worthless, 244.

Trees, combs built on, by bees, 118; apiaries should be near, 131; substitute for, 131; limbs of, need not be cut, in. hiving bees, 133; shade of, agreeable to bees, 280; honey-producing 292.

Trifolium repens (white clover) in lawns, 144 (note 2).

Trophalaxis, (food sharing) 178 (note).

Tulip (poplar, or white wood), tree yields great quantities of honey, 292.

U.

Unfertilized (drone) eggs, laying, queen's knowing how to, 38 (note 9);

Union of colonies, facilitated by giving them the same smell, 203; mode of, 203, 204; for wintering, 336.

Unbelief in revelation not prompted by true philosophy, 52.

Uncleanly persons disagreeable to bees, 313.

V.

Varnish, used by bees in place of propolis, 80.

Varro, his remark, that bees in large hives become dispirited, 208.

Ventilation, furnished to larvae by shape of cells 75; of the hive, 88-94; produced by the fanning of bees, 88; Huber on, 88; its necessity; 89; remarks on, in human dwellings, 91; provided for and easily controlled in mov. comb hive, 93, 94; artificial, must be simple to be useful, 93; *via auger holes in supers 96 (note 3);* should be attended to, after swarming, 121; ample, should be given, while bees are storing honey, 288, 366; how to give, in Winter, 338; upward, needed in Winter, 338, 340 (note), 241, 360.

Vice, effect of, on man, compared to ravages of the moth, 235.

Virgil, described the Italian bee, 318.

W.

Wagner, Samuel, *American Bee Journal, 15 (note 11);* letter of, on mov. comb hive, 17-18; theory of, on how queen determines sex of egg, 38; his account of bees building comb on a tree, 118; on the effect of soil on the quality of honey-yielding plants, 294 (note); on the Swedish white clover, for bees and stock, 295; letter of, on overstocking, 300; letter of, on the Italian bee, 317; extracts from on preserving the purity of the Italian bee, 323 (notes); states a remarkable fact concerning hybrid bees, 324 (note 2); attempt of, to import Italian bee, 328 (note); translation of Scholtz on wintering bees, 348-360.

War, how waged by different colonies, 263.

Wasps, fecundation of, 35; injure fruit, 86; should be destroyed in Spring, 87; torpid in Winter, 109.

Water necessary to be supplied for bees confined, 189 (and note; the refusal of, in Spring, by bees, indicative of a queenless colony, 219 (and note); cold, useful in checking robbery, 265; indispensable to bees when building comb, or rearing brood, 271, 342-346; bees need, in cold weather, 342-346; advantages of giving, to bees in cold Springs, 343.

Wax, scales of, wood-cuts, Pl. XII., Figs. 37 and 38; *found on hive bottom, 69 (note 1);* secreted from honey, 69, *secreted as liquid,* 69 *(note 2);* 275; pouches for, 69; wood-cut of, Pl. XIII., Fig. 38; Huber's experiments on secretion of, 69, *70 (note 3);* pollen may aid its secretion, 70; its elements, 71; large quantity of honey consumed in secretion of, 71; shavings of, used by bees, to build new comb, 72; a bad conductor of heat, 73; pollen useful in its secretion, 82, 197; *heat necessary to produce wax 99 (note 9);* origin of, discovered by Hornbostel, 204 (note); the food of the larvae of the bee-moth, 233, 247; how to render, from comb, 288; *amount procured per colony each year, 302 (note 13).*

Weather, unpleasant, delays of prevents swarming, 112.

West India honey, as bee-food, 256 (note), 270.

Wetherell; Dr. C. M., his analysis of royal jelly, 64.

Wheaton, Levi, on upward ventilation, 276 (note 1); on wintering bees, 346 (note 1).

White clover, see Clover, white.

Weigel, Rev. Mr., first recommended candy, as bee-feed, 272.

Wheeler, George, on ancient bar-hives, 210 (note).

Willow, varieties of, abound in honey and pollen, 292.

Wildman, Thomas, feats of, in handling bees, 308; states the fact that fear disposes colonies to unite, 203 (note); his approach to modern modes of taming bees, 204 (note); on the queen's odor, 226.

Winds, bees should be protected against, 103, 186, 279

Wings of queens, may be made to mark their age, 223.

Winter, wasps and hornets, but not bees, torpid in, 109, 335; quantity of honey needed by a stock in, 274; bees eat less in, when kept quiet, 335, 355, 358; bees should be protected from winds of, 337; bees in, if out of doors, should be allowed to fly, 337; how to ventilate hives in, 338; snow in, when injurious to bees, 338 (note 1); bees need water in, 342-346; when honey is candied in, bees need water, 342-344; disturbing bees in, injurious, 347, 355; fewer bees die in when hives are in clamps, than when in other special depositories, 358; temporary removal of colonies in, to a warm room, 341, 362.

Wintering bees, 335-361; objections to, in the open air, 335; bow to get honey for, in centre of hive 336; bee passages in comb for, 337 (and note 1), 339 (and note); *upper entrance for winter, 339 (note); brood rearing in winter, 340 (note 4);* in a dry vault or cellar, 348; *winter moisture vs. brood rearing water, 344 (note 6); moisture handling by bees in winter, 347 (note 7);* in special repositories, 348-360; further experiments in, needed, 360; *honey acts as winter-cold buffer, 360 (note 11);* requires caution in removing them from winter quarters, 361.

Wives a friendly word to, 220.

Wood-cuts, explanation of, 11, 371.

Women, American, suffer from bad ventilation, 92.

Worker-comb, size of the cells of, 74; all good, can be used in mov. comb hive 130; not built unless bees have a mature queen, 149.

Worker-bees, are females, with undeveloped ovaries, 29; when fertile, their progeny always drones, 36; Huber's theory concerning fertile, 37, 55 *(note 29); laying-workers, 55 (note 29);* sometimes exalted to be queens, 37; one raised from a drone egg, by Dr. Döhoff, 41; incapable of impregnation, 42; wood-cuts of, Pl. XII., Figs. 35, 36; number of, in swarm, 54 author's opinion respecting fertile, 55 fertile prefer to lay in drone cells, 55; honey-bag, 56; representation of, Pl. XVII., Fig. 54, *A.;* use of proboscis of, 56; wood-cut of proboscis of, Pl. XVI., Fig. 51; pollen basket, 56; sting, 56; wood-cut of, Pl. XVII., Fig. 53; loss of sting, fatal, 57; do all the work of the hive, 58; their age, 58; *length of life 59 (note 32);* lesson of, of industry from, 59; attention to royal cells, 62; wood-cut of abdomen of Pl. XVI. Fig. 52; two kinds of, described by Huber, 192 (note 2); differently occupied in different periods of life, 194; impulse of, to gather honey, undeveloped in early life, 195.

Worms, see Bee-moth, larvae of,

Wormwood, use of, for driving away robbing bees, 265 (note).

Wurtemberg, number of its colonies of bees, 304.

Z.

Zollickolfer, H. M., his account of bees building combs on a tree, 118.

www.ingramcontent.com/pod-product-compliance
Lightning Source LLC
Chambersburg PA
CBHW061923220426
43662CB00012B/1791